Highlights in Helioclimatology

Highlights in Helioclimatology

Cosmophysical Influences on Climate and Hurricanes

Jorge A. Pérez-Peraza
Instituto de Geofisica,
Universidad Nacional Autonoma de Mexico,
Ciudada Universitaria, Delegacion Coyoacan, 04510,
Mexico, D.F., Mexico

Igor Y. Libin
International Academy of Appraisal and Consulting,
115093, Moscow, Russia

ELSEVIER

AMSTERDAM • BOSTON • HEIDELBERG • LONDON
NEW YORK • OXFORD • PARIS • SAN DIEGO
SAN FRANCISCO • SINGAPORE • SYDNEY • TOKYO

Elsevier
The Boulevard, Langford Lane, Kidlington, Oxford OX5 1GB, UK
Radarweg 29, PO Box 211, 1000 AE Amsterdam, The Netherlands
225 Wyman Street, Waltham, MA 02451, USA

First edition 2012

Library of Congress Cataloging-in-Publication Data
Application submitted

British Library Cataloguing in Publication Data
A catalogue record for this book is available from the British Library

ISBN: 978-0-12-415977-8

For information on all Elsevier publications
visit our web site at store.elsevier.com

12 13 14 15 16 10 9 8 7 6 5 4 3 2 1

To my daughters: Sofia Isabel and Esther Cecilia.
To my sons: Nicolas, Oscar, David and Jorge.
J. Pérez-Peraza

To my son and my grandchildren.
Igor Libin

Contents

Preface

For decades the majority of the world's media has made gloomy oracles, that humanity has very few years left to live. They warn that the *greenhouse effect together with ozone holes and global warming will annihilate humanity and the Earth as well. And it is man, or perhaps more accuractly, his technological activity that is to blame for the impending tragedy.*

All these nightmares have led to formation of very authoritative and aggressive political powers in many countries. Various ecological and green movements have already become parties, in some places even dominant ones or, at least, included to the dominant coalition.

But damages caused to the world economy by extreme weather phenomena such as floods, hurricanes, earthquakes, and so on, have risen from $10 billion to $150 billion a year for the last 50 years. Natural disasters become more and more scaled, and scientists connect the growth of a number of them with climatic variations. And it is another question if they are manmade ones or not.

In addition, about 30 years ago ideas about the dominant influence of human activity on the climate appeared, and it brought to formation the main international agreements, for instance the Kyoto Protocol. Scientists understood that the problem of changing climate is really important for the humanity to survive. In 1992, representatives of the world community decided to begin practical actions during the famous meeting in Rio de Janeiro. The United Nations Framework Convention on Climate Change was an outcome of this meeting.

The Convention came into force in 1994, and 186 countries participated in it. In December 1997, members of 39 governments of the Protocol of the UN Convention met in Kyoto (Japan), where numerical obligations to reduce emissions were prescribed. The agreement was signed for most of them and began to operate in 2005.

The Kyoto Protocol is the first stage of a global ecological agreement on preventing disastrous climatic fluctuations. The main focus of the protocol is quantitative obligations of developed countries and countries with economy in transition to limit and reduce emission of greenhouse gases to the atmosphere from 2008 to 2012.

More than ten years ago a decision to form a multimillionaire fund for fighting with global warming—or with industrial and everyday emission of carbon dioxide—was taken at the OPEC conference in Riyadh, Saudi Arabia.

Moreover, the former vice-president of the U.S.A. Albert Gore got the Nobel Prize of 2007 for his contribution to fight against emission of carbon dioxide; and the International Organization Studying Climatic Variations (among which were a number of scientists of the Center of the Atmosphere of the National Autonomous University of Mexico) got the Nobel Prize together with him. Journalist Peter Obraztsov writes in the magazine *Itogi*, "The Nobel prize given to Gore is probably the last surge of activity of fighters with emission of carbon dioxide."

Who still remembers the ozone hole above the Antarctica that absorbed so much money because production of ideal coolant Freon was stopped in many countries? Such a hole, while existing since 1982, is now undergoing occlusion, but according to Alvarez and Pérez-Peraza (2005, 2007; Alvarez et al., 2007, 2009; Alvarez, 2010; Pérez-Peraza et al., 2011) one of the main factors in the depletion is the cosmo-climatologic influence.

John Coleman, an American scientist and founder of The Weather Channel, says that the global warming on the planet connected with human anthropogenic activity is a fiction invented by politicians, scientists, and businessmen exploiting it *pro domo sua*. Coleman expressed his opinion that a legal action should be brought against the former vice-president Albert Gore, a famous fighter for environmental protection. As John Coleman says, it would help to disclose "artificially invented panic about global warming." In Coleman's opinion, if this claim is noted, organizations selling greenhouse gases emission quota can also be sued.

In some scientific clubs a phenomenon of global warming is considered to be a theory that could have been falsified. After signing the Kyoto Protocol limiting emissions to atmosphere, states or separate companies can sell or buy carbon emission quotas. But it is quite another story connected with big state budgets, which should be divided correctly, if possible.

It is important to emphasize that we are talking only about years 2008 to 2012, since in 2013 new obligations, new ratification, and so on, will appear. By the way, an allowed level of emissions of greenhouse gases for years 2008 through 2012 in Russia is 100% of the level of 1990 (92% is in European Union Countries, 94% is in Japan, 93% is suggested in the U.S.A., but the U.S.A. did not sign the protocol, and unfortunately Canada retired from the signed protocol on December 13, 2011. (Canada will sign a more actualized protocol where emergent countries such as China, India, Brazil, and Mexico have more defined quotas.)

The authors of this book write in one of their works (Libin and Pérez-Peraza, 2009) "It means that in case of signing the Kyoto Protocol we have no possibilities for a free-of-charge growth of our economy (for those who do not know that the year 1990 was a failed one in the Russian economy)."

The reality is that in Kyoto a very complex treaty was reached. Though useful for many political slogans, it was very vague and contained nothing pragmatic, with very few countries able to be really concerned. Any form of international control of national emissions has yet to be established, so that its execution becomes even more doubtful. The greatest paradox is that despite the noise, riot, and clamor, the average global temperature over the last 21 years has risen very slightly and quite irregularly.

Neither representatives of meteorology nor officials should decide about the future participation of countries in the Kyoto Protocol; it should be decided on by a wide scientific society. The decision to be or not to be in the Kyoto Protocol should be taken after wide open discussions by specialists in climatic sphere and economists.

Global climate change is a problem at a planetary scale, and the whole world will have to settle it. Making a coordinated decision is as necessary and unavoidable as a common fight against terrorism. And the earlier politicians begin real actions, the less damage there will be.

We would like to understand if man is really such a self-killer, that he tries to kill himself and every living thing on the planet so passionately. From the first moment of its comparatively intelligent existence humanity has always damaged the planet to survive, since it has not had any other way to continue its existence on the Earth. All natural forces and other types of animals have always been stronger than *Homo sapiens*.

Skeptics say that technologies harmful to nature have been developed especially quickly since the industrial revolution. But nature-conserving measures have gained momentum as well. It is common sense that the existence of humanity is connected directly with development of modern technologies, or it will not support itself. But what about invocations of a future disaster in mass media? Recently, the end of one of the several cycles of the Maya calendar in 2012, has been interpreted by some media as the end of the world provoking hysteria in many places.

From time to time we listen to forecasts of helio (solar) and geomagnetic activity on the radio and television and read about it in newspapers, however, nobody thinks about the great work (of many research teams) that creates these forecasts. All of us have become consumers of forecasts and have got used to their existence, without a thought for their importance for our life. As well as everyday people lending their ear to these forecasts and planning their behavior according to them, EMERCOM specialists, operators, cosmonauts, military men, meteorologists, biologists, medical doctors, and hydrologists in many countries are also consumers of these forecasts.

All forecasts are impossible without a fundamental science on which the whole building of applied research is built. It seems that today forecasts have become a part of the world economy, because they make it possible to value and make further forecasts of expected nonanthropogenic disasters such as earthquakes, droughts, epidemics, weather cataclysms, and frequency of hurricanes. They can explain the influence of solar activity variations on processes occurring on the Earth and in the closest cosmic space.

Nobody has abolished competition in the scientific society, but the price on which the scientists are standing is very high: the normal existence of humanity in the near future. That is why the pragmatism of scientists and a desire to get results as soon as possible overpowers individual ambitions.

Worldwide globalization is widely discussed in the world but not always with respect to financial aspects. It unites different researchers' creative multinational power, and as a result, multiethnic groups have made much more during the last few years than in the previous two decades of research.

That is why I would like to introduce not only this book to the readers, but also the authors, a scientist from Mexico, Jorge Pérez-Peraza of the Institute of Geophysics of the National Autonomous University of Mexico, and from Russia, Igor Libin of the MAOK (International Academy of Appraisal and Consulting).

They are two of the first authors who have used the scientific appellation about solar influence on the climate of our planet: helioclimatology.

Eygen Treyger, Rector MAOK

Foreword

Alexander L. Chijzhevsky, an outstanding Russian scientist, was a pioneer in the history of science who, in the 1920s, paid attention to synchronism of solar activity and processes occurring on the Earth. And outstanding scientist and philosopher, he was also a talented artist, refined poet, and a musician (Chijzhevsky, 1976).

A. L. Chijzhevsky wrote, “In what way do roughness and storms occurring on the Sun influence the planet? Is our spacecraft *Earth* still sailing calmly and quietly or it is being rocked on the waves of solar cycles so much from that time to time one can hear clatter of glasses in the cabin?”

Participants of the First International Congress on Biophysics and Cosmology, which took place in September 1939 in New-York, called Chijzhevsky the Leonardo da Vinci of the twentieth century. Among the personality traits of Chijzhevsky himself, he should be defined as a person of encyclopedic learning; one of the founders of cosmic natural science (together with one more of his compatriots, V. I. Vernadsky); a founder of cosmic biology, heliobiology, aeroionofication, and electric hemodynamic; as well as a historian, poet, and artist, whose humanitarian culture let him state phenomena under study clearly.

It was noted in a special memorandum passed on that First International Congress on Biophysics and Cosmology that “the proceedings genius in idea, novelty, breadth, braveness of synthesis, and depth of analysis have put Professor Chijzhevsky at the head of the world and made him a real citizen of the world, because his proceedings are global commons of humanity.” (Chijzhevsky, 1990, 1995.)

Chijzhevsky was elected an honorary member of the First International Congress on Biophysics and Cosmology in 1939, which nominated him to the Nobel Prize. By the way, Chijzhevsky‘s bas-relief is among bas-reliefs of other great scientists of the world.

However, world glory did not save the scientist from repression; he was repressed in 1942. It is clear that Stalin, who preferred to think all cataclysms in society to be a result of class struggle, did not like Chijzhevsky‘s opinion about a possible connection between wars, revolutions, and other cataclysms in society and the number of sunspots. In prison Chijzhevsky formed a small laboratory in the camp hospital and went on working there after formal discharge to finish the next stage of testing of his curious health-improving device, which is called now Chijzhevsky‘s luster (Chijzhevsky, 1990 (9th edition)).

D. I. Blokhintsev, a famous physicist (in Yagodinsky, 1987), wrote about him and his pictures, "... maybe the main thing these pictures and poems are talking about is the following—they reveal the image of a great human being in the sense it has always been understood by humanity. A necessary and integral, obligatory characteristic of this image was not only success in this or that science but a formation of a world view. Science, poetry, art should have been only a part of the great humanist and his activity."

Chijzhevsky together with such leading figures in the world of science as V. I. Vernadsky and K. E. Tsiolkovsky initiated a new cosmic view (Yagodinsky, 1987). One of his most important contributions to modern scientific ideation is finding out the influence of cosmic factors on biological, scientific, and social processes.

Chijzhevsky defined life as a living thing's capability to let through a flood of cosmic energy. He considered that the biosphere is the place of transformation of cosmic energy, emphasizing that life is a more cosmic phenomenon than an Earthly one.

"Cosmos," 1921

Acknowledgments

We are grateful to Lev Dorman, Manuel Alvarez-Madrigal, Oleg Gulinsky, Ago Jaani, Stilian Kavlakov, Mikolas Mikalajunas, Víctor Velasco Herrera, Enrique Aspra-Romero and Konstantin Yudakhin, who at different times participated in our research.

One of us (I.Ya.L.) is grateful to the colleagues from MAOK: Evgeny Treyger, Tatiana Oleynik, and Tatiana Pustovi tova; Maxim Dzalaev (from TIMAX); and Vladimir Kuznetsov (from IZMIRAN), all of whom helped us in organizing our joint research at different times.

One of us (JPP) gives special gratitude to the PAPPIT program of the Universidad Nacional Autonóma de México (UNAM) for supporting this work through the project IN119209.

Introduction

A role of the Sun for the life on the Earth: different types of solar radiation determine thermal balance of the land, ocean, and atmosphere. Outside the Earth's atmosphere there is a little over 1.3 kilowatt of energy in each square meter of place perpendicular to solar rays. The land and waters of the Earth absorb about half of that energy, and about one fifth of it is absorbed by the atmosphere. About 30% of solar energy is reflected back to the interplanetary space, mainly by the Earth's atmosphere (what is usually called the *Albedo effect*).

Sunlight and warmth were the most important factors of appearance and development of biological living forms on our planet. Energy of wind, waterfalls, and river and ocean flow are accumulated energy of the Sun. We can say the same about fossil fuel: coal, oil, and gas.

Under the influence of electromagnetic and corpuscular radiation of the Sun, air molecules break up into separate atoms which are in their turn ionized. Charged upper layers of the Earth's atmosphere—ionosphere and ozonosphere—are formed. They draw aside or absorb ionizing and penetrating solar radiation making way to the Earth's surface for only that part of solar energy useful for a living world, and which plants and living beings have been accommodated to.

Kononovich (1967) wrote,

It is very difficult to imagine what will happen if something blocks the path of these rays to the Earth for some time. Arctic cold will begin to quickly envelope our planet. In a week the tropics will be covered with snow. Rivers will be frozen, winds will come down, and the ocean will be frozen through. Winter will come unexpectedly and everywhere. It will rain heavily but not with water but with liquid air (mainly by liquid nitrogen and oxygen). The rain will freeze quickly and cover the whole planet with a seven-meter layer. No life will be able to survive in such conditions. Fortunately, it is impossible, at least, unexpectedly or in the near future. But the picture described above shows very vividly the meaning of the Sun for the Earth.

Just a sample of that is the catastrophic impact of a huge meteorite of about 10 to 20 km in diameter opening a crater of 180 km in diameter in Chicxulub, Yucatán, in the Mexican Caribbean, about 65 million years ago, the presumed cause for the disappearance of several species (among them, the dinosaurs). This is known by scientists as the K-T event (Cretaceous-Tertiary Massive Extinction). The searing heat would have destroyed lives, and dust released into the atmosphere together with massive eruptions would have caused a global blackout (a *cosmic winter*), halting photosynthesis by lowering the temperature. Animals that feed on plant life would have died of starvation and predators would have hunted each other until they too disappeared. A real *Climatic Change* was cosmically provoked.

The current hypothesis about the influence of solar activity on Earth's seismic activity is actively discussed in scientific literature. However, it is generally reduced to the study of regularities between the active processes on the Sun and earthquakes on Earth. In Odintsov et al. [2005, 2007] a connection of global seismicity of the Earth with 11-year cyclicity of sunspots, with velocity leaps of the solar wind in the circumterrestrial space, coronal emission of solar mass is studied. It is shown that the maximum of energy of an earthquake (of magnitude greater than 5 on the Richter scale) in an 11-year cycle of sunspots falls on a decrease of the cycle and is two years delayed from the maximum of the solar cycle. It is established that the maximum number of earthquakes has a direct correlation with the solar wind velocity jumps. Possible physical mechanisms of influence of solar activity on global seismicity of the Earth are also analyzed by Odintsov.

It was found out later that climatic processes such as glaciers, warming, recurrence of typhoons, earthquakes (see figure), and precipitation are connected with the Sun. The degree of the Arctic and Antarctic ice coverage, variations of ocean levels, the Gulf Stream Pulsation, and the sea thermal regime are connected with 11-year, 22- to 30-year, and 50-year cycles.

The figure shows peak values of earthquake energy fall on time of solar activity maximums. Eleven-year cyclicity is observed in the movement of the

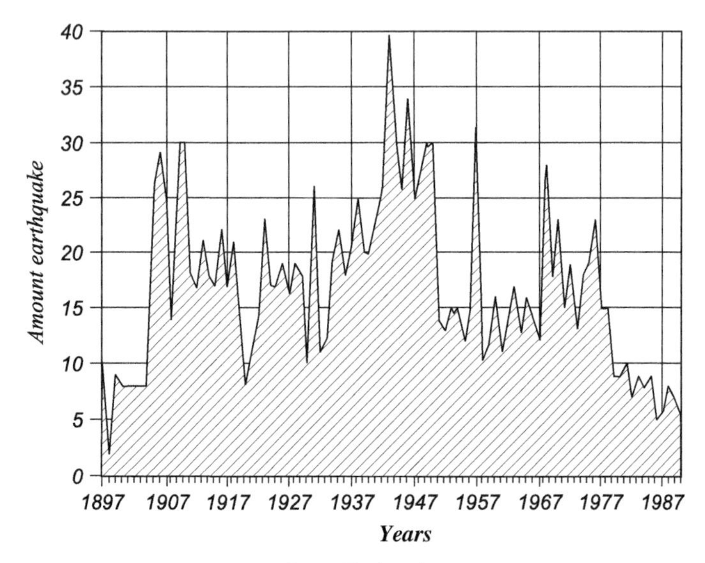

Strong Earthquakes

Earth poles, in increasing and decreasing of the Earth's rate of rotation (Russian Academy of Sciences, 2006).

In times when the Sun is changing, the intensity of the terrestrial magnetism, frequency of auroras, radioactivity of the air, the amount of ozone in the ozone layer, cyclical drought and flood, the water level in lakes and pulsation of Earth crust also change (Nelson et al., 2002; Panchenko et al., 2007; Paris et al., 2009; Volfendeyl et al., 2009).

It was proved recently that during the periods of solar flares on the Sun, blood structure is changed sharply, in such moments blood looks as if people have survived radioactive radiation. Not only do long-period cycles influence the blood (and everything in nature as well), but also year, seasonal, 24-hour, and even second-long cycles. Blood is changed constantly; the Sun is living inside us. Unfortunately, unlike representatives of exact science, biologists and medical doctors cannot get used to this, although it is already clear that flashes of epidemics and locust invasions follow the Sun (Barashkova, 2007; Elansky, 2008; Klimenko et al., 2009; Konstantinovskaya, 2001a; Obridko and Ragulskaya, 2005; Ragulskaya and Khabarova, 2001; Ragulskaya, 2004).

Tucidide, a Greek historian, wrote that epidemics are often accompanied by floods, eruptions, crop failures, and locust invasions. As solar activity variations are accompanied by atmospheric variations as well, all nations had the same systems of heaven signs.

An unusual color of the sky, cirrus clouds, aurora borealis, sunspots which one could see sometimes even with a naked eye—all these preceded the disaster.

Nikonov, 1996

The most popular and famous solar cycle is the 11-year one. But there are a great number of other cycles: 22-year, 7-year, quasibiennial-years, 1/5-year, 36-year, 80- to 90-year, 169- to 180-year, and 27-day ones. There are super long cycles, including 600-year and 1800-year, among others.

Nicolay Kondratiev, a Soviet economist who was shot in a Suzdal prison in 1938, discovered 58- to 64-year variations of economic activity which are now named after him. K. Ensen discovered a 17-year freight cycle, and Dutch shipbuilders adjusted the life time of their ships to it.

In 1964 Professor Pakkardi made a sensational report at the Leningrad conference, widely described by Shnol et al. (1998, 2000, 2001a,b; Shnol, 2009): the rate of the elemental reaction of deposition bismuth oxychloride in colloidal solution carried out by chemists of different countries made at the same time (GMT) was independent on the place of study. Even if the speed of transmission of the reaction was different for different tests, they were always the same all over the world. It means that some factor common for the whole planet influencing the reaction transmission existed. The rate of the reaction is the same as the frequency associated with solar activity.

This discovery, proved wonderfully by Shnol (2001a,b), was the last drop in collecting data about curious cyclicities. Periodicities were taken into consideration seriously: new instrumental and mathematic methods appeared, and exposure of statistically significant correlations between different processes on the Earth and in the interplanetary space began. Connection with the Sun, and with solar and magnetic activity variations was becoming clear, not only speculatively but rated.

But in what way can all these solar variations transform into changes on the Earth watched by us? Where is the mechanism of this influence?

In this book we do not only try to explain the essence of solar influence on the Earth's climate from a number of outstanding research works, but also to estimate what hazard do global climatic variations bring to humanity.

Jorge A. Pérez-Peraza
Igor Ya. Libin

Part I

Solar Terrestrial Relationships: Helioclimatology

Chapter 1

Climate and Global Warming

Chapter Outline

Climate is the weather averaged for several decades.

1.1. CLIMATIC MODEL ELEMENTS

Climate, like weather, subjects to measuring. That is why every day and every hour thousands of watchers worldwide measure weather elements such as atmospheric pressure and temperature, air moisture, wind direction and velocity, cloudiness and visibility, precipitation (quantity and type), fog, snowstorms, thunderstorms, time of sunshine, soil temperature, snow cover height, solar radiation, and so on.

Highlights in Helioclimatology. DOI: 10.1016/B978-0-12-415977-8.00001-0

The Earth climate is defined by global elements of environment: atmosphere, hydrosphere (ocean and land waters), land (continents), cryosphere (snow, ice, and permafrost regions), and biosphere.

In 1915, Alexander L. Chijzhevsky made a report about solar-biosphere connections at the Moscow Archaeological Institute, noting that the weather and climate on the Earth are tightly connected with solar activity variations. In 1918 he defended a dissertation "Research of World-wide Historical Process Periodicity" at Moscow University and was awarded his Doctor degree of Historical Sciences. Six years later he published his main work, *Physical Factors of the Historical Process*. This book laid the foundation of study into solar influence on Earth events.

It is obvious today that rising and falling temperatures on Earth are of a cyclic character and have happened lots of times without any human participation. In the past the Earth's climate has changed many times, and more essentially and radically than it does now.

Over billions of years of its existence, Earth has repeatedly experienced periods of powerful solar flares and tectonic catastrophes. The planet has survived hundreds asteroid attacks like those that presumably killed the dinosaurs. Nevertheless, the average surface temperature has never deviated below 8 °C or above 10 °C from the present.

For example, about 35 million years ago average annual temperature was 5 to 6 °C higher than now, palms and oaks grew, snakes and alligators lived on the Spitsbergen, and one could observe real tangle of tropic jungle on Taimyr.

Figure 1.1 shows the history of Northern Hemisphere temperatures during the Holocene. The modern geological period (the Holocene) has continued

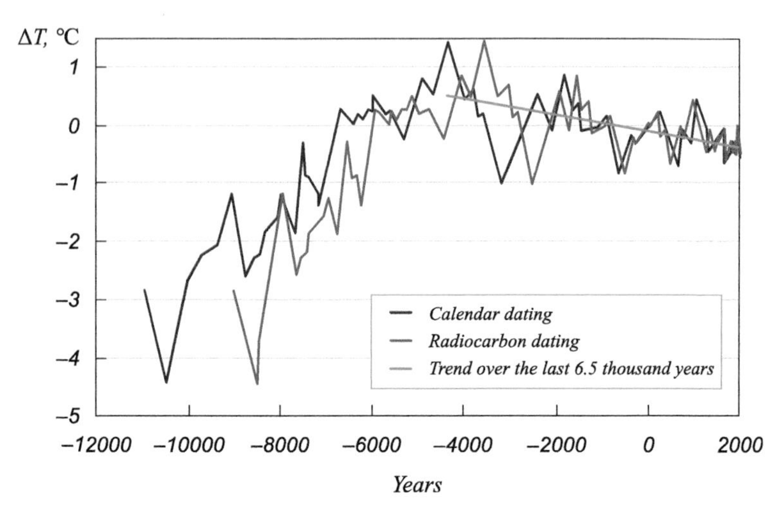

FIGURE 1.1 History of Northern Hemisphere temperatures (in deviations from the norm 1951−1980). *Source: www.polit.ru/article/2005/11/02/climate/*

for 10,000 years since the end of the last glacial period. From 9000 to 5000 years ago was the climatic maximum of the Holocene. It is extremely important that the temperature then was much higher than the modern by 1 to 1.5 °C.

Man finally began to discover the Arctic in the 1930s. Some think it was due to political will, but that is incorrect. All achievements of the 1930s were connected with enough warming in the Arctic. The ices stepped back hundreds of kilometers to the north and ships were able to achieve very high latitudes.

However, in the first two decades of the twentieth century it was significantly colder in the Arctic. Later, in 1960s and 1970s the next temperature drop happened. Glaciers stopped stepping back. Now warming has come again. According to Kotlyakov (1996, 1997):

If we have compared today's climate with former epochs, we would think that we are living in Holocene, or in the interglacial period which was observed 100 thousand years ago when temperature on the Earth was about 1.5 degrees higher than now. But 100 thousand years ago there were no men, heat stations, cars, other filth, or dirt.

In the last decades as a result of human economical activity, the content of carbon dioxide and methane has increased by one third. Tons of greenhouse gases appeared. But it has not impacted the temperature. Not long ago scientists asserted that climate warming was caused by human activity, and man's influence on the environment. But the facts tell us that temperature is raised first, and then the concentration of carbon dioxide. Gradual and inevitable variation of the climate can be explained by natural reasons. These changes are surely influenced by the human factor, but it is not a determinant, global one. Scientists admit that the mechanism of global warming is very complex and is not yet understood by science.

Many research works prove a connection of the climate with emission of charged particles of different energies from the Sun, with the interplanetary magnetic field direction, and the solar wind. The connection of solar activity with atmosphere processes has also been proved.

Atmosphere is a central element of a climatic system (see Figures 1.2 and 1.3); human beings have been changing other elements through it.

Atmosphere is in any Earth point, it is global. Other elements are local.

Ocean comprises 70.8% of the Earth's cover; land comprises the other 29.2%. Glaciers are a little more than 3% of the Earth's cover and 11% if it includes sea ices and snow cover.

In modern natural science, life problems on the Earth are united by a common term: *biosphere.* The term was coined by the Austrian geologist Seuss (1875), to denote the unity of all living organisms on our planet. The biosphere is extended globally, but irregularly on the Earth's surface; its upper border is 25 to 30 km, the lower border (in the Earth's crust) is about 2 to 3 km, and from 3 to 10 km is in the water.

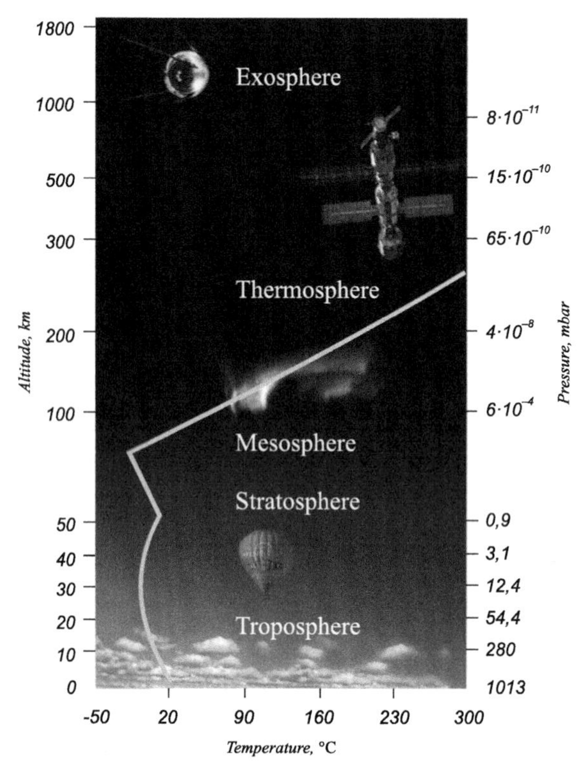

FIGURE 1.2 The structure of the Earth's atmosphere. *Sources: www.kowoma.de; www.theresilientearth.com; www.scienceofaircraft.blogspot.com*

The biosphere is a unity of objects of wildlife and inanimate nature involved in the life sphere. Two basic components of the biosphere—living organisms and their environment—constantly interact and influence each other.

The Sun is the basic source of energy for life on Earth and a great majority of processes occurring on our planet are connected with its radiation (see Figures 1.4 and 1.5). The whole biosphere is opened for cosmos and, metaphorically speaking, it takes a bath in flows of cosmic energy. While processing this energy, a living substance transforms the whole planet. In this sense we can consider the origin, formation, and functioning of the biosphere as a result of cosmic power activity.

It was discovered recently that during periods of flares on the Sun, blood structure is changed sharply. In such moments blood looks as if people have survived radioactive radiation (Shnol et al., 2000; Shnol, 2009).

Cosmic factors which influence biogeochemical processes and the Earth's climate are defined by their spatial location concerning the Sun (the tilt of the Earth axis to the Earth orbit plane), the distance from the Earth to the Sun, conditions of solar rays transmission, and, mainly, by the processes occurring

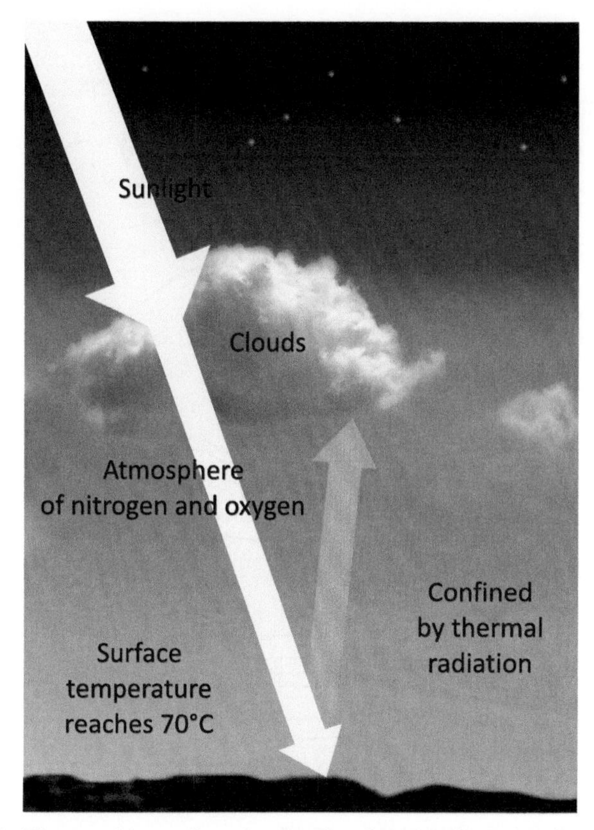

FIGURE 1.3 The greenhouse effect. *Sources: http://en.wikipedia.org/wiki/Greenhouse_effect; http://www.weatherquestions.com/What_is_the_greenhouse_effect.htm; http://www.google.com.mx/search?q=greenhouse+effect&hl=es-419&prmd=imvns&tbm=isch&tbo=u&source=univ&sa=X&ei=YPJjT86CEIeW2QWSksnbCA&sqi=2&ved=0CEwQsAQ&biw=1413&bih=705*

on the Sun. That is why the study and defining of the nature of solar-terrestrial (especially solar-biospheric) relationships have a great meaning literally for all the processes on Earth.

An *atmospheric gas* is pervasive. It is constantly exchanging with other elements of the climatic system. Constituent parts of the atmospheric gas resolve in the hydrosphere. From the hydrosphere they come to the air and penetrate into pores and lithosphere clefts. In its turn the atmosphere is filled with volcanic gases' emissions and their weak flows from the lithosphere.

Atmospheric gases are also found in ice caps. When glaciers are melting, they become free in the form of bubbles and come back to the atmosphere. The atmosphere is exchanging gases with the biosphere during the process of breathing.

The biosphere has created oxygen in the atmosphere. The atmosphere as an element of the climatic system is the most movable among all the other elements.

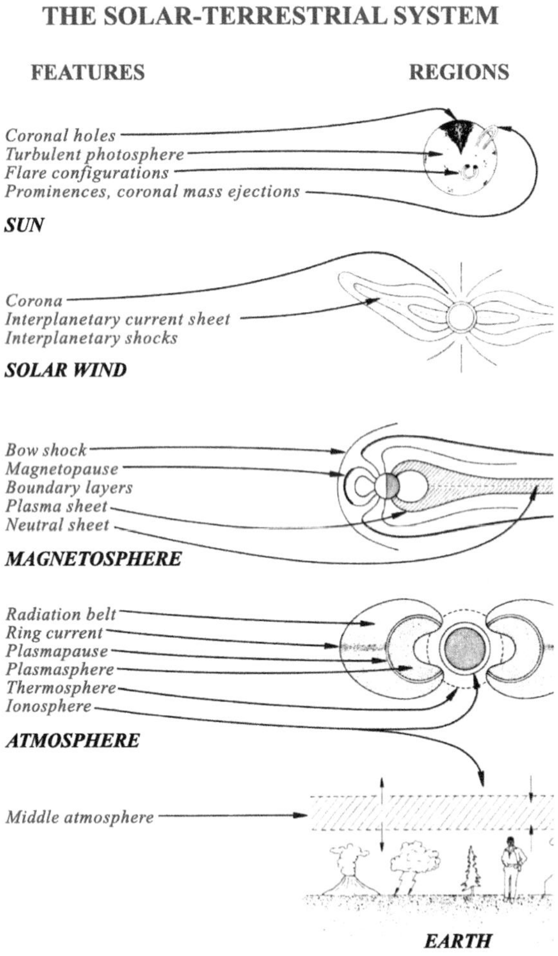

THE SOLAR-TERRESTRIAL SYSTEM

FEATURES
REGIONS

Coronal holes
Turbulent photosphere
Flare configurations
Prominences, coronal mass ejections
SUN

Corona
Interplanetary current sheet
Interplanetary shocks
SOLAR WIND

Bow shock
Magnetopause
Boundary layers
Plasma sheet
Neutral sheet
MAGNETOSPHERE

Radiation belt
Ring current
Plasmapause
Plasmasphere
Thermosphere
Ionosphere
ATMOSPHERE

Middle atmosphere

EARTH

FIGURE 1.4 The solar-terrestrial system. Source*: Program of the Scientific Committee on Solar-Terrestrial Physics (STEP), 1997.*

The *hydrosphere* of the Earth and, primarily, the world's oceans (as well as seas, lakes, and rivers) are important components of climate formation. Warmth, mass, and movement energy are given from the atmosphere to waters and back. The mutual interaction covers around two third parts of the ground surface, keeping in mind that 33.6% of the earth's surface are deserts.

Surface flows in the ocean are formed by atmospheric winds which carry a great deal of warmth. The ocean is an enormous warmth accumulator. Mass of ocean water is 258 times more than mass of the atmospheric gas. For raising the temperature of the atmospheric gas 1 °C higher, the ocean water should give the same quantity of heat energy. As a result, water temperature will fall 1/1000 °C. Such temperature changes are difficult to measure.

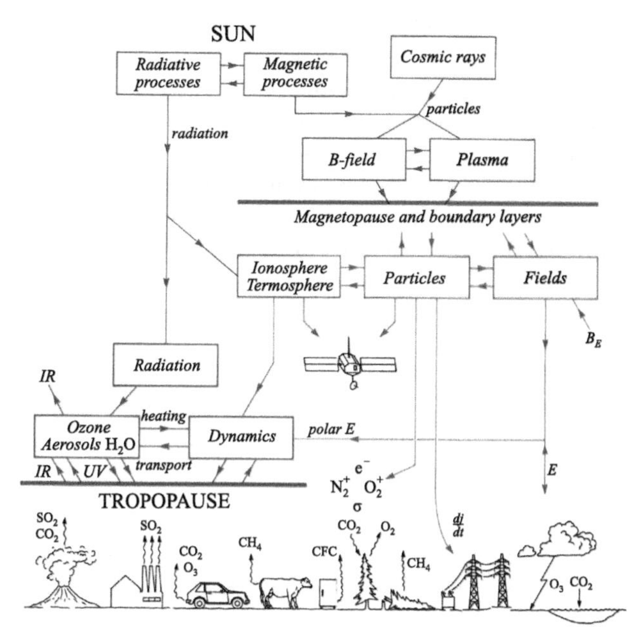

FIGURE 1.5 Basic interaction channels between the principal domains of the STR system: Sun and solar wind (red), magnetosphere and ionosphere (blue), middle atmosphere (green), troposphere and Earth (black). Source*: Program of the Scientific Committee on Solar-Terrestrial Physics (STEP), 1997.*

In the last few decades very important features of ocean water circulation have been revealed. Together with famous ocean flows, scientists have found underwater ocean whirlwinds similar to cyclones and anticyclones in the atmosphere. The diameter of these whirlwind-like ring structures can reach hundreds of kilometers. Water features inside these whirlwinds are unlike water features outside them. Underwater and surface ocean water movements have also been revealed (e.g., El-Niño in the Pacific Ocean near South America).

The hydrosphere is a moving sphere, although in comparison with the atmospheric gas its speed is tens and hundreds times less: a medium speed of ocean movement is several centimeters per a second, while the wind speed can be several hundred meters per second.

The *cryosphere* (snow and ice) also participates in formation of the climate. As ground cover, snow and ice increase the reflective capability of the Earth. As a result, about 90% of solar warming energy is reflected back to the cosmos. The basic mass of ice is concentrated in Antarctica, where 90% of the world's ice is located.

In our case, it is not the mass of ice but the area of the Earth's surface it covers that is the most important factor. The largest area on Earth is occupied by sea ice and snow cover. In summer sea, ice in the Arctic Ocean is covers and area of about 8,000,000 km^2, in winter this area grows twice as much. Snow covers on average about 60,000,000 km^2 of the ground for a year.

Land can be considered the most inertial and passive element of the climatic system. It changes little for short periods of time. It is changed under the processes of soil formation, airing, erosion, and desertification. Continent drift takes place over periods of millions of years before it fully changes the Earth image.

The biosphere is a very active component of the climatic system. During periods of growth vegetation, changes of plant communities, widening and shortening of areas occupied with plants, increasing or decreasing biomass of its influence on the climate variations are displayed in different ways, in different time scales.

If we compare the climatic system with a living organism, we can say that the role of blood in the organism is comparable to water in different states (steam, liquid, snow, ice). Water is a carrier of mass and energy in the climatic system. Specialists believe that the climatic system is a basically self-regulating system, which means that many inner and outer changes (disturbances) are put out.

1.2. BASIC THEORIES OF CLIMATE VARIATIONS

Why does climate change? Nobody can answer this question precisely. There are a lot of hypotheses that examine possible reasons of such change.

All the hypotheses about reasons of glaciations and global warming epochs can be divided into two basic groups: carbon dioxide- the greenhouse effect and solar activity, as discussed in the next two subsections (Libin and Pérez-Peraza, 2009).

1.2.1. Carbon Dioxide and Greenhouse Effect

One group of hypotheses supports the view that the reason for climate variations is the absorption of a solar energy flow by the Earth. The idea is based on measuring data that show that from time to time some conditions appear in the Earth's atmosphere in which solar energy is absorbed much more than usual and the temperature is falling greatly.

According to some scientists, it is necessary to look for the reason for this change in the amount of energy that is being absorbed: some energy is reflected back to the cosmos, some is let in to the Earth's surface, and some is used to warm up the atmosphere (Bashkirtsev and Mashnich, 2003, 2008; Bashmakov, 2009; Brand et al., 2007; IPCC, 2007).

This capability of the atmosphere depends on its structure, but it is clear that the Earth's atmospheric structure has changed radically over its history. Carbon dioxide plays the most important role in this process, even though the absolute quantity of it in the atmosphere is very small—only 0.03% of the volume. CO_2 in the atmosphere works in the same way as a film on a greenhouse, according to the principle: to let in, but not let out. As a result, 30% of solar radiation coming in is reflected from the upper atmosphere and goes back to the cosmos, but its larger part comes through the atmosphere and warms the Earth's surface.

The warming surface emits infrared radiation. Some gases entering the atmospheric structure in a small quantity (0.1%) are able to keep infrared radiation. These are called *greenhouse gases*, and the phenomenon is known as the *greenhouse effect*.

Greenhouse gases have existed in the atmosphere for almost the entire history of the Earth, and their balance has been kept on account of a natural cycle. If they were absent, the air temperature at the Earth's surface would have been 30 to 33 °C lower than now. Before the epoch of industrial development the concentration of carbon dioxide in the atmosphere was 280 ppm (particles per million), but now it has become 30% higher and reached 368 ppm.

If a natural greenhouse effect kept the Earth's atmosphere in a state of a heat balance productive for animals and plants, anthropogenic increasing of greenhouse gas concentration in the atmosphere would disturb the natural heat balance of the planet at the expense of strengthening a warming effect. The result is global warming. Adherents to the theory of anthropology influence on the climate think as follows:

Solar rays coming to the Earth's surface move through the atmosphere freely. Certainly, part of radiation disperses because of the atmosphere's relative opacity. The light energy is partly absorbed and warms the Earth. Part of the solar energy is reflected on the Earth's surface (both land and water) back to the atmosphere and then to the cosmos. The hot Earth, like any hot object, begins to radiate.

When something is given infrared or ultraviolet radiation. This radiation leaving the Earth is retained by carbon dioxide CO_2. If there were no CO_2 in the atmosphere, the average temperature on the Earth's surface would be substantially lower than it is, and conditions of the glaciations epoch would appear. If the quantity of CO_2 in the atmosphere were increased, it would lead to a global-warming epoch.

Supporters of global warming believe that a very fragile balance had appeared on the Earth between all the sources of carbon dioxide before the industrial revolution of the twentieth century. It is clear that if such a balance is disturbed, the quantity of CO_2 in the atmosphere would change and climate changes would happen, so today's global warming is connected with human anthropogenic activity.

This theory, however, has several weak points. The logic of the global warming theory supporters are as follows: since carbon dioxide leaks ultraviolet and visible solar radiation but does not leak warming and infrared radiation, the surface is warmed more than cooled, which results in global warming (Domysheva et al., 2007a, b; Kazansky et al., 2011; Klimenko, 2008).

This effect was called greenhouse because it behaves like a glass roof of a greenhouse. If it were so, it would be necessary to decrease CO_2 emissions to a minimum and shorten industry. However, some years ago the American scientist Robert Wood doubted the validity of the greenhouse effect explanation. Wood became famous for the simplicity of his experiments and coming up with witty solutions to the most complex physical problems (Seabrook, 1946). In this case, he built a table greenhouse with a roof made from the transparent crystal of common salt (salt transmits infrared and ultraviolet rays). There should not have been a greenhouse effect there, but there was! The greenhouse worked perfectly: the temperature inside was raised up to the point where a small door was opened.

In his experiment, Wood established (and all gardeners should take note) that it is warm in the greenhouse not because of radiations and surface features but because the door is closed and there is no air exchange with the atmosphere. Open the door and the greenhouse effect disappears. Certainly, this conclusion is correct for the greenhouse effect on the Earth, which, as they say, is caused by carbon dioxide. In other words, CO_2 is neither here nor there, and industrial production is not the reason for the rising temperature of the Earth atmosphere.

In the 1970s new data about climate warming of the planet appeared. It's not that scientists did not know about constantly changing average temperatures on the Earth. In fact, they knew about glacial periods (including small ones)—let us recall Brueghel's pictures with skaters on Holland's frozen canals (see Figure 1.6). They also knew had evidence for the cooling and warming in different centuries of human history.

FIGURE 1.6 *Winter Landscape*, painted in 1601 by Peter Brueghel the ELDER, showing skaters on the frozen canals of Holland (Saunders, 2009). The cold winters necessary for the canals to freeze were regular features of this time (the Little Ice Age). Skating on these canals has now become a rare event due to rising global temperatures. Source*: Figure courtesy of the Kunsthistorisches Museum, Vienna, Austria. Erich Lessing/Art Resource. (Brueghel's pictures:* http://www.ibiblio.org/wm/paint/auth/bruegel/*)*

The Vikings named Greenland a little more than 1000 years ago. They were shocked by the climate there, where oaks, ash trees, and grapes grew and a lot of birds lived. This was not millions of years ago, but fairly recently.

Roger Ravellel, a professor at Harvard University, on comparing diagrams of temperature behavior and content of carbon dioxide in the atmosphere found a very well-expressed correlation: a raise in temperature was proportional to a rise in CO_2 quantity (Revellel and Suess, 1957). Yes, during the last half-century, an evident raise of temperature under observation and CO_2 was really discovered.

The authors of this discovery found that carbon dioxide is formed first of all from burning any organic material, for example, fuel. So the more carbon dioxide that appears, the more power plants exist in the country, and the better industry is developed. That is why if we do not want the ice in the Arctic and Antarctica to melt, and cities such as Amsterdam, New York, or St Petersburg to go down under water, it is necessary to shorten industrial production or to use other ways of burning fuel that does not emit CO_2 to the atmosphere. This was the way the global-warming theory appeared, and the basic methods mapped out to fight it.

Many scientists think that the atmospheric temperature is not raised because of CO_2 emission, but vice versa, that CO_2 quantity in the atmosphere is raised because of warming.

There is a sense in this. Sea and ocean water contains CO_2, and gas solubility, as we know, is lowered when the temperature is raised, so carbon dioxide is emitted to the atmosphere. (Try to make an experiment like Robert Wood (Seabrook, 1946) did and warm a glass of gas water; a lot of bubbles will appear.)

Moreover, and importantly, the correlation of the two processes can be aroused by some common reason because of how those first two are changed in relation to each other. But what can be the reason for the climate fluctuation? Figures 1.7 and 1.8 show the behavior of solar activity in the last 500 years and temperatures over the last 1000 years (Khorozov et al., 2006). Does that give one pause for thought? Perhaps this common reason is our Sun?

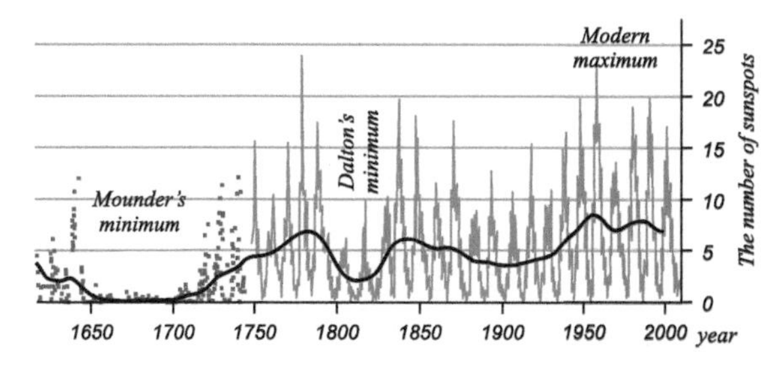

FIGURE 1.7 The behavior of solar activity in the last 500 years. Source: *Khorozov et al., 2006.*

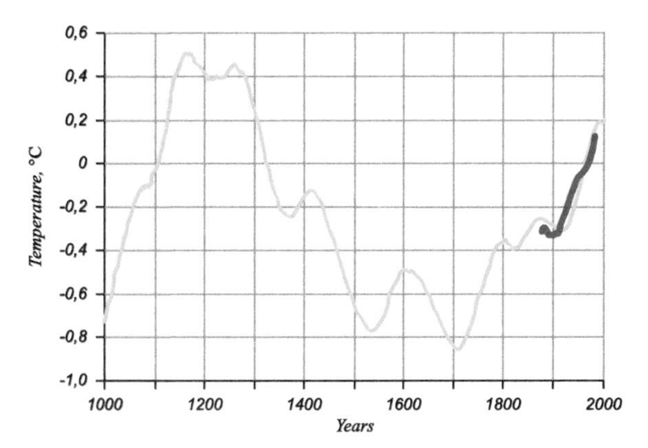

FIGURE 1.8 The behavior of temperature in the last 1000 years. Source*: Khorozov et al., 2006.*

1.2.2. Solar Activity as a Source of Climate Fluctuations

The Sun is a basic source of energy, providing the warmth that the climate depends on (Andreasen, 1993; Denkmayr, 1993; Friis-Christensen and Lassen, 1991; Libin and Pérez-Peraza, 2009). All hypotheses of this second explanation for the climate fluctuations come from the fact that the solar energy flux is considerably changing from year to year. That is why the quantity of warmth the Earth gets from the Sun is also changing. Figure 1.9 shows the general scheme of the influence of the Sun on the Earth.

Why does the Sun change energy? We know that solar processes have a certain periodicity, and the duration of such periods can be hundreds of millions of years. Solar activity is changing periodically with periods of approximately 11, 22, 35, 90, 200, 600, and 2000 years. The quantity of energy

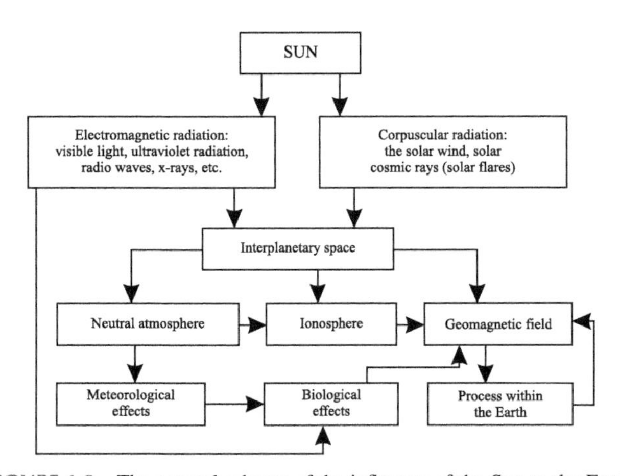

FIGURE 1.9 The general scheme of the influence of the Sun to the Earth.

that the Sun transports as solar charged particles into the circumsolar space depends on the solar activity level.

It was established earlier (Pérez-Peraza et al., 1995, 1999; Libin and Pérez-Peraza, 2009) that climatic processes, such as glaciers, warming, typhoons, earthquakes, and precipitation, are also changed periodically with periods of 11, 22 to 35, 80 to 90, and 200 to 280 years.

The degree of the Arctic and Antarctic ice coverage, variations of ocean levels, the Gulf Stream pulsation, and the sea thermal regime are connected with 11-year (see Figure 1.10), 22- to 30-year, and 150-year cycles. Peak values of earthquake energy correspond with solar activity maximums. The 11-year cycle is also observed in the movement of the Earth poles and in the increasing and decreasing of the Earth's rate of rotation.

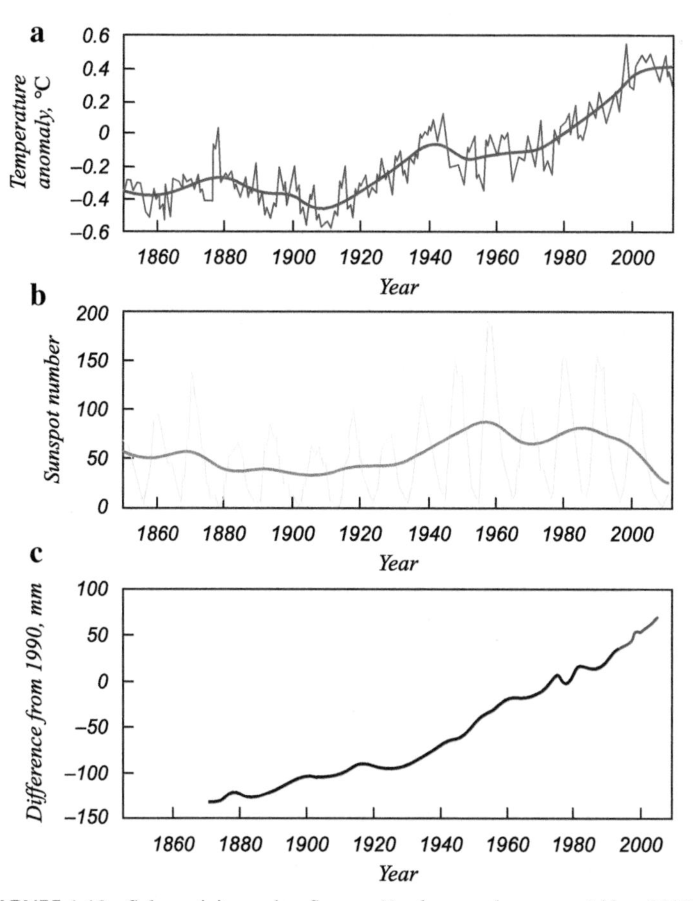

FIGURE 1.10 Solar activity cycles. Source: *Numbers on the curves, Ishkov, 2007.*

Not long ago, canals for filling the Caspian Sea (which became catastrophically shallow about four years ago) were going to be built. But now we observe a threatening rise of the Caspian Sea level and floods of coastal cities that do not have any canals.

In times when the Sun is changing the intensity of terrestrial magnetism, there are also changes in frequency of auroras, radioactivity of the air, the amount of ozone in an ozone layer, cyclical drought and flood, water level in lakes, and the pulsation of the Earth's crust. Questions about exact influence of solar activity on hydrological and climatological processes has often been raised in literature (Dorman et al., 1987a,b; Libin and Jaani, 1987, 1989, 1990; Libin et al., 1987).

An interesting theory about concrete influence of solar activity on Earth processes was suggested by Kondratyev, K.Ya., and Nikolsky, G.A. (2006). Such results were developed also in, Nikolsky (2011) and in: Nikolsky G.A. THE SUN, OCEAN AND CLIMATE. *http://vd2-777.narod.ru/files/SUN_AND_CLIMATE.pdf.* According to these authors, changeability of the solar constant is several tenths of a percent.

At the same time the solar radiation coming to the troposphere can be changed in connection with solar activity variations with several percent amplitude—the result of the influence of a stratosphere modulation mechanism. Figure 1.11 shows the location of solar and galactic cosmic rays in the Sun–Earth system.

Nikolsky was the first scientist who showed the role of galactic cosmic radiation in changing the atmospheric state and the contribution of changes of the solar constant to changes of the atmosphere and biosphere as observed on Earth. It was discovered later from the research carried out in the Danish national cosmic center that cosmic rays influence the Earth's climate fluctuations much more than specialists thought earlier (Svensmark, 2007, 2008).

In November 2008, Svensmark published the first experimental results he had got from his five years of research on the influence of cosmic rays on cloud covers (Svensmark, 2008; Svensmark et al., 2009). A full report on this work was written by Svensmark and Calder (2007).

Changes in the amount of cosmic rays in the atmosphere directly influence the increase of cloud cover of our planet. If there are a lot of clouds the Earth reflects solar radiation back to the cosmos, and in turn, the planet gets cooler.

Svensmark and Friis-Christensen (1997), Svensmark (2008), and Svensmark H. and Nigel Calder (2008) support that the Earth is today surviving a period when cloud cover is less because of a decrease in cosmic rays. They argue that this is the basic reason for today's global warming and that carbon dioxide emissions connected with anthropogenic activity influence the climate much less than scientists have thought. These results were published a week after the report of the intergovernmental expert group UNO (2007) about climate

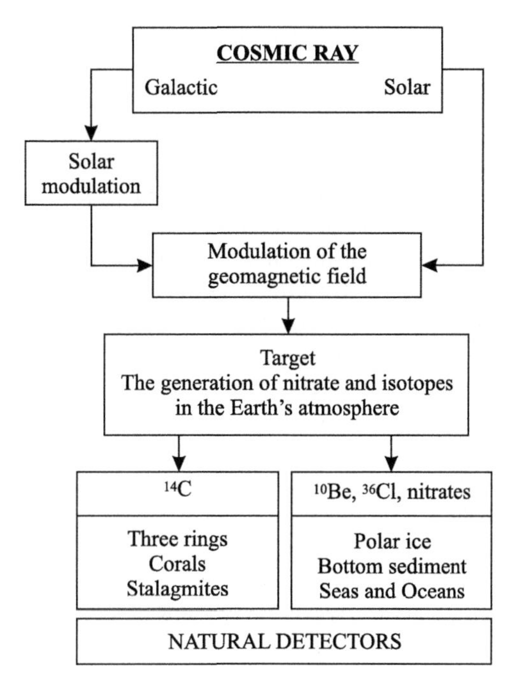

FIGURE 1.11 Cosmic rays in the Sun–climate system.

fluctuations, which concluded that carbon dioxide emissions connected with anthropogenic activity will lead to a temperature increase of 4.5 ° by the end of the century (Climate Change 2007, UNO 2007).

Svensmark (2007, 2008) wrote:

For a long time we have thought that climate fluctuation influences cloud appearance, but now we see that clouds are the dominant factor of the climate. That was not taken into consideration when creating models which help to determine the effect aroused by carbon dioxide. It can happen that CO_2 arouses less warming than we had thought. These results demonstrate that on coming to the atmosphere, cosmic rays form charged particles which attract molecules of water from the atmosphere and form the clouds.

Results of research on ice samples show that temperature and CO_2 changes had happened earlier in the past (Kocharov and Ogurtsov, 1999). In the near future a group consisting of 60 scientists from all over the world will commence a detailed experiment with the use of the Geneva particle accelerator to reproduce by the effect of incoming cosmic rays into the atmosphere. They hope that to show that cosmic radiation really influences cloud cover changes. If it is so, climatologists will possibly have to revise their conceptions of global warming mechanisms.

Incidentally, Danish scientists' results are fully confirmed by research of temperature and solar activity having being made by Russian scientists.

Regular measures of temperature of the upper atmosphere have been taken since 1860.

To understand and reveal the real reasons of global warming and the possible connection of this phenomenon with the Sun, it is necessary to have data from a longer period of time. This can be done successfully today by studying the annual rings of trees. In the process of growing, a tree is constantly affected by changes in physical and chemical state of the environment. Width of rings is a sensitive indicator of environmental changes such as temperature and moisture. By cross dating of two separate oak chronologies, Russian scientists formed the European chronological scale for North Ireland and Germany, the duration of which is 7272 years.

Ring chronologies made from European oak and pine-tree carbon fossils cover the last 11,000 thousand years, practically without any breaks. However, researchers thought this period too short for making a serious conclusion.

Besides tree rings, scientists have used corals and stalagmites in their research. These natural formations contain the isotope oxygen-18 which helps to make a conclusion about the quantity of precipitation during a period.

Glaciers can also give a lot of useful information. Their age in Greenland and Antarctica reaches sometimes hundreds of thousands of years. With the help of ice chemical structure, scientists can define cyclical climate fluctuations since the moment the icy crystals appear.

Moreover, science is capable of making out cosmogeneous isotopes in ice mass: beryllium-10, a component that is a good indicator of the number of cosmic rays coming to Earth in different epochs. The beryllium-10 mean half-life period is 1.5 million years (Dergachev, 2006, 2009; Dergachev and Dmitriev, 2005; Klimenko, 2007a, 2009a, 2009b; Raspopov et al., 2006).

It took years of hard work to gather and analyze data got from different sources. But the time to make first conclusions has come. An oxygen-18 curve which tells us if precipitation amount coincides with curves got with the help of beryllium-10. The obtained positive results indicate that there is a certain connection between cosmogeneous isotope (cosmic ray) and the climate.

It became clear, after such research, that cosmic rays coming to the atmosphere form the cosmogeneous isotopes state, mentioned above, which ionize particles weighed in the atmosphere and form a core of condensation on which drops forming thick low clouds appear. On that basis scientists could prove a direct connection between the cosmic ray flux and the area covered with clouds. In turn these clouds prevent warming of the planet surface by the Sun. Maybe because of such constant atmospheric state, the Earth gets cold up to the degree of global cooling. This new theory supporting that "cosmic rays are responsible for the process of global warming" is well accepted in several scientific environments (e.g, Stozhkov et al., 2008).

Although energy of the cosmic rays flux falling on our atmosphere is approximately 108 times less than the solar magnetic radiation flux that warms the Earth, cosmic rays are the main source of ionizing the air. They provide work of the global electric chain, formation of thunderstorm electricity, and lightning. Weakening of thunderstorm activity is a good indicator that the flux of galactic cosmic rays coming on the ground from the interstellar space is decreasing.

In the opinion of Pudovkin and Raspopov (1992a, 1992b) the amount of cosmic rays depends on changes of the Earth's magnetic field, which serves as a basic defense from cosmos influence. We know that the Earth's magnetic field is constantly changing. There is an effect that makes magnetic poles on a timeline of 2000 years bend aside by 10 degrees of latitude and longitude. A full overturn of a magnetic field can occur on a scale of approximately a million years, and then poles change their places and the intensity of the Earth's magnetic field in this period (thousands of years) becomes several times less. As a result, a screening action of the field for cosmic rays disappears and their flux comes from the interplanetary space to the Earth's atmosphere.

Pavlov (2010) wrote:

Now in laboratory conditions we model the situation when a cosmic ray flux in orbit round the Earth becomes several times higher. Preliminary results show that global disasters which have already shocked the Earth are connected only with activity of cosmic rays.

In ionizing the atmosphere, cosmic rays influence the formation of active molecules as nitric oxides (which fully confirms the conclusion of the Danish group). This is reflected in the stratosphere where the basic ozone layer is found. Nitric oxides destroy ozone (but they themselves are not consumed) and results in different unpleasant effects, for example, in slowing down the process of photosynthesis, a process that gives energy to all living organisms.

Solar flares play an important role on the Earth both long-term (tens and hundreds of years) and short-term; in Canada, in 1989, a powerful solar flare disabled an electric system of a whole region, and Quebec was left without electricity for 10 hours. According to specialists, that solar "prank" cost billions of dollars. Obridko (2011) wrote about the possibility that Moscow could be without energy for 24 hours. A solar flare breaks short distance connections, such as short radio waves, and leads to failure of radiolocation systems.

A prognosis given by scientists from IZMIRAN (the Institute of Earth magnetism, ionosphere and radiowaves propagation in Troitsk, Moscow) fixed an approximate minimum of solar activity in 2007. However, after a short break (2008 and maybe partially 2009) there was a new rising, and a new flash of solar activity by 2010, reaching approximately the same level as in 2000 (coinciding with the destruction of the Russian submarine Kursk and the fire on the Ostankino TV tower). Before that, in 1990 and 1991 similar levels coincided with putsch, the breakdown of the U.S.S.R., and the appearance of hotspots. Also, 1917 and 1937 were years of maximal solar activity.

In fact, in March 2012 an abrupt increase of solar activity took place, where two major flares ejected Coronal Mass Ejections (ionized magnetized plasma) causing geomagnetic storms at their arrival on earth.

After 2011, solar activity is expected to be very high for several years. In fact, in March 2012 an abrupt increase of solar activity took place, where two major flares ejected Coronal Mass Ejections (ionized magnetized plasma) causing geomagnetic storms at their arrival on earth. Does it mean that we should expect certain consequences on the Earth from climatic surprises, mass hysteria, or in increase in death rate from a number of accidents and disasters?

The question of a possible connection of seasonal and many-year changes of the Earth's atmospheric state with heliophysical and cosmophysical phenomena has often been discussed in the scientific literature. Nowadays, the fact that the Sun is the reason for different atmosphere alterations is beyond dispute (Libin and Pérez-Peraza, 2009; Bashkirtsev and Mashnich, 2004a,b, 2008; Mashnich, 2004a,b, 2007). The atmosphere circulation is subject to cyclical influence of changing solar activity which controls the state of geomagnetic activity (Levitin, 2006; Kovalenko, 1983; Ptitsina et al., 1998) and timely variations of galactic and solar cosmic rays (Dorman, 1982).

We may expect complex interaction between all phenomena mentioned above. Parameters of each of them have their own spectrum of seasonal and many-year periodicities (Libin and Pérez-Peraza, 2009; Chertkov 1985), because in spite of common possible mechanisms, a part of variations under observation will be characterized by cause-and-effect relations.

The fact that such connections exist follows from a simple qualitative comparison of timely variations of average monthly and average yearly Wolf number values (W), solar sunspot areas (S), geomagnetic activity (Kp-index), cosmic ray intensity (ICR), as well as storm level index (P) describing frequency of dangerous winds (with the speed $\geq$12 m/s) in the North Sea (Mikalaunas, 1973a,b, 1985; Vitinsky, 1973; Glokova, 1952).

The atmosphere variations as well as modulation of the flux of cosmic rays observed on the Earth seem to be connected by the same processes in the interplanetary space: powerful interplanetary waves, solar flares, high-speed solar wind streams, and a sector structure of the interplanetary magnetic field (IMF) (Akasofu and Chapman, 1975; Belov et al. 2005a,b.; Charakhchian, 1979; Chirkov, 1978; Chistiakov, 2000; Dorman, 1975).

One of the most remarkable peculiarities of the Sun is its almost periodical and regular configurations of the different ways it displays solar activity, as well as the observed changing phenomena (fast or slow) on the sun. These are sunspots—areas with a powerful magnetic field that results in lowered temperature and solar flares—which are massive and quickly developing explosive processes that touch the whole solar atmosphere over an active zone. They include solar filaments—plasmic formations in the solar atmosphere magnetic field that resemble drawn fiber-like structures and are up to hundreds of thousands of kilometers long. When filaments come to the visible border (limb) of the Sun it is possible to see protuberances, which are much more active yet calm formations of various forms in a complex structure.

Note also, coronal holes—areas in the solar atmosphere with a magnetic field opened to the interplanetary space. These are windows where high-speed bunches of solar-charged particles are thrown from (see Figure 1.12).

High-speed corpuscular plasma fluxes change the structure of the solar corona. When the Earth gets in such a flux, its magnetosphere deforms and a magnetic storm appears. Ionizing radiance has a large influence on upper atmosphere conditions and creates disturbs in the ionosphere. Influence on many other physical phenomena often occur.

An active zone on the Sun (AZ) is where a totality of changing structural formations in some limited area of the solar atmosphere occurs in connection with the intensification of a magnetic field in it from 10 to 20 to several (4−5) thousand times the field background. The most remarkable structural formation of an active zone in the visible light of the solar spectrum is the dark chiseled sunspots that often appear in groups. Among the many small spots are usually two large ones which form a bipolar group of spots with opposite polarity of a magnetic field in them.

Sunspots are some of the most famous phenomena on the Sun. Written sources tell us that it was Galileo Galilei who firstly noticed through a telescope and described sunspots in 1610. We do not know when and how he learned to weaken a bright sunlight, but beautiful engravings representing sunspots and published in his famous letters about sunspots in 1613 became the first systematic observations. Even earlier than that, Maya and Aztecs may have been the first to pay attention to sunspots. Deciphering symbols on a roof of a sarcophagus from Palenque and developing the theory of these symbols,

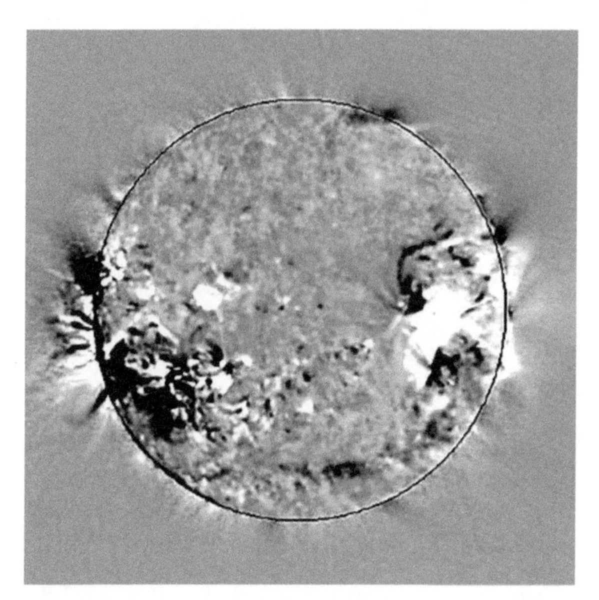

FIGURE 1.12 Sunspots and solar flares. *Source:* www.izmiran.ru/projects/space/INTERHELIOPROBE/publist

M. Cotterell discovered the connection between Maya and Aztecs' calendars (Figs. 1.13 and 1.14) and the formation of sunspots (Gilbert and Cotterell, 2000; Gilbert and Royzman, 2000).

This representation is the most famous in the world because of the name "astronaut" and similarity of the Maya King Pakal rising from the dead to the cosmonaut in the spacecraft [http://en.wikipedia.org/wiki/File:Pacal_the_Great_tomb_lid.svg] (Fig. 1.15). King Pacal himself is represented in the center, and there is a tree symbolizing the Milky Way and a mysterious bird

FIGURE 1.13 Aztec calendar.

FIGURE 1.14 Mayan calendar. Source: *I. Libin.*

FIGURE 1.15 Image of "astronaut" on the burial cover Pakal, Palenque. Source: *Wikimedia.*

pointing the way to the other world of the Cosmos near it. The tsar's headwear looks like a cosmonaut's spacesuit and surrounding things are like a control console of a spacecraft (Libin and Pérez-Peraza, 2009).

It is interesting that sunspots were seen by the Maya and Aztecs and, apparently, not again until the seventeenth century. Furthermore, at the end of the nineteenth century two observers, G. Sperer in Germany and E. Maunder in England, noted that in the 70-year period up to 1716 there were not enough spots on the solar disk (Pustilnik and Yom Din, 2004). Later, Eddy (1984) analyzed all the data and concluded that there was a solar activity decay in that period, known as the Maunder's minimum.

Wolf numbers were coined by R. Wolf, director of the Zurich observatory, who studied in detail early data of observing sunspots and organized a systematic registration of them. He coined a special index to characterize the solar spot formation activity proportional to the sum $f + 10g$, where f is the number of all separate spots noticed on the disc of the Sun and g is the number of groups formed by them. Later this index began to be called the Wolf relative number $W = k\,(f + 10g)$.

Wolf's "Zurich system" was generally accepted. It appeared that maximum and minimum interchange of the Wolf figure row is not strictly periodical but cyclic in timely intervals varying from 8 to 15 years. Cyclicity of solar activity (SA) was firstly discovered by Henrich Schwabe, a chemist from Dessau, Germany who had regularly observed the Sun while researching for an unknown planet, and registered the number of sunspots he noticed over a period of 43 years, since 1826 (see http://ossfoundation.us/ *projects/environment/global-warming/solar*; Libin and Pérez-Peraza, 2009). Having been persuaded that this number was changing periodically, Schwabe made the first announcement in 1843 and published in 1844. In 1851 Humboldt published his data in *Cosmos* and drew scientists' attention to Schwabe's discovery. There are daily data since 1749, but before that are only separate observations made by chance.

Today, in addition to Schwabe and Wolf's 11-year cycle of solar activity discovered in the middle of the nineteenth century, other cycles have been researched and discovered: Gansky's cycle of 72 years (two 36-year ones) was discovered at the end of the nineteenth century and Rubashev's cycle of more than 600 years was discovered in the middle of the twentieth century. Many studies have observed modulation of the 100-year and 11-year cycle of the solar activity and climatic parameters, relating to temperature (Figure 1.16).

Scientists of Kislovodsk observatory (Dr. Gnevyshev and Dr. Ol) have argued that the well known Hale Cycle (connected with the change of sunspot polarity every 22 years) may be described as a pair of 11 years cycles. They even determined for the decade of the 1996's that the second cycle of such a pair is 1.4 times higher than in the previous one (Ol', 1973).

Almost all indexes of solar activity indicate changes that repeat every 11 years on average, and because of that they are called the 11-year solar cyclicity.

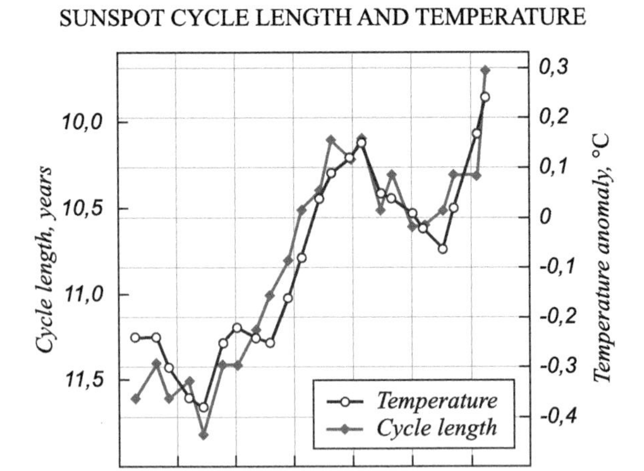

FIGURE 1.16 Sunspot cycle length and temperature. *Source: Friis-Christensen and Lassen, 1992.*

A lot of processes occurring on the Earth are connected with them to a greater or lesser degree.

The following are indexes of helio- and geophysical activity:

- Wolf relative numbers, W
- Sunspot areas, S
- Area of calcium floccules, K
- Radio radiation flux, 10.7 sm (2800 MHz) F10.7
- Index of flares, 1,2m...
- Irradiance (solar constant), Q
- K-index (the observatory geomagnetic field vector variation averaged over the thee directions and 3-hour intervals)
- Kp planetary K-index, averaged to 12 observatories; variants ap, Ap, and others are also used

Basic peculiarities of the 11-year cyclicity include the following:

- Cycle duration is from 7 to 17 years (Figure 1.17).
- Growth phase is from 2 to 5 years, and decay phase from 5 to 12 years.
- Amplitudes of successive cycles are changed gradually from values W ~50 (low) to W ~200 (high cycles).
- Succession of magnetic polarity of basic spots in groups is kept during the cycle, but the polarity is opposite in both hemispheres. In the next cycle the polarity is changed to opposite.
- Spot formation zone during the cycle moves from middle latitudes (30–35°) to 5° at the end of the cycle (Sperer's law; see *http://sunphys.ru/sun/sun-exploring-history*).

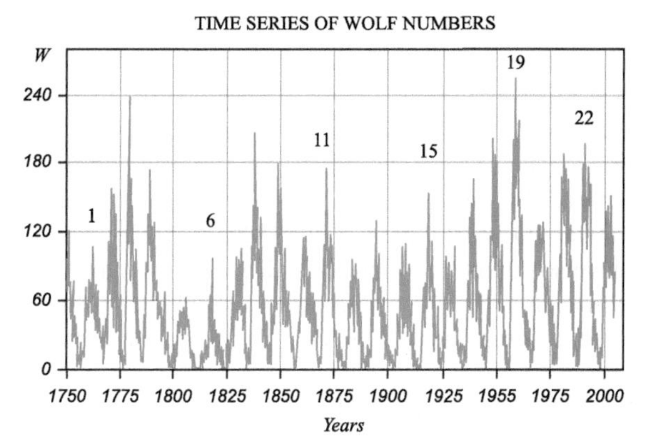

FIGURE 1.17 Cyclicity of solar activity for cycles 1–23. *Source: http://sunphys.ru/sun/sun-exploring-history.*

The main peculiarity of a solar activity cycle is the law of spot magnetic polarity changing: during each 11-year cycle all leader sunspots of bipolar groups have the same polarity in the Northern Hemisphere and opposite in the Southern Hemisphere. The same is true for trailer sunspots, the polarity of which is always opposite to the leader sunspot polarity. In the next cycle, polarity of leader and trailer sunspots is changed to the opposite. Together with polarity of a general magnetic field of the Sun, the poles that are located near rotation poles are changed. That is why it is more correct to say about 22-year solar activity cycle, but not 11-year one. This cycle was discover by Hale et al. (1919) and as we mentioned before it is known as the Hale cycle.

1.3. SOLAR-TERRESTRIAL PHYSICS AND FUNDAMENTAL COSMIC RESEARCH

The Sun and solar-terrestrial relations are studied with the help of terrestrial and cosmic watch facilities. The most important results have been achieved within the last few years due to cosmic research. Most types of observations are possible only from the cosmos, and this explains an indispensable role of cosmic research in solar-terrestrial physics (Kuznetsov and Oraevsky, 2002a,b; Krainev and Kalinin, 2011, Krainev and Webber, 2005a; Kuznetsov and Zeleny, 2008).

Growing understanding of the influence of cosmic weather factors on the climate, the Earth's environment, and different spheres of human activity leads to the determination of the practical value of research in this sphere. More discussion of many fundamental scientific problems of solar and stellar physics and plasma astrophysics with the help of observing the Sun is expected.

Solar influence resources on cosmic weather is composed of several fundamental processes. These often disturb calmness on the circumterrestrial cosmic space what is translated on the Earth. About 37,000 flares happen during an 11-year cycle of solar activity (according to data for 22-year cycle of solar activity from 1986 to 1996). In the maximum of a solar cycle, on average 1 flare happens every 1 to 2 hours; at the very minimum 1 to 2 flares occur per a day. Other powerful manifestations of solar activity include coronal mass emissions, which occur on average 5 to 10 times per day in the maximum of a cycle. A fraction of them spread to the direction of the Earth and trigger geomagnetic storms (Oraevsky et al., 1998, 2002a, 2002b; Krainev and Webber, 2005b; Kuznetsov and Oraevsky, 2002a,b). Moreover, about 500 magnetic storms happen in each solar cycle, and these can influence people's health and lead to dangerous, sometimes disastrous, influence on different technical systems on the Earth.

At IZMIRAN there is a geophysical forecasting center which observes solar activity state based on terrestrial and cosmic data and provides interested organizations and offices with information about the Earth's magnetic field state and magnetic storms. Observations made there in cooperation with medical institutions showed that during magnetic storms the adrenalin rate in human organisms is increased and bloodstream nature is changed in microcapillaries such that bloodstream becomes interrupted, which leads to changes in pulsation and blood pressure. As adrenalin discharge happens also during stress, magnetic storms can be regarded as a stress influence on the human organism, the consequences of which are well known, and they are especially noticeable and essential for people with cardiovascular diseases. In these periods of perturbations the numbers of hospitalizations of people with cardiovascular diseases are raised, another indication of the a high degree of correlation with magnetic storms.

Observing the Sun from the cosmos has great value for fundamental astrophysics and practical purposes, and leads us to attempt to answer questions that have been considered controversial for many decades. These include questions concerning the following:

- The origin of mass coronal emissions and solar flares
- The warming of a solar corona and fastening of the solar wind
- The mechanisms of a solar cycle
- The global structure of the heliosphere and perturbation coming from the Sun
- The forecasting of solar phenomena geoeffectiveness

Active and workable cosmic projects from all around the world are being used to solve these and many other problems. As an example, we will look at on one of the Russian cosmic programs.

1.3.1. CORONAS Program and CORONAS-F Project

For researching the Sun and solar-terrestrial relations on different phases of the 11-year solar cycle, IZMIRAN worked out and put into practice an international program CORONAS (COCOSA: Complex Orbital Circumterrestrial Observations of Solar Activity). Within this program the first satellite CORONAS-I (launched in 1994) observed the Sun near the minimum of its activity. The second satellite CORONAS–F (see Figure 1.18), launched in July 2001 was addressed to study solar activity in the previous solar cycle (23d). The maximum of the 23d solar activity cycle was achieved in April 2000, and a high level of activity was kept over 2 to 3 years. The maximum of the current 24th solar cycle is expected in 2013 (Oraevsky et al., 2002a; Kuznetsov and Oraevsky, 2002a,b; Kuznetsov and Zeleny, 2008).

The orbit of the satellite CORONAS -F (orbit inclination is 82.49°, minimal removal from the Earth's surface is 500.9 km, maximal removal from the Earth's surface is 548.5 km, period of satellite is 94.859 min) provides periodical changing periods of continuous observations of the Sun with duration of

FIGURE 1.18 CORONAS-F satellite in orbit. Source*: Oraevsky et al., 2002a,b.*

about 20 days, which is very important for patrolling solar phenomena and flares for registration of solar global variations.

The basic scientific problems of the project CORONAS-F are observations of solar global variations and on the basis of these, studying the depth and inner structure of the Sun. Questions of complex researches of active zones, flares, plasma emissions in a wide range of wavelengths, studies of solar cosmic rays accelerated during active phenomena on the Sun, conditions for their occurrence, spreading in the interplanetary magnetic field, and influence on the Earth magnetosphere are some of the goals to be solved.

A large number of scientists, engineers, and specialists of a wide cooperation of Russian, Ukrainian, and foreign organizations and agencies under the IZMIRAN participate in realization of scientific experiments of the project CORONAS-F, including PIAS (Physical Institute of AS), PTI (Physical and Technical Institute), IAG (Institute of Applied Geophysics), RINP MSU (Research Institute of Nuclear Physics of the Moscow State University), ISR PAS (Institute of Space Research of the RAS), and MEPI (Moscow Engineering and Physical Institute).

1.3.2. Global Variations of the Sun and Its Inner Structure

Observed solar activity is a reflection of processes occurring in its depth. Thermonuclear reactions occur in the Sun's core, a discharged energy of them are brought to the outer layers giving rise to the complex structure and dynamics of these layers: a convective zone, photosphere, chromospheres, corona, and solar wind. So, studying the inner structure of the Sun creates an understanding of the solar activity nature. Characteristics of the inner layers of the Sun such as density, temperature, their depth distribution, dependence of angular velocity on a radius and width, depth of a convective zone, and so on, are important.

Helioseismology is the science of the inner structure of the Sun. It is one of the most workable modern methods of studying the inner structure of the Sun, and deals with the natural oscillations of the Sun. It began in the 1960s when 5-minute oscillations covering the whole surface of the Sun were discovered (with periods of 3–10 of minutes). With the help of helioseismology it is possible to research the solar structure from the convective zone down to the core. In the spectrum of natural oscillations there is information about temperature, pressure, magnetic fields, and rotational velocity and their dependence on the depth.

1.3.3. Cosmic (extra-atmospheric) Helioseismology

Terrestrial observations of global oscillations meet a variety of difficulties. A wish to achieve maximally possible spatial and temporal resolution of the global oscillation spectrum requires carrying out observations regularly over a period of at least two weeks because frequency resolution is inversely proportional to time of observation.

Observing from the Earth's surface it is possible only if there are several points of observation distributed along the terrestrial longitude and which have the same equipment, or if observations are done from polar spheres. Even in these cases, getting long rows of data depends on good weather. Besides, instability of the Earth's atmosphere and presence of natural oscillations lower the signal-to-noise correlation, and observations in some parts of the spectrum, in ultraviolet, for example, become impossible because of strong absorption of solar radiation by the atmosphere.

That is why it is better to observe global oscillations of the Sun from the cosmos. And solar-synchronic orbit of the satellite CORONAS-F providing observations without any breaks for 20 days allows us to watch the dynamics of different modes of global oscillations, namely growth, saturation, and amplitude decreasing phases, which have characteristic times from several hours to several days.

Data of solar radiation intensity over a wide spectrum obtained on the spacecraft allows us to see the variation nature of the solar constant and to collect information about sunspots, faculae, chromospheric network, and other expressions of solar activity to these variations. It also enables study of the dependence on the 11-year cycle of solar activity of the parameters of global oscillations under study.

1.3.3.1. Solar Flares

Solar flares are the most powerful expression of solar activity. When flares occur, accelerated particles and perturbation of the solar wind reach the Earth; they influence the magnetosphere and atmosphere, raise radiation hazard in the circumterrestrial cosmic space, and lead to many effects that have become the subject of cosmic weather research. A complex of the scientific equipment consisting of three devices makes complex research of solar cosmic rays (SCR) and their manifestations in the circumterrestrial cosmic space.

1.3.3.2. Representing X-ray Spectroscopy of the Sun

Getting images of the Sun representing the most characteristic traits of its face has become the main purpose of modern solar cosmic projects. However, a new direction in solar astrophysics has been realized on the CORONAS-F satellite in the network of the experiment SPIRIT (FIAN). It performs X-ray spectroscopy of the Sun which permits the reestablishment of three-dimensional structure based on monochromatic pictures of the Sun and research dynamics of plasma formations of solar atmosphere over a wide range of temperatures existing on the Sun (from 5×10^4 °K to 5×10^7 °K). More than 200 Sun images are registered every day.

The project CORONAS-F is a constituent part of a perspective program of solar research worked out by IZMIRAN under the aegis of Rosaviacosmos and RAS in cooperation with other Russian and foreign organizations. This program includes the project Interhelioprobe for researching the Sun from close distances and the project Polar-Ecliptic Patrol for global vision of the Sun and controlling the cosmic weather.

1.3.3.3. Project Interhelioprobe

In the project Interhelioprobe a spacecraft located on the heliocentric orbit performs many gravity-assist maneuvers near the Venus and come close to the Sun at the account of attraction of the Venus along a torsile trajectory. As a result, a spacecraft will hover above a certain area of the Sun's surface over the course of a week, and it will establish important direct correlations for solar-terrestrial relations of phenomena on the Sun and in the interplanetary space.

Circling around the Sun for approximately a third of the year, the Interhelioprobe will take different positions relative to the Sun–Earth line, locating and observing the Sun from the side of that line and the reverse side invisible from the Earth. The project answers the basic questions of solar-terrestrial physics and astrophysics concerning the solar corona warming mechanisms, origin and acceleration of the solar wind, and origin of the most powerful expressions of solar activity—solar flares and coronal mass emissions.

1.3.3.4. Project Polar-Ecliptic Patrol

In the framework of the Project Polar-Ecliptic Patrol created by IZMIRAN continuous monitoring of the solar activity and the solar wind is planned, including solar emissions and heliospheric perturbations moving to the direction of the Earth and observing polar areas and the back side of the Sun (Kuznetsov and Oraevsky, 2002a, 2002b).

Two small spacecraft are located on polar (or tilted at an angle of 45 ° to the plane of the ecliptic) heliocentric orbits at the distance 0.5 a.u. so that their orbit planes are mutually transverse to each other, and on the orbits the equipment is allocated for quarter of the period (about 130 days).

Within such an orbital scheme the control for a Sun–Earth line is constantly provided from one of the spacecraft and from both spacecraft during a long time. When one of the spacecraft is located in the plane of the ecliptic, the other one is located above one of the solar poles, and when one of the spacecraft moves away from the plane of the ecliptic, the other one moves closer to it. So, simultaneous monitoring is carried out in ecliptic and in polar areas. It allows the study of regularly low- and high-speed solar wind, 3D images of a solar corona, and solar emissions.

At different times, one of the spacecraft will be located in the other hemisphere of the Earth relative to a Sun–Earth line, and so this spacecraft will observe the reverse side of the Sun invisible from the Earth.

FIGURE 1.19 Neil Armstrong at the meeting of COSPAR, Leningrad, 1971.

In 1966, S.P. Korolyov, a general constructor of soviet cosmos, wrote in the article "Steps to the Future" (see Korolyov, 1980),

There is no other branch in Soviet science which is developing so quickly as cosmic research. A little more than 8 years have passed since a man-made cosmic body—the first Soviet artificial Earth satellite—firstly appeared in the Universe. The history of astronautics is only about three thousand days, but it is so rich in events the most important for the humanity, that it is possible to separate out whole epochs in it.

Nowadays continuous cosmic research of the Sun, solar wind, interplanetary space at the distance of tens a.u. from the Earth, and solar-terrestrial relations have become commonplace. Spacecraft Voyager-1 and Pioneer-10, Cassini and Galileo, Genesis and Soho, CORONAS and Interheliaprobe, SOHO, MKS (International Space Station), and others work in the cosmos.

Incidentally, the first direct results of measuring the solar wind were sent to the Earth from the Moon by the spacecraft Apollo. (The Moon, which has no atmosphere and no magnetic field, has appeared to be a good place for gathering such information.)

On the July 20, 1969, at 20:17 (GMT) the American spacecraft Apollo-11 landed on the Moon's surface. For the first time, man realized his dream and set his foot on another celestial body. Setting foot on the Moon's surface, American cosmonaut Neil Armstrong (Figure 1.19) said, "This small step of a human is a gigantic step of the humanity."

Chapter 2

Solar Activity Influence on Cosmophysical and Geophysical Processes

Chapter Outline

B.M. Kuzhevsky (Kuzhevsky et al., 2002; Kuzhevsky, 2005) writes in his work *A Research Object is the Sun*,

Everything that happens with the Earth as a planet—geological and climatic processes in it, conception of life, evolution of it from a primitive level to Homo sapiens, spiritual and psychical life of a separate human being and the whole nations—is connected to the Sun.

Figure 2.1 shows a scheme of the influence of cosmic processes on Earth.

In some sense it is possible to assert that we live in the Sun's atmosphere. That is why wide research of our heavenly body is very important, especially the study of processes in its atmosphere.

Highlights in Helioclimatology. DOI: 10.1016/B978-0-12-415977-8.00002-2

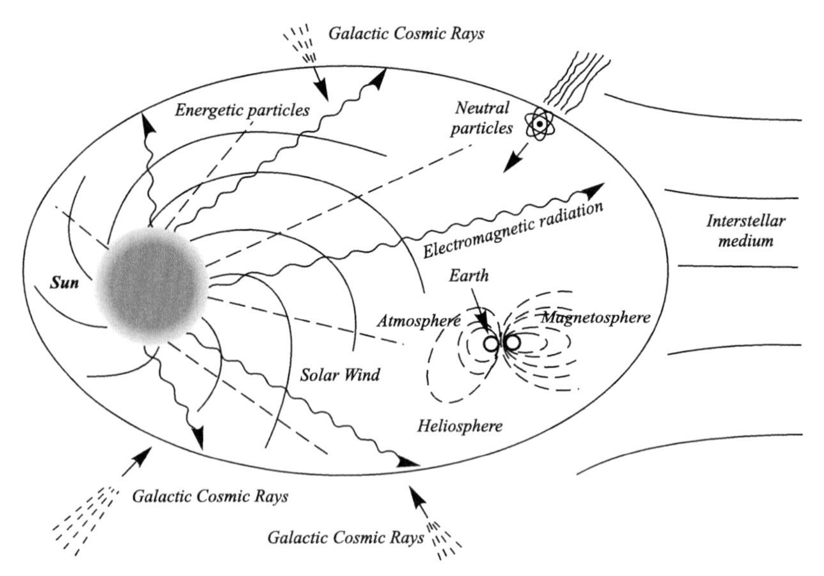

FIGURE 2.1 Scheme of the influence of cosmic processes on Earth. *Source: Thurber et al., 2007; Dorman, 2009.*

Powerful short-time energy releases from spatially-local regions of the atmosphere are translated in the generation and emissions of neutral and charged particles, as well as electromagnetic radiation of all wavelengths, which pierce through the interplanetary space influencing its objects, including the Earth. That is why physical processes taking place in the solar atmosphere are very important for humankind.

2.1. ACTIVE PROCESSES ON THE SUN

From the ancient times people noticed dark spots of different sizes and shapes on the surface of the Sun, which appeared and disappeared. Today we know that sunspots are external manifestations of subsurface active processes on the Sun. Observing them allows us to see different cyclicities in solar activity, and these become apparent during changes of the different parameters that characterize the sunspots.

One more impressive phenomenon, which was discovered more of 100 years ago, is chromospheric flare, a sporadic release of a great deal of energy equal to a synchronic explosion of 100 million 1-kiloton nuclear bombs ($\sim 10^{32}$ ergs) in that part of solar atmosphere, the chromosphere.

Sometimes an unexpected explosion of a small volume of solar plasma can be observed in a well-developed active region. This powerful manifestation of solar activity is called a solar flare. Solar flares appear in the area of the magnetic field polarity change where powerful opposite magnetic poles collide in a small region. As a result, their structure is essentially changed.

A solar flare is usually characterized by a fast rising (up to 10 minutes) and a slow decay (20–100 min.). During a flare, radiation in practically all ranges of the electromagnetic spectrum is raised. A chromospheric (solar) flare spreads far not only the chromosphere, but also covers the lower (solar photosphere) and higher (solar corona) regions.

About a half of common flare energy is removed by powerful emissions of plasma which come through the solar corona and reach the orbit of the Earth as corpuscular fluxes interacting with the Earth's magnetosphere, sometimes leading to auroras and magnetic storms.

Powerful energy is released in flares in different ways, as gas-dynamic movement of solar atmosphere plasma and as electromagnetic radiation over a wide range. High energy particles are also produced, with energies from some tenths of millions of electron-volts (MeV) up to tens of billions of electron-volts (GeV), the so-called *solar cosmic rays* (SCR).

Carrington and Hodgson were the first to observed a flare, independently, in England on September 1, 1859. Observations in those days were made in white light of the electromagnetic spectrum.

The easiest way to observe solar flares is in the hydrogen red line radiated by the chromosphere. In the radio-frequency range, and increase of radio brightness in active areas can be so large that a full radio-wave energy flux moving from the Sun becomes tens or even a thousand times larger. These phenomena are called solar radio bursts.

The role of the magnetic field is the fundamental factor: when the magnetic field (MF) density is increased in some area of the chromosphere or the corona (for example, on account of a merging of a new magnetic flux from a convective zone) the topology of the magnetic field lines distribution is changed.

When old and new emerging magnetic fields approach each other in areas where field lines are of an opposite direction, when they meet neutral points and lines appear where the MF becomes zero. In the vicinity of neutral points and neutral lines, MF fluxes should be reallocated and their common structure should be changed. In the course of these topology changes an instability state is reached producing electric currents, and then fast thermal energy release. Following such MF annihilation, sometimes intense electric fields are produced with the subsequent acceleration of local particles up to SCR energies.

As a whole the process can be globally described as a powerful explosion accompanied by plasma particle acceleration up to high energies and plasma cloud emissions into the interplanetary space with a velocity of thousands of kilometers per second, the so-called *coronal mass ejections* (CME).

Solar Flares are characterized by several indexes, according to different parameters: for instance, the intensity in X-rays radiation, their area, their duration, etc. An illustrative example of one of such flare indexes is shown in Table 2.1

TABLE 2.1 Flare Index

Class	Average Duration (min)	Area in Solar Hemisphere Fractions	New Classification (Area in 1/1 min Arc of Solar Hemisphere Area Fractions)
1–(S)	Subflare	$<10^{-4}$	<100
1 (1)	20	$(1–3)\cdot 10^{-4}$	100–250
2 (2)	33	$(3–8)\cdot 10^{-4}$	250–600
3 (3)	62	$(8–15)\cdot 10^{-4}$	600–1200
3+ (4)	–	$>15\cdot 10^{-4}$	>1200

2.1.1. Electromagnetic and Corpuscular Flare Radiation

During a flare radiation, almost all ranges of the spectrum are raised. In the visible wavelength this increase is relatively small—not more then 1.5 to 2 times during the most powerful flares observed even sometimes in the white light of the background of the bright photosphere. But in the far ultraviolet and X-ray spectrum regions and especially in the radio-region on meter waves this increase is very substantial. Sometimes one can also observe gamma ray bursts.

Proton flares (Figure 2.2) are powerful flares that usually produce high-energy charged particles or protons with tens and hundreds of mega electron-volt energies. They are accompanied by electron fluxes with more than 40 kilo-electron-volt (KeV) energies, and sometimes even relativistic electron fluxes with more than ten mega electron-volt energies (MeV). Also heavy nuclei of some KeV-MeV are accelerated in some of these flares.

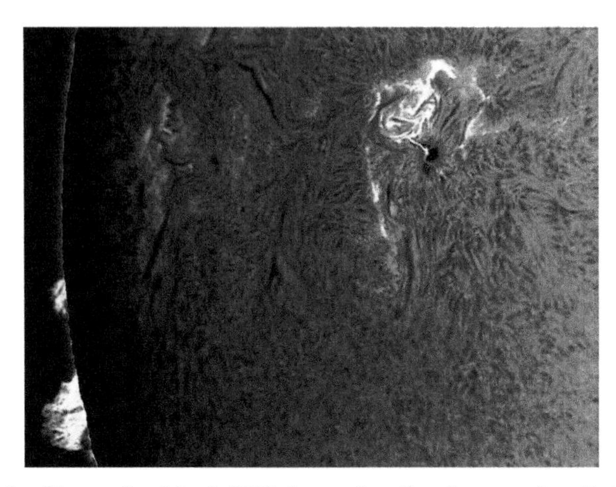

FIGURE 2.2 Flare on Sun May 2, 2006. *Source: http://apod.nasa.gov/apod/ap060502.html*

Particle energy fluxes from proton flares are dangerous for cosmonauts' life and health in the space. They can arouse failures and degradation of onboard computers and other devices. One can see powerful flares even in a white light on the background of a bright photosphere, but it is very rare.

It is clear that the absolute energy particle flux and a SCR energetic spectrum range are connected with flare power. It was discovered that particles of relatively low energies are registered before a SCR flare over the previous 24 hours, on average. This phenomenon is known as preflare increase.

It is possible to use observation of low-energy preflare increase to raise reliability of radiation environment forecasting in the interplanetary space. The first works in this direction gave essential improvement of the forecast; the probability of exact forecasting has become 85 to 90% higher than other methods for forecasting.

Discovering the low-energy particle preflare increase phenomenon has confirmed many researcher's opinion that it's possible to observe processes of particle acceleration in the Sun's atmosphere continuously; only the power of the active processes is changing. That is why the terms *calm* or *active Sun* are relative and do not reflect the real life of the Sun.

Instrumental observations in cosmic condition have allowed a better understanding of the phenomena of solar activity and solar cosmic rays. Solar cosmic rays are fluxes of high-energy-charged particles accelerated in the corona and chromosphere which appear during solar flares. They are registered on the Earth's surface by neutron and meson monitors as sudden and sharp increases of cosmic ray density on the background of the more high-energy particles, and are called *galactic cosmic rays.*

A conditionally accepted lower limit of solar cosmic ray energy is $5 \times 10^4 - 10^6$ eV. If energies are lower, then particle flux possesses qualities of plasma which cannot neglect electromagnetic interaction of particles with each other and the interplanetary magnetic field.

Energy and charge distribution of solar cosmic rays on the Earth is determined by the particle acceleration mechanism in the source (the solar flare), peculiarities of particle yield from the acceleration area, and the conditions of their spreading in the interplanetary space. Particle acceleration is tightly connected with solar flare origin and the development mechanism. (Gallegos-Cruz and Pérez-Peraza, 1995; Pérez-Peraza and Gallegos-Cruz, 1994, 1998).

Leaving their acceleration source in the Sun, solar cosmic ray particles propagate in the interplanetary magnetic field for many hours, and by scattering step-by-step on its uniformities they move to the periphery of the solar system. A part of them penetrate into the Earth's atmosphere arousing additional ionization of atmospheric gases (mostly in the sphere of polar caps).

If it has been possible since 1942 to register and study particles with energy of hundreds of thousands of electron-volt and higher ($\geq$ 500 MeV) based on observations from the Earth's surface (Neutron Monitors), at present with scientific equipment brought to the interplanetary space by spacecrafts and satellites it has become possible to register particles of low energies from the Sun down to some units and tens of thousands of electron-volt (« 500 MeV). Fluxes of such relatively low energy particles from the Sun are registered much more often than high-energy particle fluxes.

2.2. MODERN CONCEPTIONS ABOUT THE SOLAR CORONA STRUCTURE

A solar corona is a key to understanding processes occurring on the Sun and an important precursor and indicator of future events in the heliosphere. Experimental corona research methods consist of observation of separate corona line radiation or parts of its radiation spectrum.

Observations in the 1.5-meter wavelength were done for the first time on May 20, 1947 in Brazil, with a help of equipment fixed on board the Griboyedov ship (by means of a synphased antenna fixed on the ship board). Vitaly L. Ginzburg (a future Nobel prize-winner) and other astronomers and physicists such as I.S. Shklovsky (State Astronomical Institute after P. Sternberg—SAIS) and Y.L. Alpert (IZMIRAN), took part in the expedition (Figure 2.3).

After the Second World War (1939–1945) academic N.D. Papalexi asked V.L. Ginzburg to estimate conditions of the reflection of meter and decimeter range radio waves from the Sun. This problem appeared in connection with N.D. Papalexi's ideas [www.prao.ru/] about the possibility of locating not only the Moon and planets, but also the Sun. At that time Prof. Ginzburg had

FIGURE 2.3 A group of participants of the Soviet expedition ship Griboyedov. *First row*: First on the right is S. Khaikin; fourth from the right is G. Ushakov. *Second row*: Fourth from the left is Vitaly Ginzburg; ninth from the left is B. Chihachev. *Third row*: Second from the right is I. Shklovsky.

a developed a theory of radio wave transmission in plasma, and he came quickly to an uncommon conclusion that radio waves would be absorbed in the corona and chromosphere.

An interesting conclusion followed that the source of solar radio-frequency radiation is not the photosphere, like in the optics, but the upper chromosphere and a solar corona, the temperature of which is about 10^5–$10^{7\circ}$ Kelvin for longer waves of the meter range.

Consequently, V.L. Ginzburg wrote two radio astronomical reviews in the journal *Success in Physical Science* (Ginzburg, 1947, 1948). In these works he discussed a question about radio-wave diffraction on the lunar limb which permits an increase in angular discrimination of details on the Sun during a solar eclipse. These works initiated more work on the theory of synchrotron cosmic radio-frequency radiation and its connection with the cosmic ray origin problem and high-energy astrophysics. According to Prof. Ginzburg, the link between radio astronomy and cosmic rays led to the formation of new directions in astronomy: cosmic ray astrophysics and high-energy astrophysics or gamma- and X-ray astronomy.

The basic method of research of a corona fine structure and corona dynamics are observations of the so-called *white corona*, which make it possible to research coronal processes. Processes in the corona are conventionally called quasi-stationary (for the time >24 hours) and sporadic (for the time <24 hours). In the absence of sporadic processes (or if they are weak) the corona is quasi-stationary.

According to Eselevich (2002a,b,c), research of a quasi-stationary corona in the white light "is first of all studying the brightest constituent part of it—the belt of corona streamers" (Figure 2.4). A cross section of streamers on corona

FIGURE 2.4 Solar corona streamers.

representations is regarded as a highlight of the helmet turning into a pencil beam as it goes away from the Sun.

All preceding research on this subject can be divided into two large periods: before the launch at the end of 1995 of the spacecraft SOHO (Solar and Heliospheric Observatory) with an instrument LASCO (Large Angle Spectrometric Coronagraph) on board, and after launching it. LASCO consists of three parallel coronagraphs with concentric and overlapping fields of view.

When observed in the white light, streamers in the corona are regarded as ray-like structures of highlight reflecting distribution peculiarities of the magnetic field forming them. Because of their global characteristics, the unity of them in the space forms a streamer belt several degrees thick covering the Sun (surface) inside of which a slow solar wind is flowing with a high density plasma several times higher than the surrounding plasma density. A streamer belt in the corona divides regions with opposite polarity of the Sun's radial magnetic field (or magnetic flux tubes of open field lines with opposite polarity coming from neighbor coronal holes). This means that a neutral line of a radial magnetic field goes along the belt, the position of which comes from the estimation of the magnetic field in the corona by a technique known as the *potential approximation.*

2.3. PHYSICAL BASIS OF DISTURBANCE FORECASTING

2.3.1. Solar Wind (Parker, 1965, 1982)

Due to the solar wind, the Sun loses about one million tons of matter per second. The solar wind consists of electrons, protons, and helium nuclei (alpha-particles); nuclei of other elements; and non-ionized particles (electrically neutral) in a very small quantity.

Although the solar wind comes from the outer layer of the Sun, it does not reflect real element composition in this layer; as result of *differentiation processes* the contents of some elements are increased and some are decreased (the so called FIP-effect).

Intensity of the solar wind depends on solar activity fluctuations and their sources. Depending on the velocity, solar wind fluxes are divided into two classes: slow (about 300–400 km/s at the level of the Earth orbit) and fast (about 600–700 km/s about the Earth orbit). There are also sporadic high-speed (up to 1200 km/s) short-time fluxes.

A slow solar wind is born in the quiescent regions of the solar corona during its gas-dynamic expansion; when the temperature is about 2×10^6 K, the corona cannot be in hydrostatic balance conditions and this expansion should lead to coronal substance acceleration up to supersonic speed in boundary conditions.

Warming of the solar corona up to such a temperature is the result of the convective nature of heat transport in the photosphere of the Sun. Development of convective turbulence in plasma is carried out by intensive magneto sonic wave generation, and in their turn sonic waves are transformed to shock ones when distributing to a direction of decreasing density of solar atmosphere, and shock waves are absorbed effectively by corona matter and warmed to a temperature of 1 to 3×10^6 K.

Though parameters of the solar wind evolve with time during different periods of the solar cycle, mainly during CME, and may have drastic changes due to interplanetary disturbances, however an average picture their values is illustrated in Table 2.2.

Fluxes of a recurrent fast solar wind are emitted by the Sun during several months and have a frequency period of 27 days if it is observed from the Earth (the Sun rotation period). These fluxes are associated with *coronal holes*—corona areas with relatively low temperature (about $0{,}8 \times 10^6$ °K), low plasma density (one forth density of the surrounding corona), and radial magnetic field relative to the Sun.

Moving in space full of plasma, slow solar wind sporadic fluxes compact plasma behind its front, forming a shock wave that moves together with it. It was suggested at first that such fluxes were induced by solar flares, but nowadays it is regarded that sporadic high-speed fluxes in the solar wind are conditioned by coronal mass emissions. It is also necessary to note that solar flares and coronal mass emissions are connected with the same active regions on the Sun and there is statistic dependence between them.

The solar wind defines the heliosphere as the cavity where interstellar and solar wind magnetic fields meet, and due to that it prevents the interstellar gas penetrating into the solar system. Besides, the solar wind gives birth to such phenomena as auroras and planet radiation belts on the planets of the solar system possessing a magnetic field.

TABLE 2.2 Parameters of Solar Wind

Parameter	Average	Slow Solar Wind	Rapid Solar Wind
Density n, cm^{-3}	8,7	11,9	3,9
Velocity V, km/s	468	327	702
nV, $cm^{-2} \cdot s^{-1}$	$3{,}8 \cdot 10^8$	$3{,}9 \cdot 10^8$	$2{,}7 \cdot 10^8$
The proton temperature T_p, K	$7 \cdot 10^4$	$3{,}4 \cdot 10^4$	$2{,}3 \cdot 10^5$
The electron temperature (T_e °K)	$1{,}4 \cdot 10^5$	$1{,}3 \cdot 10^5$	$1{,}0 \cdot 10^5$
T_e / T_p	1,9	4,4	0,45

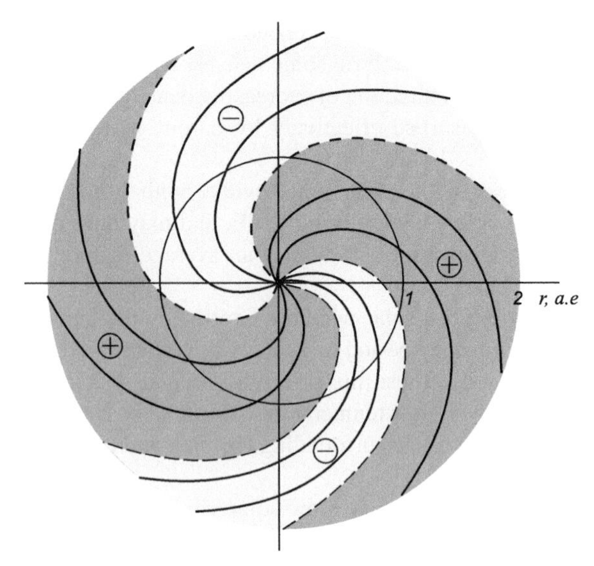

FIGURE 2.5 Schematic picture of magnetic field lines in the solar wind, showing a possible structure of four-sectors. *Source: http://www.ipages.ru/index.php?ref_item_id=1284&ref_dl=1; http://club.radioscanner.ru/topic643.html.*

The solar wind stream nonuniformity (far from planets) gives birth to the interplanetary magnetic field (Svalgaard et al., 1974; Krainev M.B and Kalinin, 2011).

A scheme picture of magnetic force lines in a solar wind representing a possible four-sector structure is shown in Figure 2.5.

2.3.2. Disturbance Forecasting in a Circumterrestrial Environment

The main types of energetic fluxes (particles and electromagnetic radiation) from the Sun the influence of which leads to some disturbance in a circumterrestrial environment (magnetosphere, ionosphere and atmosphere of the Earth) are:

a. Fluxes of comparatively dense ($n \sim 1-70\ cm^{-3}$ on the Earth orbit) quasi-neutral and low-energy ($E < 10$ KeV) solar wind plasma which arouse magnetospheric and ionospheric storms lasting 24 hours and more

b. Fluxes of energetic ($E \sim 10-100$ MeV) flare protons of a low density ($n \sim 10^{-10}-10^{-7}\ cm^{-3}$) lasting several hours and arousing the phenomenon called *polar cap absorption* (PCA),

c. Outbursts of ultraviolet radiation fluxes from solar flares which arouse concentration change in different layers of the ionosphere, with a characteristic time of about 1 hour

d. Outbursts of soft and hard X-ray radiation fluxes from flares which arouse sudden ionospheric disturbance in D-layers of the ionosphere, with characteristic time of several minutes

Solar wind fluxes induce an inconsiderable rebuilding of the magnetosphere and ionosphere. That is why basic attention is paid to their study. It has been established from (a) that fluxes can be divided into two main classes: quasi-stationary solar wind (SW) fluxes with a source lifetime of more than 24 hours, and SW sporadic fluxes, with a source lifetime of less than 24 hours. In its turn the quasi-stationary solar wind is divided into two types: fast SW which flows from the area of coronal holes and reaches the Earth on the orbit at V ~ 400–800 km/s and a slow SW which flows in the streamer belt or streamer chains at V ~ 250–400 km/s.

Knowledge about researches of SW fluxes and their characteristics on the Sun allows us to estimate and forecast SW parameters at the distance 1 a.u. (astronomical unit) and geomagnetic activity indexes Kp(t) and Ap(t) evolution in time. Knowledge of Kp(t) and Ap(t) allows us to define positions of the most important spatial structures using the models of disturbed magnetosphere and ionosphere: plasma sheet boundary, plasma layer boundary, place and time of substorm beginning, and position of the main ionospheric trough. Figure 2.6 shows the Van Allen belts, which are the natural protection of the Earth.

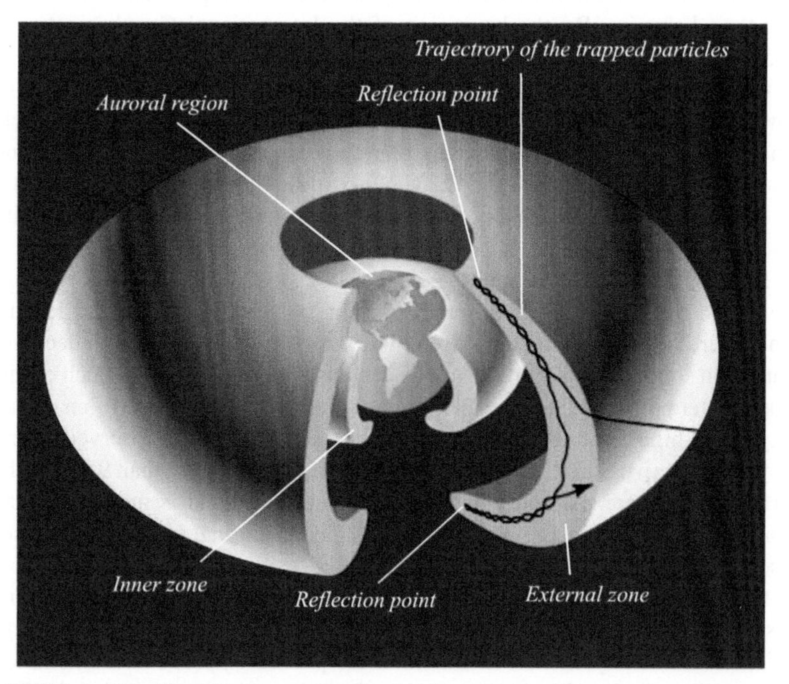

FIGURE 2.6 Van Allen belts. *Source: Reprinted with permission of The Aerospace Corporation.*

2.4. PRECURSORS OF COSMIC STORMS

In 1896 Olaf Birkelund, a Norwegian physicist argued that besides light the Sun also emits corpuscular radiation, and its speed is almost one thousand times less than the light speed. On reaching the Earth this radiation induces magnetic storms and auroras. Birkelund supposed also that auroras can be created by electrical charged particles (*corpuscular rays*) which are released from the Sun and taken in by the Earth's magnetic field near poles.

Cosmic rays in the Earth's atmosphere. The main source of CR inside the galaxy is explosions of supernova stars. Cosmic rays are accelerated on shock waves that are formed in these explosions (e.g., Stozhkov, 2007; Stozhkov et al., 2008). Maximal energy that particles can reach in such processes is $E_{max} \sim 10^{21}$ eV, though there is some controversy about whether cosmic rays of the highest energies ($>10^{16}$ eV) are formed or not in the methagalaxy.

According to their origin CR can be divided into the following groups.

- CR of galactic origin (GCR)
- CR of presumable methagalactic origin (with energies from $E > 10^{16}$ eV to $E \sim 10^{21}$ eV)
- Solar CR (SCR) generated on the Sun during solar flares. ($< 2 \times 10^{10}$ eV)

Constant observation of CR is being done to study peculiarities of their long-time behavior. By the beginning of the International Geophysical year (1957) a worldwide network of CR stations had been formed.

Long-time measurement of CR fluxes led to the discovery of a number of new phenomena. First of all, in CR one can observe an 11-year cycle conditioned by the 11-year cycle of solar activity. When the Sun is calm and solar activity is at its minimum, the CR flux in the heliosphere and on Earth, the orbit reaches maximal values. If the Sun is in periods of maximal activity the CR flux is minimal.

In CR one can observe sporadic changes of their intensity called *Forbush decreases*. Suddenly during several hours or less, a CR flux in the Earth's atmosphere or on satellites registered by terrestrial stations begins to decrease sharply. In some cases amplitude of this decrease is as high as 10%. Such events happen after powerful flares on the Sun. As solar flares occur more often during the years of high solar activity, a Forbush decrease is also often observed during the years of active Sun. CR loses the majority of energy (more than 95%) in the Earth's atmosphere. Although this energy is not large and is much less than solar energy falling on Earth, the role of CR is primordial in many processes observed in the Earth's atmosphere.

Cosmic rays discovery history. In 1912 the Austrian physicist Victor Franz Hess (see Figure 2.7) made a flight by balloon with equipment for cosmic rays registration to study cosmic rays nature. He proved by visual demonstration that cosmic radiation flux is increased as the balloon increases its height. The

FIGURE 2.7 Victor Franz Hess after a flight on August 7, 1912.

next proof of solar corpuscular radiation appeared during the study of another phenomenon—magnetic storms—which, as it was discovered, are connected with the Earth magnetic field fluctuations. As they usually appear two days after a solar flare the explanation of English physicist Sidney Chapman about solar corpuscular radiation has become widely accepted.

Finally, in the 1940s the American researcher Scott Forbush (Forbush, 1946, 1954, 1958) discovered that intensity of cosmic radiation reaching the Earth in the period of high solar activity was low and was decreased sharply during magnetic storms. In other words, the higher solar activity there was, the less cosmic radiation particles reached the Earth surface. Forbush wrote,

It seems that something in solar emissions must hinder penetrating of cosmic rays into the solar system, and this hindrance is increased when the Sun becomes very active.

Eugene Parker, an American astrophysicist, suggested this radiation (which hinders cosmic rays to reach the Earth) be called *solar wind* Parker (1958). The SW blows constantly from the Sun into the solar system. As stated before, it flies past the Earth at speeds of 300 to 500 km/s and determines the weather in cosmic space near the Earth.

Parker (1982), wrote,

A powerful hydrogen wind is constantly blowing in the solar system. Appearing on the Sun it flies past the Earth at speed 400 km/s, reaches remote planets and goes away into the interstellar space. Like a broom it sweeps out gases flowing from planets and comets, small particles of meteoritic dust and even cosmic rays. This wind is responsible for outer regions of the Earth radiation belts, for auroras in the Earth's atmosphere and for geomagnetic storms. It can even participate in formation of a general weather picture on the Earth.

The solar wind is the main reason for such events as geomagnetic storms and their influence on human life and the world economy. In this connection systematic observation of a solar wind is necessary for understanding astro-geophysical phenomena and estimating perspectives of human activity in the immediate future as well.

Among the populations of the interplanetary medium which feel the modulating action of the Solar Wind are cosmic rays – research of turbulence in the solar wind in connection with shock-wave transmission is one of the central problems of modern astrophysics. Shock-wave transmission in a turbulent cosmic medium conditions many physical processes, for example, acceleration of cosmic rays. The most convenient object to study the solar wind is the interplanetary medium accessible today by modern experimental methods.

CR fluctuations are precursors of cosmic disturbances. A widespread method of study of the turbulence of the interplanetary medium is to study fluctuations. Short-period changes of the interplanetary magnetic field and fluctuations of cosmic rays are connected since they appear on account of cosmic ray charged particle scattering on random nonuniformities of the interplanetary magnetic field (Libin, 1983a; Libin and Dorman, 1985; Libin et al., 1996c).

The first fluctuations were discovered at the end of 1960s by Dhanju and Sarabhai (1967, 1970), with the help of a gigantic (at that time) 60 square meter detector array at the Chacaltaya station in Bolivia. In more recent years there have been such detector arrays in almost all cosmic laboratories on the Earth.

Although research fluctuations in the cosmic rays takes its place today among basic instruments of modern cosmophysics, fluctuation existence as an effect of displaying influence of the interplanetary medium on cosmic rays is still in some cases disputable. What is more, in such cases whether by defects of mathematical methods of revealing fluctuations, equipment shortcomings, or researchers' bad luck, even negative results have allowed for conclusions about fluctuation nature.

Absence of meaningful fluctuations during observations on the equator and their simultaneous presence during observations in high latitudes are persuasive evidence that Earth magnetic field changes can not be the source of fluctuations under study, because such changes on the equator are minimal. It is necessary to note that fluctuation revealing is a very delicate and difficult matter. Methods of research of spatial-temporal flux changes, cosmic rays spectrum, and composition which have existed until today have met serious difficulties during the study of fluctuations, because to reveal these fluctuations one had to operate heliophysical and geophysical nonstationary processes which are constantly changing amplitude, phase, and frequency, and to study a number of correlation connections between fluctuations themselves.

Usual correlation spectral methods appeared to be uninformative because data array were used very little. Scientists had to use special spectral methods based on autoregressive analysis. Advantages of autoregressive spectral analysis are in controlling comparison of different notes, as functions of the same parameter—frequency—are compared, and no serious limitations are imposed on this process. Applying autoregressive spectral analysis has allowed the estimation not only of the presence of fluctuations with different periods in observing data but verification of them as well.

Studying fluctuation behavior dynamics, their origin, and development coupled with different heliophysical processes has revealed some curious regularities: a sharp increase in fluctuation amplitude is connected with solar flares and shock waves in the interplanetary medium near the Earth, and such an increase is observed on the Earth at least 10 to 12 hours before beginning the Earth disturbance, and is weakened 12 to 20 hours after the maximal disturbance. So, fluctuation increase is an indicator of expected disturbances on the Earth.

The effect of fluctuation advance of expected disturbances can be easily explained. Disturbance transmission velocity in the interplanetary medium is 300 to 500 km/s, and cosmic ray charged particles' velocity is a bit less than the light velocity, in this case the distance L at which cosmic rays feel approaching disturbance is $L = R/300B$, where B is intensity of the interplanetary magnetic field, R is particle rigidity, and L is sensitivity distance or Larmor particle radius.

Consequently, it is very easy to take into consideration differences in time when cosmic rays have felt disturbances (and their fluctuations) and disturbances themselves reach the Earth. As this distance is determined only by particle energy (field density for each concrete case is constant) it is possible to probe a cosmic space at different distances from the Earth by registering fluctuations of cosmic particles of different energies on the Earth. Under these circumstances, many experiments and calculation methods of shock wave short-period forecasting and diagnostics were worked out, with the help of observing terrestrial fluctuations of cosmic rays with periods from 12−15 to 420 minutes which feel effectively the shock wave approaching the Earth. Figures 2.8 and 2.9 show the scintillation telescope at IZMIRAN (Dorman et al., 1979, 1983; Belov et al., 2011a,b,c,d).

As mentioned above, the effect of cosmic rays' advance of magnetic field disturbance can be easily explained: cosmic rays feel magnetic field irregularities at the distance of transfer range particle scattering which neutron supermonitors (Figures 2.10 through 2.14) and scintillation telescopes (Figures 2.7 through 2.9) register on the Earth surface. It is important to take into consideration that cosmic ray velocity is about one thousand times higher than the velocity of the disturbances moving toward the Earth, and cosmic rays will bring the information about approaching disturbances almost instantly, but disturbance will travel to the Earth for several hours.

FIGURE 2.8 View of one of four horizontal arrays of the Baksan scintillation telescope. *Source: Courtesy of A.V. Voevodsky.*

As energies of cosmic rays registered on the Earth are in a wide range, distances at which cosmic rays will feel disturbances approaching the Earth can be essentially different; registering particles of different energies one can find nonuniformities up to several a.u. It is necessary to note that possibilities to study fluctuations are not limited by that. Research of the power spectrum of cosmic ray fluctuation has shown a connection with the level of storminess of the interplanetary medium of not only fluctuations of specific frequencies but the whole spectrum. Furthermore, inclination of the cosmic ray fluctuation spectrum increases step by step to its maximal value several hours before approaching of the interplanetary medium disturbance to the Earth, stays minimal during the disturbance, and decreases after it passes by the Earth.

The aforementioned authors discovered that the spectrum power increases on different frequencies when the interplanetary magnetic field spectrum boundary passes by the Earth, and it seems that a high-speed flux boundary also should be noticed by observations of cosmic ray fluctuations.

We connect further perspectives of studying cosmic ray fluctuations and using them for diagnostics and forecasting with, first of all, further development of theoretical research (kinetic theory of fluctuations, approached in the framework of isotropic and anisotropic diffusion models; theory of fluctuation of atmospheric and geomagnetic origin; theory of fluctuation appearance in the generation process of solar cosmic rays and their propagation in the corona, in the interplanetary space and in the Earth's magnetosphere).

The widening of experimental works will be of great value in forecasting IMF disturbances. For example, in the registration of cosmic ray fluctuations with the help of precision instruments on the Earth's surface from different asymptotic directions using a world network neutron monitors as a united

(a)

(b)

Top view of the installation SSTIS

NaI NaI
NaI NaI

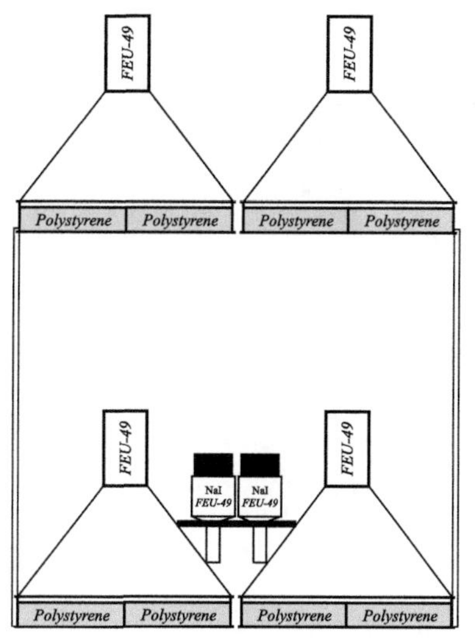

FIGURE 2.9 (a) One of the sections of the supertelescopes at IZMIRAN. (b) Scintillation telescope (top view); scintillation telescope (side view). Photomultiplier-49 (PMT-49) and polystyrene scintillator.

planetary multiway super device; in the registration of cosmic ray fluctuations with the help of multidirectional super telescopes on different levels of depth under the Earth; and wider use of observing cosmic ray fluctuations from balloons, satellites, and spacecrafts.

FIGURE 2.10 Cosmic ray station at IZMIRAN.

FIGURE 2.11 Cosmic ray station at IZMIRAN: neutron monitor.

It will soon become possible to solve important problems of quantitative diagnostics and forecasting of important events for cosmic research and its terrestrial applications, such as powerful solar flares, interplanetary shock waves, and disturbances propagating in the interplanetary space. Very soon we will get estimations of near space state around the Earth, as regularly as cosmic rays registration. The first steps have been already made.

The centenary of Cosmic Rays discovered by Victor Hess on August 7, 1912 will be celebrated on the proscenium of different scientific forums during 2012. Could Victor Franz Hess imagine at the start of his famous flight by

FIGURE 2.12 Center forecasts of space weather at IZMIRAN.

FIGURE 2.13 Neutron monitor station (NMS) of the Instituto de Geofísica (IG) at the Universidad Nacional Autónoma de México (UNAM).

a balloon, that he will open the door to modern space physics?, that the rays which he had discovered are the precursors of cosmic storms?

2.5. SOLAR ACTIVITY AND COSMIC RAYS

2.5.1. Long-Period Variations of Solar Activity and Cosmic Rays

Changes in magnetic field structure occurring in deep layers of the Sun and reaching its surface determine configuration of coronal magnetic fields. In their

FIGURE 2.14 Neutron monitor detector (NMD) of the (IG) at UNAM.

turn these fields (whose form is changed periodically which one can see quite well in white light during solar eclipses) form conditions of flowing out of the solar wind, which influences intensity of cosmic rays by modulating processes. Such changes of solar magnetism occur every 11 years and are expressed in various solar activity manifestations. That is why different characteristics can be used as indexes of solar activity, depending on the problem.

To solve problems of solar-space connections the following tools are commonly used: data of slowly changing solar activity components (number and area of sunspots, coronal indexes, density of a solar radio-frequency radiation flux, total areas of photospheric and chromospheric faculae and quiet protuberances) and data of quickly changing solar activity components (solar flares, radio-frequency radiation outbursts, active protuberance). For very long time periods when there is not trustable data, some indexes, for instance the Carbon-14 (Figure 2.15), are used. Under other circumstances ice cores and tree rings are also employed. These are known as *proxies* of some absent data.

Most of parameters of slowly changing solar activity components are interconnected and characterized by 11-year cyclicity. Among all solar activity characteristics mentioned above, sunspot figures (W) are the classical ones and have become the most frequently used index when studying solar-atmospheric and solar-space connections. Analysis of Wolf figure temporal changes and cosmic ray intensity has helped researchers to reveal 11-year and 22-year variations of cosmic rays (Dorman, 1989; Dorman and Libin, 1985).

However, as it is noted in Dorman (1989), sunspot area (S) is a more objective index (in comparison with W), as in this characteristic definition there are no mistakes connected with the observer's individual peculiarities and the method of observation. Sunspot area and Wolf figures are connected with each other, however, a coefficient connecting these characteristics is

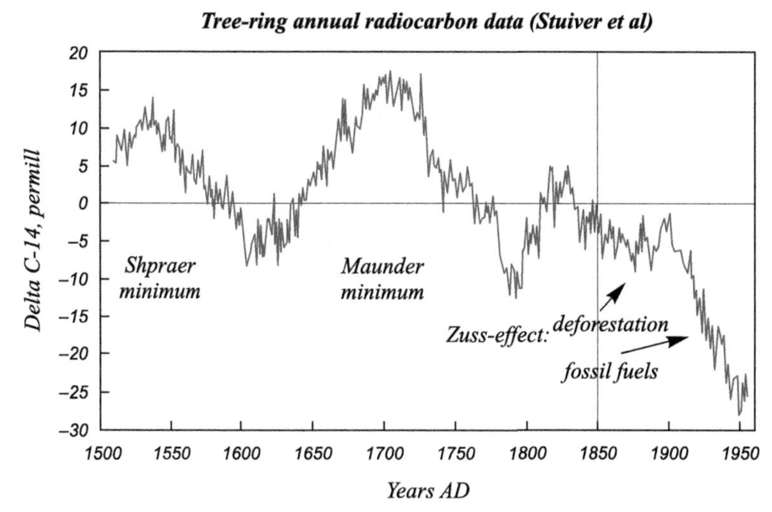

FIGURE 2.15 The content of C-14 from 1500 to 1950. *Source: Solanki et al., 2004.*

changed when a solar cycle is developed, indicating some independence among both parameters.

During researching long-period variations of cosmic rays, a change of sunspot area as a solar activity index is preferable, because it reflects dependence of transfer distance for particle scattering on density of magnetic irregularities brought by the solar wind and reflecting sunspot area (Novikov and Panasiuk (2007); Belov et al., 2004; Guschina et al., 1968, 1970; Guschina, 1983).

On making a diagram of cosmic ray intensity—sunspot area for an 11-year period, one observes a characteristic hysteresis loop. The loop is the result of delay of electromagnetic condition change in the interplanetary medium relative to processes that give birth to them on the Sun. Retardation time is estimated from hysteresis, $\tau_{ret} \approx 1$ year, and, consequently, the modulating heliosphere size is very large ($\approx 80 - 100$ a.u.) when it is supposed that the solar wind velocity is constant ($U \approx 400$ km/s). It was shown later that to reveal long-period variations in cosmic rays it is necessary to use data retrieved with the help of a 6- or 12-month moving average, but in smoothing data in general one reveals comparatively short-period variations for one month.

Retardation value is different in different periods of solar cycle. The largest one is in the solar activity decay period. Let us note that in the same period variations of cosmic rays of one year are most clearly defined. Moreover, τ_{ret} depends on cosmic ray particle energy and decreases while energy is raised.

It has been discovered the absence, (for some periods) of high correlation of cosmic ray temporal changes and sunspot numbers, when considering the whole solar disk (Libin and Perez Peraza, 2009). This absence was regarded as a result of an unsuccessful choice of the solar activity parameters as given by

(Dorman, 1967; Simpson, 1963). Later, it was suggested to use intensity of coronal radiation (with a wavelength λ 5303 Å) as a measure of effluent plasma. Intensity of this coronal line is a good indicator of solar cyclic activity (Guschina et al., 1968), changing essentially from solar maximum to minimum.

Some peculiarities of an 11-year cycle of the Sun, important for studying solar activity and long-period variations of solar rays, were firstly noted by analyzing data of green coronal line intensity. So, Gnevyshev (1963) researched temporal changes of this index and established the presence of the index of the second maximum, which did not yield to the first one energetically in the 11-year cycle. Later it was revealed for other characteristics of solar activity.

A choice of intensity of the coronal index as a solar activity index was made in terms of Parker's correlation between a solar wind velocity and temperature of the inner corona (Parker, 1982), which in turn is connected with intensity of the green coronal line (I_λ) by correlation dependence.

Further research of coronal radiation revealed large-scale emissive structures in the corona conditioned by a sector structure of the solar magnetic field (Parker, 1965). This interconnection of I_λ and structure peculiarities of the solar magnetic field and, consequently an effluent solar wind confirmed the validity of coronal activity as a solar activity index.

Time dependence of cosmic and coronal radiation intensity for 11 years reveals the same cosmic ray retardation effect relative to solar activity manifestations as the index that characterizes sunspot area. Scientists tried to explain an abnormal behavior of cosmic rays in solar maximum periods noted in Stozhkov (2007), Stozhkov et al. (2008), and Veselovsky et al. (2007) in the following way: heliolatitude distribution on solar active regions is changed as an 11-year cycle is developed. However, taking this factor into consideration, this anomaly is left in cosmic ray behavior. The answer can appear only if considering the change of the sign of the solar global field, which is going on in solar maximum (Valdes-Galicia et al., 2005; Krainev and Webber, 2005a, 2005b; Krainev and Kalinin, 2011).

Using the results (indexes) characterizing different manifestations of changing solar activity (SA) cyclicity, such as coronal radiation intensity, radiofrequency radiation, and total sunspot area, scientists noted a breach in correlation of cosmic ray intensity (CRI) with SA in inversion periods of large-scale solar magnetic fields.

What is the connection between different values characterizing solar activity with the solar magnetic field, the source that induces this activity? Babcock (1959) and Leighton (1964) developed a persuasive model to explain SA cyclicity. On the basis of this model there is the idea that all irregular activity manifestations on the Sun surface and above it are induced by changes of poloidal and toroidal components of the magnetic field during differential rotation of the Sun.

In general, the solar wind wafts away just the poloidal magnetic field of the Sun into the interplanetary space (see Figure 2.16). It is possible to observe

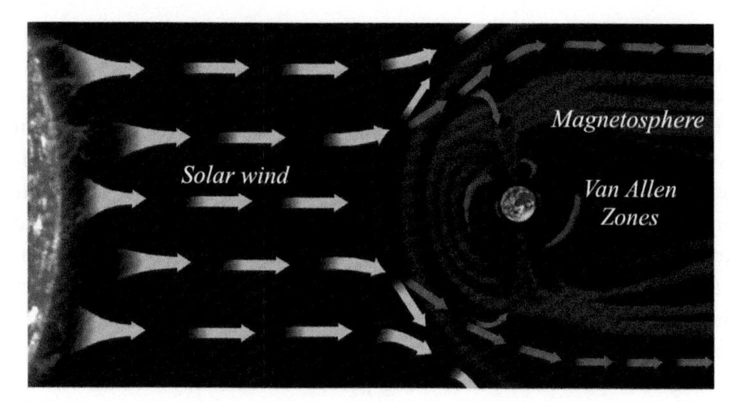

FIGURE 2.16 Artistic design of the emergence of solar wind into the interplanetary medium. *Source: Compilation of www.sickscent.com and www.cmarchesin.blogspot.com.*

a magnetic cycle of a poloidal component on the basis of coronal activity changes (Parker, 1982). Particularly, according to Dorman (2005), brightness of the green coronal line is a measure of magnetic activity and corona warming. Fields of active areas existing on the Sun surface are regions of a toroidal azimuth field emersion which form the famous latitudinally elongated biopolar magnetic regions. Therefore, poloidal and toroidal solar magnetic fields and their changes during an 11-year cycle are connected with each other and generated by the same dynamo-process.

Polarity of sunspot magnetic fields as well as magnetic fields of polar regions are characterized by a 22-year cycle (Vasilieva, 1984), and polarity of preceding and following spots in biopolar groups in both hemispheres of the Sun changes while transmitting from one 11-year cycle to another, or in solar minimum (and this time one observes maximum of the poloidal field). The magnetic field sign in polar regions changes near maximum of the 11-year cycle, and the process of inversion of these fields is rather long. Latitude change of the sign of common solar magnetic field is also revealed during development of an SA cycle in latitude distribution of its active regions (Vitinsky, 1983). All this confirms the validity of the choice of total sunspot area and coronal radiation intensity as basic SA indexes for helio-climatological problems.

Since the beginning of the 1950s when 11-year modulation of cosmic ray intensity by solar activity was revealed, scientists have suggested many different modulation models to explain this fact. However, nowadays Parker's model (Parker, 1982) is the fullest and generally accepted. On the basis of it there is the idea that cosmic ray modulation and propagation in the interplanetary space is determined by features of a plasma flux flowing out from the Sun and "carrying a frozen magnetic field" and nonuniformities characterizing it, and the density spectrum is changed together with the 11-year cycle (Libin and Pérez-Peraza, 2009).

Later, Parker's idea was successfully confirmed with many experiments in the interplanetary space including that on the International Cosmic Station (ICS) at the beginning of this century.

In Parker's differential equation for cosmic ray density, cosmic ray diffusion process and simultaneous convective particle transmission for a spherically symmetrical case with consideration of particle energy changes in interaction of them with the solar wind is included. In solving equations of cosmic ray anisotropic diffusion, joint data of observing coronal radiation and sunspot area for definition transfer distance for particle scattering took into consideration peculiarities of solar wind modulating influence in different epochs of a 22-year solar magnetic cycle.

Finding out peculiarities of long-period cosmic ray variations during periods of solar maximum or periods when a solar magnetic field sign in polar zones is changed results in discovering 22-year variations of intensity of cosmic rays themselves. It demands a more detailed approach to solving the problems of distribution of opposite particle charge in large-scaled interplanetary magnetic fields (IMF) which change direction with a particular regularity.

For a theoretical description of a 22-year cycle in cosmic rays, a model of occurrence of drift flows changing direction subject to solar common field sign is generally accepted. (Direct changes of cosmic ray intensity in the space during the last 25 years confirm appraisals of modulating heliosphere size—which were got earlier empirically from comparison CRI and SA—as 50–100 a.u.) Common characteristics of solar cycles and their energy have been described, for instance, in Ishkov and Shibaev (2005) and Ishkov (2007, 2008).

An average duration of solar cycles is 10.81 years, and for the last 8 cycles there was a decreasing tendency of its duration to 10.44 years. An estimation of average energy released during a cycle is determined by the full kinetic energy of rotation and the full energy of electromagnetic radiation during a cycle. Other phenomena release much less energy, although they can determine separate processes during solar cycle development.

Mathematical images of even and uneven cycles give a possibility of forming a closed structure of the solar physical magnetic field. The low statistic of trustable data of solar cycles (14) and the lack of a physical model of SA cycle development do not allow a true cycle forecast before a cycles begins. But the situation changes with the beginning of a new cycle. In 18 to 24 months it is possible to reveal its height, the dates of its maximum, and the possible duration of the current SA cycle.

Besides the above-mentioned long-period 11-year and 22-year cycles, other periodicities connected with solar activity are observed in the intensity of a cosmic ray flux variation: fluctuations with periods of 4 to 5 years, 2 years, 1 year, and several months are observed in intensity of a cosmic ray flux (Libin and Jaani, 1989; Libin et al., 1998).

Nature of extra-atmospheric short-period cosmic ray variations were discovered more than 40 years ago when measures of a cosmic ray muonic component were analyzed (e.g., Pai and Sarabhai (1963)); they are explained by changes of the Earth position in the space relative to the helioequator.

It is important to note that studying short-period variations and fluctuations of cosmic rays and solar activity is of common interest, as comparison of periodicity in both processes can give information about their interaction. So, making calculations of power spectrums in the interval of 50 to 1000 days (10-day averaged data of CR registration with the neutron monitor of Deep River, sunspot numbers, data about a radio-frequency radiation in the frequency of 2800 MHz, and data about measuring parameters of the Sun acceleration relative to a common solar system mass center were used), scientists established oscillations with periods 650 to 680, 350, 238, 170, and 75 days (Stozhkov and Charakhchian, 1969) modulated by an 11-year wave that confirms the solar origin of some of them (2 years, 1 year, 1 to 2 months).

Model research of 11-year solar activity changes and cosmic ray intensity made by Libin and Jaani (1989) and Libin et al. (1998) have showed that a great part of variations (2-year, 1 year, and 1 to 2 month ones) are harmonic components of 11-year and 22-year cycles.

Because the high interest for research in solar-terrestrial connections of the 2-year (or *quasi-biennial*) cosmic ray variation, Charakhchian (1979) has quickly undertaken this research. In her works she showed that the energetic spectrum of the 2-year cosmic ray periodicity is close to the spectrum of 11-year variations, and relations of their amplitudes is of steady character, but on average equal to $A_{11}/A_2 = 5.3$ which is much higher than if 2-year variations were harmonic components of 11-year ones.

Moreover, a 2-year variation is observed in solar and magnetic activity and also in climatological processes. It was found that 2-year variations have an extraterrestrial origin and are connected with physical conditions in the region of the interplanetary space closest to the Sun (Libin and Pérez-Peraza, 2009).

Quasi-biennial variations of cosmic rays are manifested in alteration of solar-daily anisotropy phases subject to the position of the Earth relative to the helioequator. While the poles of the common solar magnetic field are altered, an alteration order of anisotropy phases is changed to the opposite correlation with behavior of quasi-biennial variations (QBV) of north—south asymmetry of the solar wind. The reasons for quasi-biennial variations of cosmic ray intensity are cyclic changes of solar wind characteristics in the whole heliosphere and appropriate variations of the behavior of different atmospheric processes.

Quasi-biennial variations (QBV) are found in many solar and terrestrial processes including a low-latitude stratospheric wind (LSW), the solar magnetic field on the surface (B_r component), and the Earth rotation speed V. QBV behave differently on the Sun and on Earth. In LSW they dominate, but in many processes on the Sun, QBV are found only after observation filtration.

A statistic analysis of 36-year tables of filtered values U^*, B_a^*, and V^* was made to reveal solar-stratospheric connections (Ivanov-Kholodny and Chertoprud, 2005) where U is the averaged velocity of LSW, B_a is a module B_r

average in all solar latitudes, and symbol * means filtration according to the scheme used in (Ivanov-Kholodny and Chertoprud, 2005). An important correlation between values of U^* and B_a^* ($r = -0.58 \pm 0.08$) and a linear connection U^* with B_a^* and V^* characterized by a coefficient of multiple correlation $R = 0.68$ are found. So, it is proved that there is a certain connection between stratospheric and solar QBV.

As for all the other short-periodic variations mentioned above, revealing and analyzing them (as a manifestation of solar activity) demands a certain care because of the complex behavior of them from cycle to cycle and a lack of reliable correlation connections with solar activity (Libin and Gulinsky, 1979), although much research by several authors in each case show only a rather stationary character of them during short periods of time (3–4 years).

Periodical cosmic rays variations in the time interval 1953 to 2004 have been researched with the help of modified methods of spectral analysis. An initial data table of monthly cosmic ray flux has been synthesized from measures made in different time intervals throughout this period, with 100 terrestrial neutron monitors distributed about the whole surface of the Earth.

These measuring station data are barometrically corrected and are on the server http://spidr.ngdc.noaa.gov/spidr/query.do?group=CRI4096& in format 4096. The method of formation of a joint synthesized data row is based on principles of processing unequal temporal rows with a prescribed balance or mean-square measure errors.

It is found out that weak quasi-periodic components appear in periods of 0.75, 1.0, and 2.0 years and more powerful ones appear in periods of 1.5, 1.75, 2.6, 3.0, 4.1, 9.0, and 23.0 years in a spectral periodogram of synthesized data. On the basis of synthesized data values, a cosmic ray activity index has been coined, and interaction of it with solar activity indexes in X-rays, radio, and optic ranges has been studied (Dergachev and Dmitriev, 2005).

One of the mechanisms of variation appearance in cosmic rays in periods that are not connected with cyclic solar activity are heliosphere self-oscillations, which are of a clear quasi-stationary character. A period of such oscillations which appear because of the disturbing action of large-scaled nonuniformities of the interplanetary medium (Attolini et al., 1983, 1984, 1985) and from the solar wind can be found from 2 to 8 to 9 years (Libin and Pérez-Peraza, 2007).

What is the practical appliance of such regularities? The IZMIRAN made a forecast of the cosmic ray flux for the next solar cycle and reestablished CR behavior in 17 to 20 centuries on the basis of the model connecting cosmic ray (CR) modulation with solar activity indexes (Belov et al., 2004). A forecast of a cosmic ray (CR) flux was made on the basis of the forecast of the main solar magnetic field characteristics (see forecast station in Figure 2.12). To reestablish CR behavior in the past, sunspot numbers and geomagnetic activity indexes were used.

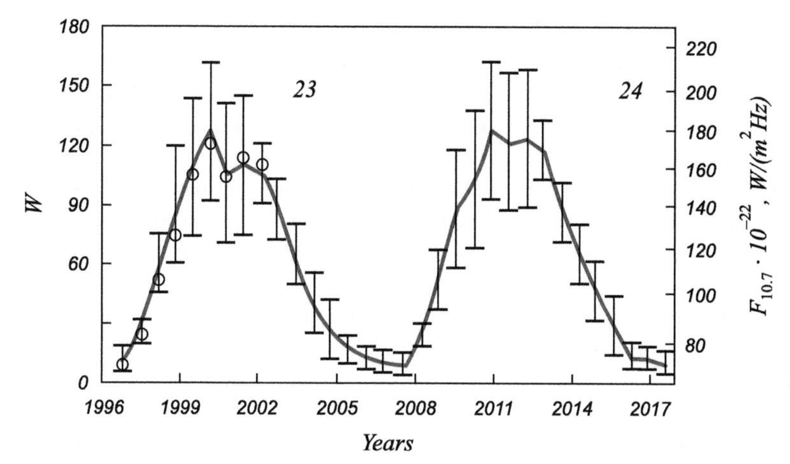

FIGURE 2.17 Forecast of solar activity before 2017 [MGU]. *Source: Khramova et al., 2002.*

For forecasting, scientists needed reliable data about CR variations in the network of CR stations as well as data about changing solar global magnetic field characteristics (inclination of a heliosphere current sheet, average intensity of the solar magnetic field on the surface at the source of a solar wind and field polarity). A worked multiparametric model of CR modulation in the heliosphere allows the precise estimation of CR variations under study enables a long-time forecast.

Such a forecast before 2017 was made in the MGU (RINP and SAIS, see Figure 2.17) with a method of phase mean-values for Wolf figures and a radiofrequency radiation flux F10.7 based on the information for the first 22 cycles (from 1 to 22, see the picture at the top of the figure). The circles are observed data and the error bars are confidence intervals corresponding to 99% of probability.

2.5.1.1. Solar Activity Forecasting

Solar activity is the reflection of solar dynamic processes as a nonlinear dynamic system. Connection between all the processes occurring on the Sun is complex and indirect, which is why on observing and fixing values of some black box outputs (the Sun as we see it nowadays) it is possible to expect that only certain regularities such as cyclicity will be revealed (Lychak, 2002, 2004, 2006; Makhov and Posashkov, 2007; Solntsev and Nekrutkin, 2003).

In his works, Lychak (2002, 2004, 2006) suggested the use of a set-theoretical approach to building up a mathematical model of signals on outputs of a dynamic system characterized by chaotic oscillations. He argues that chaotic events and processes on consideration may have a qualitative difference from a realistic approach.

New methods of presentation of a mathematic uncertainty model lead to new approaches to the solution of frequent problems of manipulating measurable (observational) results that are received with some level of mistakes. Particularly the necessity to know the interval characteristics of uncontrolled mistakes, or to work out methods of experimental determination of them. Also to get not only close values of finite data, but interval estimations of possible mistakes that follow from them.

Lychak showed a cycle synchronism to working out methods of statistic processing of solar activity indicators (Wolf figures, solar radio-frequency flux values). Based on both solar activity factors, he obtained interval estimations of periods of its cyclicity manifestation. On the basis of concepts about solar processes as chaotic oscillations, a forecast was made of solar activity till the end of the last cycle (which had not been finished yet) based on both indicators.

For statistical data manipulation he used data of daily values of the Wolf number from January 1818 to May 2003 and also monthly average values of it from January 1700 (in fact, from 1700 to 1748 there were yearly average values) to December 1984 (Libin and Pérez-Peraza, 2009). Averaging out monthly values by using daily Wolf number values after December 1984 resulted in monthly average values of it from January 1700 to May 2003.

To illustrate his prognosis method, Lychak did the following: He considered the Wolf numbers for the whole period 1700 through 2003 as 49.96. Monthly average values of a Wolf number were smoothed out with a 13-monthly rectangular window, and beginnings and ends of standard (11-year) solar activity cycles are determined basing on these smoothing values. Dates of beginnings and ends and, correspondently, duration of these cycles have been obtained (Lychak, 2002, 2004, 2006). The obtained figures of alternative Lychak cycles for the period mentioned above is 26, and the last (the 27th cycle) has not been finished yet (the date of the beginning is September of 1996). In this case, if you do not take into consideration the numbering of Lychak (2006), their allocated cycles are in good agreement with conventional cycles.

It should be emphasized that conventional standard cycle values assume that we are living the 24th solar cycle since January 10, 2008, which will reach its maximum in 2013 (that is, there is a diphase of 3 cycles). Therefore, following Lychak the average value of the cycle duration is 131.19 months (10.9 years). A maximal duration was observed in the 8th cycle (from September 1784 to April 1798): 163 months (13.6 years). Then in the 9th cycle there were 152 months (12.7 years). A minimal duration was observed in the 6th cycle (from June 1766 to June 1775): 108 months (9 years). Then in the 7th cycle there were 111 months (9.25 years), and in the 12th cycle there were 113 months (9.4 years).

As an alternative estimation of the cycle average duration, the average distance between absolute maximums of Wolf number values on the neighboring cycles were used on the basis of the data. It is equal to 131.08 months (10.9 years) or almost the same when estimating with a standard method. It is interesting to note that in general, duration of cycles with larger values of a maximal Wolf number on the cycle is shorter.

On manipulating data of an average Wolf number with a wider 396-month (33-year) sliding window the so-called *century cycles* were established among which the last one is not finished yet.

There are three cycle groups (with duration of about 100 years): 1 to 9 (with duration of 1182 months, or 98.5 years), 10 to 17 (with duration of 1088 months, or 90.7 years), 18 to 26 (with duration of 1141 months, or 95 years).

The solar-terrestrial physics data of a solar radio-frequency radiation flux on a wave of 10.7 sm for the period from February 1947 to May 2003 were used. Its average value for the whole period equal to 119.63 has been found. Over this period one can observe 4 full cycles of solar radio-frequency radiation flux variations on a fixed wave of 10.7 sm.

The cycle duration in months is max = 144; min = 116; mid = 127. A timely interval between cycle absolute maximums in months is max = 148; min = 109; mid = 132.

Friis-Christensen and Svensmark (Friis-Christensen and Svensmark, 2003) also identified the range of variations of the SA, including the well-observed frequency of a century (Figures 2.18 through 2.20).

An average value of the cycle duration (average on 4 full cycles) is 126.75 months (10.6 years). An average Wolf number of the 4 full cycles (19–22) gives the same value. It means that average Wolf numbers in solar activity cycle

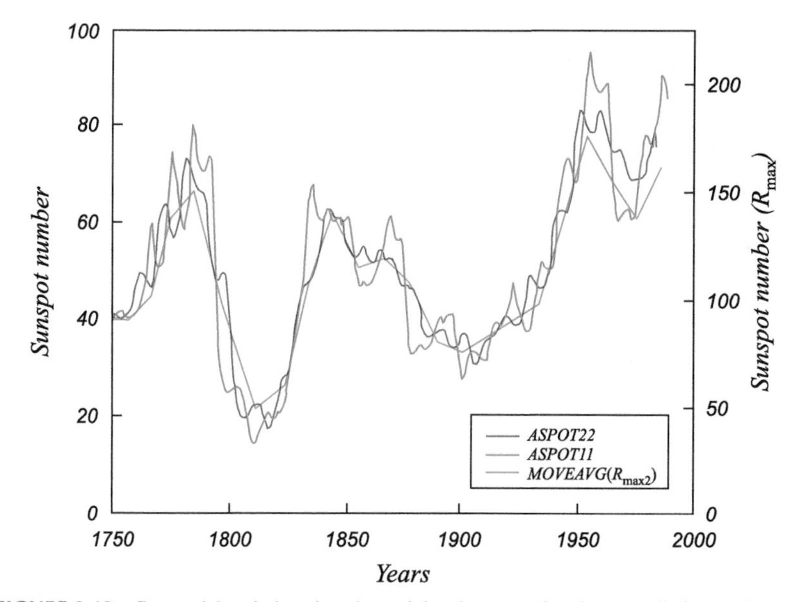

FIGURE 2.18 Centennial variations in solar activity. *Source: What do we really know about the Sun-climate connection? Solar-Terrestrial Physics Division, 1997, Pg 914.*

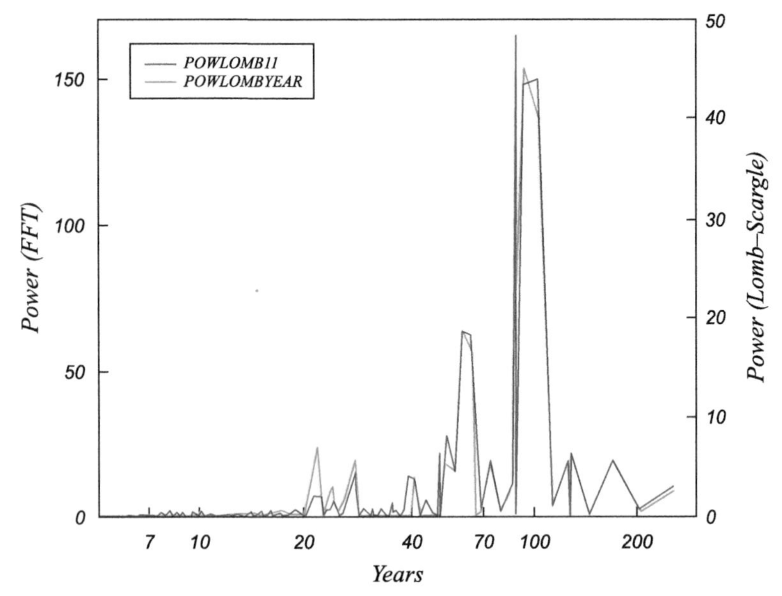

FIGURE 2.19 FTT power spectra for different window functions are calculated for 11-year running of mean sunspots numbers, spanning from 1732 to 1987 ($N = 256$). The columns in the frequencies $Vk = k/N$ ($k = 0., 1, 2, \ldots N\backslash 2$) on the logarithmic abscissa scale are slightly shifted. *Left*: Parzen window. *Right*: Welch window. The curves are Lomb-Scargle periodograms for the same data set as the FTT spectra (fully drawn) and for annual R values from 1732 to 1987 (dash, only drawn for period > 20 years), respectively. *Source: What do we really know about the Sun-climate connection? Solar-Terrestrial Physics Division, 1997, Pg 914.*

duration and values of a radio-frequency radiation flux coincide during the synchronic measuring of both factors.

Comparison of values (see Figure 2.21) of a radio-frequency radiation flux (the upper blue line) with relevant monthly average Wolf numbers for the same period (the lower red line) shows a synchronism of cycles with regard to both solar activity factors.

Lychak made a forecast of solar activity relative to monthly average Wolf numbers till the end of his 27th cycle. For that, he compared its observed part with relevant parts of the preceding 26 cycles normalized with multiplication by a coefficient equal to the relation of an average Wolf number of the 27-cycle observed part to an average value of the cycle.

Such cycle (according to mean-square deviation minimum) is the 24th cycle. The forecast was made by joining a relevant part of 24th cycle to the part of the 27th cycle that corresponds to the solar activity recession phase. Satisfied results of the forecast conformation were obtained for both solar activity factors during the first 7 months.

Several research techniques for estimating solar activity factors—Wolf numbers and a level radio-frequency radiation flux on wave 10.7 sm—were worked out. (A new mathematic model of time changes of these factors as

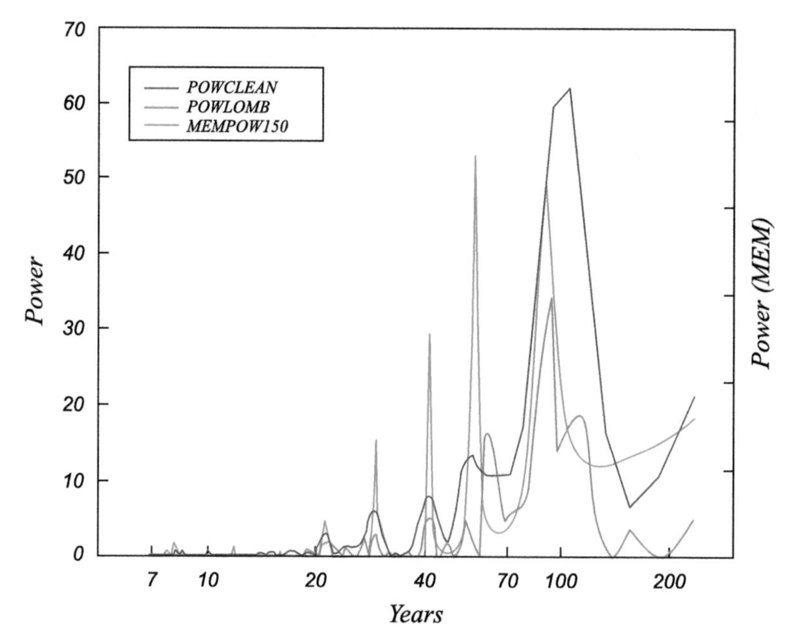

FIGURE 2.20 Comparison of power spectra for different methods. Spectral analysis of the centenary of solar activity variations. *Source: What do we really know about the Sun-climate connection? Solar-Terrestrial Physics Division, 1997, Pg 914.*

a chaotic process with some interval characteristics was used.) On the basis of the presentation of solar activity manifestations as a result of some oscillations of solar dynamics, a forecast of the last solar activity cycle for Wolf's numbers, particularly of duration of it, and the averaged Wolf

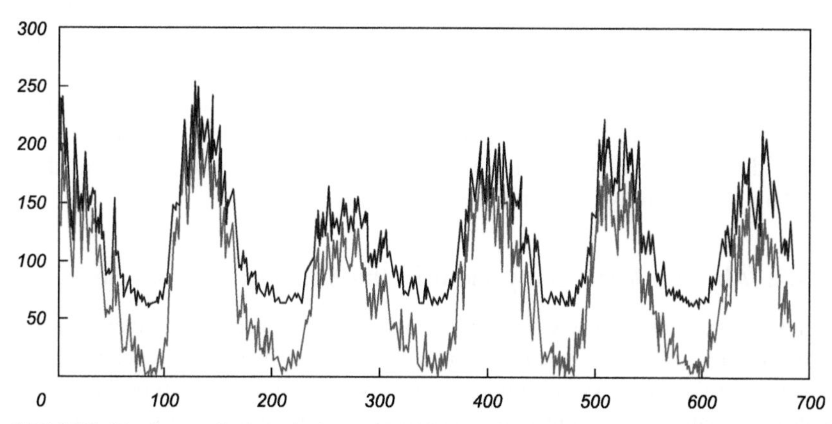

FIGURE 2.21 Syncronization of solar cycles with regard to both the radio flux (blue line) and the monthly average Wolf numbers (red line) in 21–26 cycles. Vertical and horizontal axis are given in reltive values and months respectively (Lychak, 2002).

numbers on the whole cycle was made. An analogical forecast was made relative to the other solar activity factor—a level of radio-frequency radiation flux on wave 10.7 sm.

Comparison of forecasted values with real ones (the forecast was made beginning from May 2003) showed a rather good coincidence of oscillation form of monthly average values during the first six forecasting months. (After that period the difference of forecasted and real values increases sharply.) Though the majority of forecasts of the 24th solar activity cycle were made in supposition that the minimum would not last for long, it was possible to expect that some active regions forming a new solar activity cycle would appear on the Sun in 2008. However, although rebuilding of the main magnetic field was noted, as an indicator of a new cycle, the beginning of solar activity is still very low. Moreover, calmness during the last year has been recorded. There has been no such long solar pause in preceding activity cycles.

Forecasters do not know what to think and have to review not only the minimum duration, but the beginning moment of the whole active part of the cycle. It is possible to expect that telescopes TESIS fixed onboard the CORONAS-FOTON spacecraft will be able to watch the formation of new activity zones, although there were no sunspots on the solar surface, and the Sun seemed calm as seen from the Earth. It is the same configuration that two active areas registered with telescopes in the solar corona on March 27, 2009, and observed for 4 days without a break. Both areas are rather far from the equator, and this fact says much to astrophysicists. Just here, on high heliocentric latitudes, powerful magnetic fields of a new cycle that have been formed in the depth of the Sun for several years should come to the top. This zone will come closer to the equator for several years, increasing activity and forming something that is called a solar cycle.

A new high-latitude activity zone is usually formed together with destruction of an old equatorial zone which is left from the preceding cycle. Thereby activity cycles collide. Besides, as a rule two symmetrical activity zones—in the Northern and Southern Hemispheres—are formed on the Sun.

In the last few years, besides the lack of the equatorial zone there was a lack of the Southern Hemisphere zone. Time shows how long such asymmetry will stay and if it is a peculiarity of the new solar cycle. So, we know in this way that the 24th solar activity cycle initiates on January 10, 2008.

2.5.2. About a Year Component in the Solar Activity Cycle

In research of cosmic ray effects connected with peculiarities of solar activity heliolatitude distribution and the slope of the ecliptic plane to the helioequator plane, a question about a year component in the solar activity cycle, its peculiarities, appearance, and disappearance dynamics becomes important (Dorman, 1998).

To find oscillations in SA close to 1 year, sunspot areas for 1 solar turnover of the Sun as a whole and of the Northern and Southern hemispheres separately were used by Kachanov and Kozlov, (2008) and Libin and Pérez-Peraza (2009). Oscillations with periods of 1.2 years were found with methods of a sliding average and epoch syncopation. This periodicity does not depend on an 11-year cycle number and forms a continuous series from 1878 to 2007.

A similar research for cosmic rays was made with the use of the HL-index of solar activity (Guschina et al., 1970), because as there is heliolatitude dependence of SA and solar wind velocity, a year cosmic ray variation has to be observed because of misalignment of the solar equator and the Earth orbit planes. Such variation (with amplitude 0.34% and maximum in January) was found according to monthly average values of the cosmic ray neutral component intensity (data of cosmic ray stations of Inuvik and McMurdo were corrected for the temperature effect and noncyclic variations). A character of 1-year cosmic ray variation supposes the presence of a transverse gradient of cosmic ray density of 8% per 1 a.u, relative to the solar equator plane, which was later confirmed in satellite experiments. Long-term changes of a year (and half-year) cosmic ray variations revealed a large changeability, but year variation amplitude was always 2 to 3 times higher than that of the half-year one. A lack of anomalies in a year (and half-year) wave in solar activity and in cosmic rays shows that their amplitude and phase changeability is conditioned by broadband noise.

Intensity spectra of cosmic ray and solar activity variations (in a range from 3×10^{-8} to 3×10^{-7} Hz) calculated by the authors for each 11 years during the period 1960 to 2008 are well approximated by the function $P(f) \sim A\,f^{-2.4}$. Herewith a half-year wave exceeds only a 90% confidence interval, when a year one exceeds 95%.

A more complex correlation connection (changing from cycle to cycle and depending on a solar activity phase) is observed between cosmic ray intensity and geomagnetic activity and atmospheric processes.

According to calculation data for the period 1960 to 2008, the authors showed a good correspondence between short-periodic variations of CRI, solar radiation, and closed water system levels for, at least, a half of observed periods.

2.5.3. Connection of Solar Activity with Geomagnetic Processes

Solar activity influence on the Earth's magnetosphere is manifested by two basic types of magnetic storms: those that begin suddenly and those form gradually. (As a rule magnetic disturbances with a sudden beginning are connected with chromospheric flares and change in the 11-year cycle in phase with solar activity.) A reason for a sudden beginning magnetic storm is that a front of a shock hydrodynamic wave appearing on the Sun (during a sudden particle emission from a chromospheric flare) induces a sharp compression of the Earth's magnetosphere. Twenty-seven day repetition and the most powerful development in solar minima (for 1–3 years) are characteristics of gradual magnetic storms. This type appears when the Earth crosses sector boundaries of the interplanetary magnetic field (IMF) which rotates together with the Sun. According to Vasilieva (1984), the common solar magnetic field is at a maximum 1 to 2 years before a solar minimum, which is why magnetic fields in sector boundaries are the largest just in this period.

Geoeffective boundaries are those boundaries between sectors where a vertical component of the interplanetary magnetic field is changed from a north direction to a south one (in the ecliptic coordinate system). IMF influences the geomagnetic field because the IMF component directed to the south allows interplanetary field lines to reconnect with geomagnetic lines of force on a magnetosphere daily which leads to forced transmission of these field lines to the magnetospheric tail. It also promotes field line reconnections in the tail which trigger magnetospheric disturbances.

Long-term changes of geomagnetic disturbances reflect 11-year, 22-year, and century (80–90 year) solar activity cycles (Rivin and Zvereva, 1983). In some works (Rivin, 1985; Libin and Pérez-Peraza, 2009) it is shown that flare magnetic storms typical for 11-year cycle maximum are better developed in uneven cycles, and recurrent disturbances, developed on a cycle recession curve, are much stronger in even cycles. That is why it is more correct to connect geomagnetic disturbances with a 22-year cycle of solar magnetic activity.

The Earth climate oscillations are also of a polycyclic character, as previously mentioned. These are cycles which last 2 to 3 years (quasi-biennial), 4 to 7, 10 to 12, 20 to 23, 80 to 90, and 380 to 450 years. For the first time a spectral analysis of a 1000-year row of a correlation index of deuterium content to hydrogen content was made in the work (Ol', 1973; Vitinsky et al., 1976; Vitinsky et al., 1986) (in tree rings this index changes are proportional to changes of the atmospheric temperature). As a result of the analysis, a 22.26-year period close to a 22-year solar activity cycle was determined.

2.6. A ROLE OF THE INTERPLANETARY MAGNETIC FIELD IN ATMOSPHERIC PROCESSES

In Ol' (1973) it is shown that an 11-year cycle is expressed in meteorological processes as much weaker than a 22-year one, although an 11-year cycle (which is the basic one in sunspot formation activity) has a larger amplitude than a 22-year one. Ol' confirmed the existence of variations with periods 7, 11 to 12, 17, and 30 months in meteorological processes, and also nonstationary (or quasi-stationary) variations with periods 27, 9 to 14, and 6 to 7 days which are observed synchronically in almost all meteorological indications and in disturbance characteristics of the Earth geomagnetic field (Hamilton and Evans, 1983; Pérez-Peraza et al., 1997; Oraevsky et al., 2002b; Libin and Jaani, 1990; Libin et al., 1990).

Ol' supposed in his work that 6- and 9-day rhythms in the Earth's atmosphere are connected with the interplanetary magnetic field structure. The 9-day one corresponds to the existence of 6 sectors with three geoactive boundaries in the IMF, and the 6-day one corresponds to eight sectors with four geoactive boundaries. So, existence of common solar rhythms in atmospheric processes and geomagnetic disturbance tells us about a common solar source connected with a sector structure of the IMF.

In many works a connection of the lower atmosphere with the interplanetary magnetic field and a solar wind was discovered. Herewith, a correlation sign between atmospheric processes, the solar wind, and the IMF was changed to the opposite, when the Earth moved from one sector of the IMF to another. The IMF sector structure induces changes of the vorticity index (determined as the area where the atmosphere circulation per area unit reaches the value 20×10^{-5} s^{-1}), corresponding to a well-formed cyclone on isobaric surface of 300 and 500 MB in the Northern Hemisphere.

Area of hollow low pressure regions in the Northern Hemisphere in winter can become minimal 24 hours after the Earth crosses the IMF sector boundary, and a vorticity minimum value is higher percentagewise for regions in the troposphere that are characterized with more intensive circulation.

Processing of observing results made by Wilcox's group (Wilcox and Hundhausen 1983) in the 1980s showed that the process of passing sector boundaries is accompanied by proton fluxes with energy equal to tens of MeV. Vorticity index minimum connected with sector boundaries which were followed by proton fluxes is almost twice deeper than the index minimum connected with usual boundaries. Wilcox's conclusions supporting that boundaries with proton fluxes were accompanied with a large increase of IMF density and a deep vorticity index have been accepted until recently.

According to observing data of cosmic ray intensity, IMF and data of vorticity areas in the troposphere (index VAI) with the time epoch syncopation method (a day when the Earth crosses a sector boundaries is considered to be a zero day) the following results were discovered.

- Crossing an IMF sector boundary induces a stable effect in VAI (about 20%), while the effect is unstable in cosmic rays.
- When there are 21 crossings of IMF boundaries one can observe a 0.5% increase for a cosmic ray flux three days before crossing, with further 1.0 to 1.5% decrease in the same period of time.
- When there are 28 crossings of sector boundaries, effects in cosmic rays have not been observed.
- When there are 17 crossings effects in cosmic rays have been observed but they have an opposite character compared with the first case.

So, after research of sector boundary influence on the troposphere vorticity index it was concluded that the effect cannot be observed directly through cosmic rays.

By studying high-speed solar wind flux influence on the atmosphere circulation, geomagnetic activity, and galactic cosmic ray intensity it was proven that sudden recession of galactic cosmic ray intensity begins 1 to 2 days before velocity maximum in the solar plasma flux reaches a minimum in the first day and is rebuilt to the initial value on the 4th or 5th day.

Looking at the time dependence of a geomagnetic activity index *Kp* a clear peak is developed in the day of plasma flux velocity maximum. In the middle and upper tropospheres of the Northern Hemisphere in middle latitudes in the moment when the Earth falls in a solar plasma high-speed flux, a sudden decrease of areas occupied by deep cyclones is noticed. Obtained results are fully confirmed by Loginov (1978) and Loginov et al. (1975, 1980), who show that an increase of solar wind velocity leads to a decrease of cyclical activity in the troposphere. A decrease of cyclical activity in the troposphere seems to be connected with intensity decreasing of galactic cosmic rays which play a certain role in disturbances of tropospheric circulation (Svensmark, 2008).

This effect is also well observed in studying influence of solar flares on the Earth's atmosphere. Flares lead to changes in atmosphere circulations in middle and high latitudes after 12 hours of observing them in the optical window.

During the last few years there were some attempts to check connection between cyclonic disturbance development and atmospheric vorticity, on one side, and on the other side the Earth's crossing the interplanetary magnetic field sector boundary, solar wind high-speed fluxes, and solar flares. Research over the last 20 years lead to the following conclusions.

1. The Earth moves through sector boundaries and high-speed fluxes of the solar wind and it induces a decrease of vorticity which happens in the same time as geomagnetic and electromagnetic disturbances (Artekha et al., 2003; Artekha and Erokhin, 2004, 2005) and the rebuilding of cosmic ray fluctuation spectrum (Libin and Pérez-Peraza, 2009).

It is true that the role of all electromagnetic phenomena in cyclone formation and other crisis atmospheric processes can clear up the facts mentioned above. Herewith a low (4, 8 km) negative area exerts the largest influence on cyclonic rotation of a whirl storm, and anticyclone movement is determined by a high (10, 16 km) positive area.

If appearing, charged areas really play a big role in formation, support, and movement of rotating atmospheric units, and a number of other factors become clearer. For example, cyclones appear more often than anticyclones, because it is easier for a denser (as it is located lower) negative area to support rotation in the whole atmospheric region in a cyclone system than for a positive area which is not so dense (as it is located higher) to rotate an anticyclone system, and because to organize a system of a smaller size is always easier. For the same reason, average sizes of an anticyclone are larger than average sizes of a cyclone, because the charged subsystem size threshold for supporting rotation is different.

Artekha and Erokhin (2005) describe a plasma model of a large-scale whirl storm to describe formation; further quasi-stationary phase of vorticity was suggested. A process of organization of a powerful cloudy structure which accompanies powerful vorticities (where a large number of charges concentrate) is also of a great interest. Here a big role can be given to forces of electromagnetic nature (e.g., dielectrophores), when a particle is influenced by the force $F = 0{,}5(\varepsilon_1\text{-}\varepsilon_2)\partial E/\partial r$ transmitting the particle with dielectric penetrability ε_1 in the environment with penetrability ε_2 to the region of higher density of the electric field.

As dielectric penetrability ε_1 of water steam, *a fortiori*, water and ice is different from dielectric penetrability of the air, the mentioned force should play a big role in the process of increasing the local moisture, cloud crowding to a charged zone, and keeping clouds in the united system.

As a result of the action of all these mechanisms for vorticities an extended structure of charged regions between the positive Earth surface and a negative layer of the tropopause nearby is formed.

2. Vorticity amplification comes after powerful solar flares in the interval of heliographical longitudes 0° to 44 °E.
3. Substantial amplification of vorticities can be connected with flares, which produce powerful geomagnetic disturbances. As a rule, these flares are located in the eastern part of the solar disk and appear serially.

So, we can observe complex interconnection between solar activity, geomagnetic disturbance, cosmic ray intensity, and atmospheric processes, and the character of this connection between all mentioned phenomena is changed from time to time and also can be different for different regions. This makes the process of studying mechanisms of solar activity influence on the climate very difficult.

It is true, in some regions, especially in the Northern Atlantics, after geomagnetic disturbance dispersion the surface pressure variability increases. It reflects a level of conversion of a potential energy to a kinetic one that inevitably has to appear in a wind field. As a result of a statistic analysis of 90-year observation of surface pressure, it is observed that the atmosphere instability increases in middle latitudes in the Northern Hemisphere after powerful geomagnetic disturbance (Mustel et al., 1981; Veretenenko et al., 2005). Because of this, research of possible connections between solar activity, the atmosphere vorticity, wind velocity, precipitation, temperature, pressure, cosmic radiation fluxes, geomagnetic, and electromagnetic activity is becoming the most important factor in understanding the mechanisms of solar-terrestrial connections, to form a helioclimatology basis.

Meanwhile, it is now clear that modulation of a cosmic ray flux and changes of climatological parameters are connected with the same processes: powerful shock waves in the interplanetary space, geomagnetic disturbances, solar flares, and so on. This is why during working out any prognostic and climatological models it is important and even necessary (as Danish, Mexican, and Russian scientists' latest results show) to take into consideration changes of cosmic ray intensity besides solar, geomagnetic, and electromagnetic activity (Libin and Pérez-Peraza, 2009).

Chapter 3

About Possible Analysis of Solar Activity Influence on Terrestrial Processes

Chapter Outline

Highlights in Helioclimatology. DOI: 10.1016/B978-0-12-415977-8.00003-4

3.1. METHODS OF JOINT ANALYSIS OF COSMOPHYSICAL AND CLIMATOLOGICAL PROCESSES

3.1.1. Spectral Analysis of Researched Processes

The analysis of parameters of solar activity (Wolf numbers W, sunspot area S, HL-index, solar radiation level on the Earth), geomagnetic disturbances (Kp-index), and atmospheric processes (pressure, temperature, precipitation, level of closed water system, ice area) was made in general with methods of correlation and/or spectral analysis.

Practical techniques of spectrum estimation always consist of several stages: preliminary analysis, calculation of sample covariance, correlation functions and spectral estimations, calculation of mutual correlation functions and mutual spectral estimations, and interpretation of the obtained results.

Preliminary analysis consists of research of temporal series to reveal stationarity and leading them, if necessary, to a stationary or quasi-stationary state to learn apparent trends and periodicities in the data set under research (important for solving problems about data filtration in a test analysis, the *pilot analysis*) (Blackman and Tukey, 1959; Danilov and Solntsev, 1997; Djenkins and Watts, 1972; Goliyandina et al., 1997, 2001).

If it becomes clear in the result of a preliminary analysis, that a larger part of estimated spectrum power concentrates on one or several specific frequencies, data filtration or conversion of each of initial data x_t and y_t to a certain data set $x't$ and $y't$ with the help of various linear and quasi-linear correlations is necessary.

To solve a problem about the use of initial or filtered data for the analysis, correlation functions are calculated to choose a width of the window, with the help of which series under research are seen during the analysis sample.

$$C_{xx}(\tau) = (1/(\mathrm{N}-1)) \sum_{L_{\max}-1}^{\mathrm{N-k}} (x_t - \overline{\mathrm{x}})(x_{t+\tau} - \overline{\mathrm{x}})$$

for values $\tau = 0, 1, 2, \ldots, L_{\max}$. (A cut-off point quantity $L_{\max}$ is chosen from the criterion of sample correlation minimum. A cut-off point $L_{\max}$ is reached when sample correlations $C_{xx}(\tau)$ become equal to 0.05 to 0.1. A problem about the use of initial or filtered data rows is solved on the basis of the condition that correlation functions of them are reduced to zero.)

After solving of a problem about the use of initial or filtered data rows and choosing a cut-off point sample, spectral densities $S_{xx}(f)$ are calculated. (To avoid terminological confusion, spectral density of the process under research is the Fourier transform of its correlation or covariance functions.)

$$S_{xx}(\mathrm{f}) = 2\Delta\left[C_{xx}(0) + 2 \sum_{\tau=1}^{L_{\max}-1} \mathrm{C}(\tau)W(\tau)\cos 2\pi \mathrm{f}\tau\Delta\right] \tag{3.1}$$

where Δ is the series sampling interval, $W(\tau)$ is the lag window with a cut-off M chosen from the correlation $M = \tau\, L_{\max}$ (there are a great deal of lag windows each of which has its own advantages and disadvantages relative to a given problem).

To find out correlation connections between different processes, a mutual spectral analysis is used. It consists of calculation of mutual correlation functions which reveal the delay time of one of the processes under study, relative to another, on the basis of the maximal coefficient of mutual correlation or shift quantity r between the maximum position of a mutual correlation function and zero value r. (Delay time estimation is of a great importance especially during analysis of connections between solar and magnetic activities and processes occurring on the Earth, and can be used by scientists in their attempts to forecast processes occurring on the Earth basing on changes of solar and interplanetary activity.)

Formulas for discreet estimation of mutual spectrums (when two or more data series are estimated simultaneously) are analogical to formulas for estimation of spectral densities of univariate analysis of processes (when only one data series is analyzed), but mutual analysis gives the possibility to get information about phase difference between analyzed processes on each frequency (the phase spectrum) and degree of correlation of processes between each other on different frequencies (the coherence spectrum).

As it is shown in many works about spectral analysis, if two series in some frequency interval are moved in relation to one another to some timely interval or they can be presented as

$$x_{2t} = z_{2t} + \beta_1 z_{1t-d}, x_{1t} = z_{1t}, x_{2t} = z_{2t} + \beta_1 x_{1t-d}$$

A phase spectrum in this frequency range is regarded as a linear function of frequency. It means that a cosine wave of frequency *f* Hz makes *fd* oscillations for a delay time *d* and, consequently, the phase delay is 2*nfd* rad. So, on the basis of a phase spectrum showing phase differences between processes it is possible to find the delay time of one series relative to another for each frequency, and it can be of a great importance during forecasting of different duration (short-term and long-term).

Analysis of parameters of solar activity, geomagnetic disturbances, solar radiation, global and local atmospheric processes, and cosmic ray intensity can be made with the help of several methods among which are univariate analysis and mutual spectral analysis, two-dimensional spectral analysis Hissa, multiple correlation conversion, and methods of periodically correlated sudden process theory.

The analysis that follows was made on the basis of data of monthly average and yearly average solar activity values (Wolf numbers *W*, sunspot area *S*, HL-index for 1880–2008), cosmic ray intensity (registration data of cosmic ray stations at Kiel, Troitsk, Appatiny, Mac-Merdo, etc., for 1963–2008), geomagnetic activity (Kp and Ap-indexes for 1945–2001), the atmosphere circulation (storm degree index *P* for 1950–2000), temperature *T* °C, pressure and precipitation (registration data of meteorological stations of Russia, Estonia, Mexico, Sweden, Lithuania, Canada for 1920–2008), levels of closed water systems (Lake Pátzcuaro in Mexico, Lake Chudskoye and Lake Baikal in Russia, the Caspian Sea for 1880–2007), solar radiation (registration data of meteorological stations of Russia, Mexico, Cuba, Denmark, and Canada for 1960–2008), and ice area in the Baltic and White Seas (registration data of meteorological stations of Russia, Estonia, Denmark, and Sweden for 1920–2008) (Abuzyarov, 2003; Babkin, 2005, 2008; Dorman et al., 1986b; Dorn, 2008; Filatova et al., 2003; Leyva-Contreras et al., 1996a,b; Doganovsky, 1982, 1990; Libin et al., 1996a,b,d,e; Marsh, 2003; Mikalajunas, 1973a,b).

Spectral estimations were calculated on the basis of all intervals as a whole and 72-month (or 144-month) realizations with 12-month sliding shift (to receive a time-detailed picture of behavior of correlation and spectral estimations during 18–24 solar activity cycles). The accuracy of obtained results was controlled with simultaneous appliance of different special methods and procedures (data filtration, closing spectral windows, etc.) as well as constant use of test programs.

It is possible to get detailed information about spectral estimation calculations from many texts, but it is necessary to take into consideration that the preliminary stage of any mutual spectrum calculations estimations of displacement between processes must be done with the help of calculation the of maximal value of mutual shift correlation (Max, 1983),

$$K_{xy}(\tau) = \sum_{t=1}^{N-\tau}(x_t - \overline{x})(y_{t+\tau} - \overline{y})/\sigma_x^2\sigma_y^2$$

or, on the basis of a shift of τ between mutual correlations function maximum and zero. An example of these calculations is shown in Figure 3.1. Here we see a cross-correlation between solar activity and temperature. The magnitude of the shift is 36 months.

All tests of spectral analysis programs allow the use of different methods in different conditions for concrete events and data series. We can test all used programs of dynamic spectral analysis with the help of the function $S(f) = \sin(f_0 + \Delta f) + \sin(f_1) + \sin(f_2)$. It is possible to be sure that if the position of density spectrum peaks is not changed on fixed frequencies, on frequency $f_0 + \Delta f$ a peak is transformed slowly to the region of higher frequencies tracing the change Δf_i.

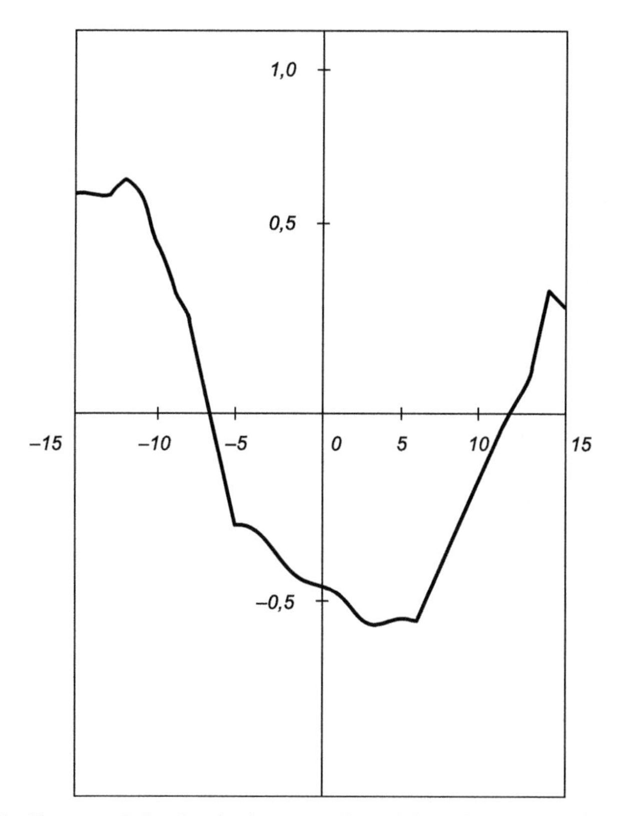

FIGURE 3.1 Cross-correlation function between solar activity and temperature for years 1950 to 2010. *Source: Pérez-Peraza et al., 2005.*

It is necessary to note that practically all spectral methods are useful for stationary processes. When all the analyzed processes are not stationary, data filtration precedes calculation of spectrums, especially the more dynamic ones.

3.1.2. Autoregressive Methods of Analysis (ARMA) for Estimation of Connections between Cosmophysical and Meteorological Parameters

Together with the use of spectral methods the use of independent periodicity methods with properties different from the others is reasonable. (This is important for mutual control of obtained results.) Statistic characteristics of the series of analyzed processes may be rebuilt. In the case of nonstationary processes the classical concept of spectrum is not well defined, such that data rebuilding based on the fast Fourier and Blackman–Tukey transform methods often give incorrect results (Max, 1983).

A usual way in such situations is the separation of quasi-stationary parts, but this usually meets some difficulties. Such parts (if they exist) can be short, and the Fourier method based on a small quantity of data gives bad results and does not allow separate close frequencies. At the same time the frequency separation problem is one of the main points of the problem under study, as each of these frequencies can be connected with different physical mechanisms of interaction of cosmophysical and meteorological parameters.

To separate frequencies on short data sections, autoregressive methods have been used. The essence of them consists of an additional supposition that each researched process can be described with an autoregressive model of an unknown order p (for $t = 0, 1, 2, \ldots$).

$$x_{t+1} = \sum_{i=0}^{p} a_{i+1} x_{t+1} \tag{3.2}$$

In this supposition, autoregression coefficients are estimated with this or that method, the best order is searched, and the spectrum is calculated on the basis of these coefficients.

This approach with the use of different algorithms such as Berg (1938), Levinson–Derbin (Bendat and Pirsol, 2008), Pisarenko (Key and Marple, 1981), and Prony (Key and Marple, 1981) methods and their modifications (Gulinsky et al., 1992; Yudakhin et al., 1991) give good results in some cases.

To get over the difficulties the authors (together with Gulinsky et al., 1992; and Yudakhin et al., 1994) suggested the following approach: It was supposed that the process could be described with an autoregressive model in which coefficients themselves were changing in time.

$$x_{t+1} = \sum_{i=0}^{p} a_{i+1}(t) x_{t+1}, \quad t = 0, 1, 2, \ldots. \tag{3.3}$$

It is clear that such a process is not stationary. Each coefficient is then presented as a given full function system series $\{\varphi_K\}$

$$a_i(t) = \sum_{k=1}^{N} C_{iK}\varphi_K(t) \tag{3.4}$$

with unknown coefficients $\{C_K\}$. A power series $\{1, x, x^2, \ldots\}$ can be particularly chosen as a function system.

Then coefficients C_k are calculated with the least area technique for the elected decomposition number N (4) and the model order p (3). The model order p and term number N in decomposition can be chosen optimally in some sense. Such approach allows us to enter the term instantaneous spectrum for the nonstationary process.

In each given moment t^* the parameters $\{C_K\}$ must correspond to an autoregressive model with known constant coefficients.

$$a_i(t^*) = \sum_{k=1}^{N} C_{iK}\varphi_K(t^*) \tag{3.5}$$

Such a process is stopped in the moment t^* (it can be continued to the eternity). This process is a stationary and some t^*-instantaneous spectrum which is calculated analytically based on coefficients C_K corresponds to it.

By building an instantaneous spectrum succession in time we can study dynamically the rebuilding of the process.

Together with classical spectral methods of analysis, other modern methods of time series nonlinear analysis and nonlinear forecast were used to control the results.

Reconstruction of universe dynamic system model in the Euclidean space of convenient dimension can be the basis of modern approaches to predict time series (Makarenko, 2005).

1. System trajectories fill an attracting set of a small dimension d—an attractor—in the phase space
2. The first coordinate trajectory protections are continuous nonlinear functions of phase coordinates
3. Protection values in discreet time becomes an observed time series
4. There is an ergodic invariant measure which can be estimated as a dwelling time of a point in the elementary phase volume

Under these assumptions, it is possible to build up an embedding of time series enclosing

$$\{x_i\}, i=1, N \text{ in } R^m, m > 2\Delta$$

which will be a topological copy of the real attractor.

A copy dynamics is specified by a nonlinear regression equation $x_i + \tau = \Phi(x_i)$, $x_i = (x_i, x_{i-\tau}, \ldots, x_{i-(m+1)\tau})$ which is just a predictor model. Function $\Phi \in C1$ is even differentiable but specified by a finite pair set $\{x_{i+\tau}, \Phi(x_i)\}$ on the series history. That is why, any approximation problem in L_2-metrics is not correct.

Possible approaches to the solution on the level of technical rigor are divided into local and global methods. The first are based on the Lorenz analog method and add up to approximation Φ in local vicinity of each reconstruction point. Polynomials of a small degree are usually used for that. Global methods approximate Φ in all points at once.

Radial basis functions (RBF) and artificial neuron networks (ANN) serve as instruments for that. The latter represents the m variable function as a superposition of functions, only from one variable, using one nonlinear standard function of a formal neuron.

There are two important problems. The first is connected with estimation of a series predictability horizon which is determined by a maximal Luapunov exponent of a universal model. The existence of such exponent is connected with recession of close trajectories of chaotic dynamics and leads on practice to exponential growth of the prediction mistake.

The situation can be improved with the help of a vector prediction scheme. The second problem is connected with a model mistake. Such mistake appears from noise in data and unknown latent parameters which Φ can depend on.

Changing of parameters leads to rebuilding the dynamic regimes that are observed as nonstationarity of time series. Makarenko (2005) discusses two approaches to decrease the model mistake. The first is based on the idea of nonuniform enclosing and meant for time series which have several correlation lengths. Using the principle of Minimal Description Length it is possible to find RBF parameters to improve the model. The second approach leads to corrections of the first prediction made with the help of ANN.

Incorrectness of an approximation problem leads to a large number of possible variants of the prediction or forecasting.

3.1.3. Modeling a Mechanism of Interaction of Heliophysical and Geophysical Processes

Modeling with the help of autoregressive models is based on the supposition that a predicted value is a weighted sum p of previous readings (one-parameter representation) or weighted sums p, q,... of previous readings of different time series (multi-parameter representation). Researches made for different meteorological, solar activity, geophysical processes and galactic cosmic rays have shown that any meteorological parameters $P(t)$ measured in some periods of time in the framework of the autoregression model can be represented as

$$P(t) = \sum_{i=1}^{q} a_i\, P(t-i) + \sum_{i=1}^{s} b_i W(t-i-w) + \sum_{i=1}^{r} c_i\, K_p(t-i-k) + \sum_{i=1}^{m} d_i\, I(t-i-i) + \xi_t \quad (3.6)$$

where a, b, c, d are coefficients calculated with known values P, W, K_p, and I. In Equation (3.6) W is Wolf number; K_p is geomagnetic activity index value; I is cosmic ray intensity for periods preceding the time t; q, r, s, and m are an order of multi-parameter autoregression used for building up a prediction model; ξ_t is a succession of independent random vectors; w, k, and i are delays between W, K_p, and I and the researched (predicted) process P.

On the basis of accumulated sets of atmospheric data, solar activity, geomagnetic activity, and cosmic ray intensity data with dimension N_0, one can create a matrix for a system of linear equations (Equation 3.6) the solutions of which determine vectors {a},{b},{c},{d}. It is necessary to take it into consideration in this case that sets of autoregression coefficients can be determined almost for each period of time. So, using monthly average data for the period 1950 to 2000 we get about 600 equations. Consequently, if we take yearly averaged values, the number of equations is decreased up to 50k where k is the maximal value of any values q, s, r, m; so, the number of equation becomes about 40.

Figure 3.2 shows the results of dynamic spectral analysis of the SA and the temperature. Similar results were obtained by the authors for the other meteorological parameters.

The aim of autoregression parameter estimation is to solve the system of linear equations like Yule-Worker one (Key and Marple, 1981). Using model

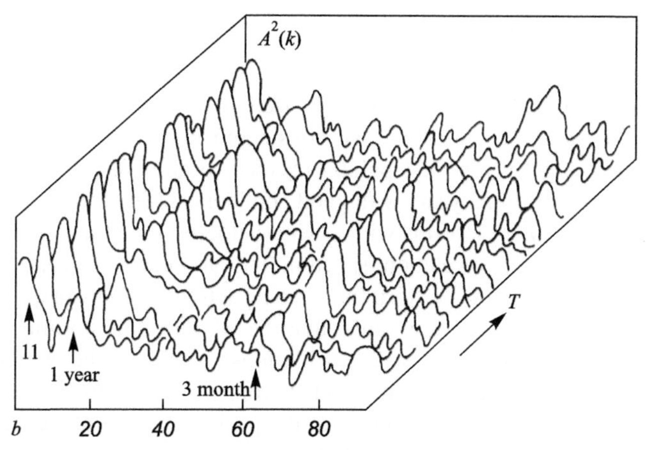

FIGURE 3.2 Results of dynamic spectral analysis of variations in solar activity and temperature. *Source: Pérez- Peraza et al., 1997.*

orders for atmospheric processes, solar activity, geomagnetic activity, and cosmic ray (influencing the atmosphere transparence (Pudovkin, 1995)) it is possible to predict average values (or any other characteristics of the Earth) of the atmosphere one step forward, and the distance of this step will be determined only by data discretization level.

One should note that calculation of autoregression coefficients can help to evaluate a contribution of this or that process to the predicted process. So, during analysis of atmospheric processes, autoregression coefficients for geomagnetic activity were very small (i.e., contribution of it into atmospheric circulation is small), but coefficients for cosmic rays observed on the Earth are rather big (Libin et al., 1996a), which later was explained in Pudovkin-Raspopov's model of solar activity influence on atmospheric processes (Pudovkin and Raspopov, 1992a,b).

So, using standard autoregressive models to work out methods of forecasting space and geophysical processes of the atmosphere changeability produces good results (Pavlishkina and Mikhaylov, 2002; Shiklomanov et al., 2003).

A result of the first steps in this direction was the yearly forecast of the level of closed water systems in Mexico (Lake Pátzcuaro) and Russia (Lake Chudskoye) based on solar activity, cosmic ray intensity, and preceding values of lake levels. Veracity of such forecast for 2003 was 9%; that for 2004 was 8%; 8.5% for 2005; 12% for 2006; and 11% for 2007. 2008 to 2011 are in the same order.

3.2. STUDY OF SPECTRAL CHARACTERISTICS OF SOLAR ACTIVITY AND CLIMATOLOGICAL PROCESSES

3.2.1. Study of Correlational and Spectral Characteristics of Space Processes and Storminess

Several years ago the authors made spectral and autoregressive circulation research in regions of the North Atlantic Oscillation (NAO; storminess in the Northern Sea) for 1950 to 2002. We considered continual sets of monthly averaged data of 52 years (624 points) to study oscillations with cyclicities from several months to 22 years with good confidence. (*Storminess* is an index of regularity of powerful storms with a wind velocity more than 17 m/s.)

As a result of joint correlation, spectral, and autoregressive analysis (with study of dynamic spectra) of storminess, solar activity, geomagnetic activity, and cosmic ray intensity, it was found that the whole set of storminess oscillations are tightly connected with oscillations of helio- and geomagnetic activity, especially in the frequency range from 2×10^{-9} to 8×10^{-8} Hz. (Model research of a spectrum show that the peak amplitude at the frequency 3×10^{-9} Hz does not provide peak amplitudes on other frequencies, and modeling slow trends on frequencies more than $5-6 \times 10^{-9}$ Hz does not lead to any spectra influence. Sudden emissions intensify only the low-frequency part of the spectrum which tells on quantity B in an essential degree and does not tell on the value of γ, if the spectrum is of the form $P(f) \sim B f^{\gamma}$.)

The authors created a data-processing system in which some secondary procedures were realized: high and low frequency filtration, elimination of regular changes, estimation of process imbalance, or appearance of non-stationarity and calculation of basic statistic characteristic of the process.

The analysis showed that processes of storminess and solar activity are, firstly, opposite in phase and are, secondly, shifted by 3.5 to 4.0 years relative to each other. (Update of this estimation based on less data intervals leads to close estimations: storminess P retardation relative to solar activity W for periods, for example, 1950 to 1963 gives a 4-year value, for 1961 to 1976 a 3-year value, for 1977 to 1994 a 3-year value, and for 1993 to 2002 a 4-year value. For the whole period of researched data—52 years—the total retardation is about 45 months.)

Analogical calculations of mutual correlation functions between solar activity (W and S) and geomagnetic activity (K_p) parameters lead to a 1.5 year; Kp-index retardation relative to solar activity, and both processes are in phase almost in the whole 52-year period.

Together with it, joint correlation analysis between geomagnetic activity (K_p) and storminess (P) leads to a 1.9 to 2.5 year storminess retardation relative to geomagnetic activity (the processes are opposite in phase) which corresponds well with the results mentioned above. Obtained results correspond well with analysis of a cosmic ray intensity retardation effect relative to a helio-latitude index of solar activity HL where retardation quantity is changed from 6 to 20 months depending on the solar activity cycle (Mikisha and Smirnov, 1999; Gulinsky et al., 2002; Mikalajunas, 2010).

Calculations of mutual correlation functions between cosmic ray intensity I (stations Apatite and Moscow) and storminess P show that cosmic rays pass ahead of storminess by 3 years which coincides with the results of calculation of P retardation relative to W if it takes pairs W-I and HL-I analysis into consideration.

Let us note that dependence of atmospheric phenomena on solar activity factors does not add up to only adequate image of processes, growth, or decay of their parameter quantities in branches of solar activity cycles (during analysis it is enough time to watch dynamics of spectral estimation behavior), but also appears to be a process of set amplitude and phase rebuilding relative to small-scale oscillations with periods 3 to 4, 6, and 12 months, and large-scale ones with periods 2, 9 to 15, 20 to 28, and even 80 to 90 years. In this case relatively constant retardation between storminess and solar activity becomes essential. It allows us to get a mode of approximate estimation of an average level of storminess 1 to 2 years before.

Research showed that one can separate four cycles in time series of storminess in the Northern Sea (and to a less degree in the Baltic Sea), as follows.

1. One of 24 hours connected with differences in heat capacity of an underlying surface
2. A synoptical one appearing in the result of cyclone and anticyclone activity
3. A yearly one connected with season intensity pulsation of energoactive zones of the Northern Atlantics;
4. One of many years (climatological) conditioned by different geophysical, heliophysical, and tropospheric factors

Similar conclusions can be made for the climatological cycle (on the basis of joint analysis of all four processes in the framework of a multiparametrical model)—W, P, I, and K_p for the whole 52-year period contain not only clearly separated (with more than 95% of veracity) 11-year and quasi-biennial variations of all processes but a precise structure in the region of higher frequencies (22-year and century variations).

Obtained spectral estimations agree with results of spectral analysis of energy exchange ocean-atmosphere in Northern Atlantics (Ariel et al., 1986) which demonstrate temporal scales of climatological and within-year variability that coincide with spectral estimations of storminess, solar activity, and cosmic ray intensity.

3.2.2. Tendency of Changes of the Baltic Sea Ice Area (glaciation) with Climate Change Estimation

3.2.2.1. Study of Glaciations with Method Track

To identify the source of fluctuations in different hydrological and climatic data, it is interesting to compare the basic oscillation periods in many year periods of the yearly average series of the level of Lake Chudskoye (see Section 3.2.4), with the oscillation periods of the series of total afflux to lakes Pesvo and Udomlya (which are located close to Tchudskoye Lake). Cross analysis shows that the detected oscillations with periods from several months to 4–5 years for the two objects indirectly connected from each other (are observed differences in the periods and amplitudes of the oscillations). Oscillations with periods of more than 11 years are the same for all the studied lakes and appears to have a common origin. Nevertheless, these objects are too intertwined. Needed other data's to compare the detected fluctuations of various meteorological series and identify their possible common source.

In the capacity of such a series, the glaciations series of maximal area of the Baltic Sea ice can be used (Libin, 1996a). Waters of Lake Chudskoye (Shpindler and Zengbush, 1896) and lakes Pesvo and Udomlya flow into this sea, and one can expect that any oscillations revealed in these lakes to be reflected in oscillations of its glaciations. There is a reason to suppose that the leading 28.5-year period in oscillations of the first two series will become apparent in the glaciations series as well.

Jaani (Jaani, 1973; Solntsev, 1997; Solntsev and Filatova, 1999; Solntsev and Nekrutkin, 2003; Libin and Pérez-Peraza, 2009) researched 273-year series of observations characterizing maximal area of the Baltic Sea ice from 1720 to 1992. It is clear that there is no single standard solution to the problem of separating the trend from long time series. The choice of the solution method depends on *a priori* information about the model of the process under study and on the behavior of the real statistic data describing this process.

The peculiarity of their data is the distribution, which is very far from normal. So, when the average value of ice areas is 218,000 km^2, the minimal one is 52,000 km^2 and the maximal one is 420,000 km^2—a standard deviation of them is rather big, 114,000 km^2.

A sample histogram shows that its distribution has almost no tails; it looks somewhat deformed (there is a noticeable positive asymmetry). A lack of tails is determined by the fact that the maximal value of ice area is limited from above by the Baltic Sea area (420,000 km^2) and from below by ice area of shallow gulfs which are frozen even in the warmest winters.

In this case it is difficult to expect that traditional procedures like a sliding average will give clear, direct information about the presence or absence of some searched slow trend. But there was an attempt to apply such a procedure—polynomial approximation—with a least-area method. The best approximation was received for a third-degree polynomial. However, the confidence belt for a regression line is too wide to talk about veracity of this solution. A result of polynomials of higher degrees is even worse.

In connection with inadequacy of the obtained results, an initiate series of ice area was researched with the method track: $M = 40$ taken as the parameter. Studying the basic components shows that high-frequency constituents dominate at a high degree. Among them the largest part of dispersion falls on oscillation with a period 5.4 years (it becomes apparent in the first 3 basic components and explains the 14% of dispersion), with a period of 7.8 years (4 and 5 components, 7.8% of dispersion) and with a period of about 3 years (7 and 7 components, 7.5% of dispersion). A slow trend is weakly presented in the first and the third components.

To suppress high-frequency components of the process, different procedures of series filtration were used. A sliding average is the simplest linear filter of this kind. However, now when the character of a signal hindering components is known it is possible to manage width of an averaging interval, find a filter characteristic to suppress these harmonic components as much as possible. We know that a sliding average suppresses wonderfully the harmonic components of which period coincides with the average interval width or keeps within this interval integrally (it appears to be a common multiple for these periods).

In our case 15-point summing up was to suppress components with periods of about 3, 5, and 7.5 years, three most powerful components. After 15-point smoothing the first 7 and the last 7 points were excepted from the obtained series (as they are calculated on the basis of shorter series segments and can induce distortions in the further analysis), and an analysis of the smoothing series with the program track was made. Then the track length was chosen to be 30, which gave a more precise reproduction of a slow trend. In this case the first main component containing more than 60% variations and picking out a slow trend is sharply distinguished.

To control the results, the analogical analysis of the series obtained by smoothing the initiate data with a sliding average method was made on the basis of seven points with further exception of the first three and the last three. Herewith the weight of the first main component corresponding to a slow trend became half as much.

This fact confirms doubts that the use of the initiate time series linear smoothing procedures is optimal, which is connected with the above-mentioned difference in distribution of researched ice area values from normal.

Such data (Blackman and Tukey, 1959; Bendat and Pirsol, 2008) suggested to apply the so-called resistant (steady) smoothing, using not an arithmetical mean but a median average or weight averaging and special procedures of flat segment exception.

Those authors made five different smoothing procedures (the procedures were realized in the statistic packet STATGRAPHICS). Each procedure uses not more than five series points coming one by one. Obtained results appeared to be rather close to each other and much different from linear smoothing results. For the further analysis, a series was obtained with the average of the five mentioned smoothing variants and worked up with the track method. In this case a slow trend becomes apparent in the first main component, explaining 26.5% of dispersion; the second and the third components (28% of dispersion) release oscillations with a period of about 19 years.

The most important result of this stage of research may be the following: the three smoothing procedures gave almost coinciding curves of a slow trend looking like a sinusoid segment with a period of about 300 years or a cubical polynomial with a curve bend part. Thus it is proved that the hypothesis about the climate warming in Northern Europe is not confirmed by the researched material. The opposite supposition about a cooling period supports that it began in the 1950s because slow periodical climate oscillations can be expressed. Moreover, it is necessary to note periods when failures happen in the process dynamics. Such periods are visible on diagrams reestablished on the basis of separate components of the process values. These periods correspond to time intervals 1790 to 1800 and 1910 to 1920. These intervals on the diagram of a slow trend were reestablished, on the basis of the first main component, and correspond to the points of minimal and maximal ice areas.

In the process of data analysis with the track method, a great number of periodic components with periods from 2 to 40 years were found. Thus, in main components with big number oscillations, periods 5.9 years and about 20 years are also found.

To control these periodic components, a periodogram of the initiate series was also calculated. Although exactness of process component period definition is small, it is possible to say that the basic components received with the track method on the previous analysis stages are found in this case as well. Thus the following components were found in a descending order—about 300 years, 90 to 100 years, about 46 years, 27 to 30 years, about 20 years, 14 to 15 years, 10 to 13 years, 8 to 9 years, and some others.

3.2.2.2. Study of Glaciation with Method of Multivariate Spectral Analysis

The works of Wald (1960) (a consecutive analysis and a general statistical decision theory), Blackman and Tukey (1959) (methods of jack knife and multiple comparisons), Efroimson (2002) (step-by-step procedures), Huber (1981) (robust procedures), Efron (1979) (bootstrap method), Rao (1971) (common multivariate linear models), and many others led to a step-by-step review of a statistic analysis common technique which was applied 20 to 30 years ago. That is why together with the research on the track method based on the 282nd series of observations characterizing maximal ice areas of the Baltic Sea since 1720 to 2002 and the methods of Blackman and Tukey (1959), a spectral and autoregressive analysis was made and analysis was carried out separately for even and uneven solar activity cycles. The results of the autoregressive analysis showed the existence of Baltic Sea area ice oscillations with periods of 80 to 90 years, 20 to 22 years, 9 to 13 years, and 4 to 7 years.

It is necessary to note that mutual autoregressive analysis for the whole data set and separately for even and uneven solar activity cycles shows an interesting picture. In uneven cycles, one can observe well-expressed 4- to 7-year, 10- to 12-year, and 80- to 90-year area ice variations (on the background of weakly expressed 300-year variations), in even cycles 20- to 30-year, 89- to 90-year, and 300-year variations prevail.

It is natural to suppose that the existence of 11-year area ice variation in uneven cycles and the absence of it in even cycles only intensify 22-year variation. Another important aspect is that obtained results correlate well with any solar activity indexes.

There is still one more conclusion made by the authors from those calculations. During filtration of the initiate series with a sliding average of periods 50, 75, 100, and 150 years and further joint auto-regressive analysis with analogical solar activity data, a secular trend with period longer than the prevailing found variations was discovered (>700–1000 years). This trend behavior indicates a total (very large) common warming of the climate in the Northern Hemisphere especially in the last 100 to 200 years.

Obtained results fully coincide with the results of many years of research made by V.V. Betin and Y.V. Preobrazhensky, who studied glaciations of the

Baltic and winter severity in Europe for the period between 1770 and 1950 (Clayton, 1933; Shnitnikov, 1949; Klimenko, 2008, 2009a,b; Lukin et al., 2009; Pokrovsky, 2005). Their aim was to make a forecast of these measures for the next 30 years, up to 1980.

The predicted change of the Baltic glaciations justified itself; the Baltic maximal glaciations predicted for 1959 to 1960 really took place, and later, after 1960, as it had been predicted, the Baltic Sea glaciations began to decrease.

Research showed that the Baltic Sea glaciations changed with different periods, such as 80 to 90 years, 28, 22–20, 15–11, 6–5 years, and even 3–2 years (Dergachev and Raspopov, 2000; Klimenko et al., 2007, Libin et al., 1996a; Libin and Pérez-Peraza, 2009; Sitnov, 2009).

Air temperature and river flows are changed almost in the same way (researchers used temperature data measured in Helsinki and river flow measuring results). In his work Babkin (2008) describes the results of researching the Volga flow oscillations in the late Pleistocene. It is shown that the Volga flow in separate periods of the late Pleistocene was changed in the range 600 to 120 km^3 per year. Structures of many year flow oscillations of the upper Volga, Oka, Don, and Dnepr for the period of hydrological research lasting 4 to 5, 10 to 12, 20 to 24, and 80 to 90 years were found (Vakulenko et al, 2003).

Figure 3.3 shows the results of the mutual spectral analysis of the ice cover of the North Atlantic Oscillation (NAO) and solar activity (for the years 1950–2005). Similar results were obtained by the authors for the other meteorological parameters.

3.2.3. Modulation of Solar Radiation Observed on the Earth and Its Possible Connection with Solar Activity Changes

The solar radiation spectrum is close to the spectrum of a black body heated to the temperature 5770 °K with radiated energy deficit in the range of close ultraviolet radiation. Solar radiation intensity in far ultraviolet radiation and in the X-ray range is several orders more than corresponding radiation of an absolutely black body.

The difference in solar and black-body spectrums is explained as follows: short-wave radiation in different ranges of wavelengths are generated in different regions of the Sun's atmosphere. Particularly radiation with a wavelength $\lambda <$ 1500 Å is generated in the chromosphere and the solar corona, or in the regions where the temperature is much higher than the temperature of the photosphere.

At the same time we know that the chromosphere and corona parameters are very changeable and depend greatly on the solar radiation level, which is why it is not surprising that short-wave solar radiation intensity also changes together with solar activity level from day to day.

A relative value of solar radiation intensity cyclical variations reaches 10% if $\lambda = 300–500$ Å and decreases sharply if $\lambda > 2000$ Å. Consequently the solar constant is not subject to variations of more than 0.1%.

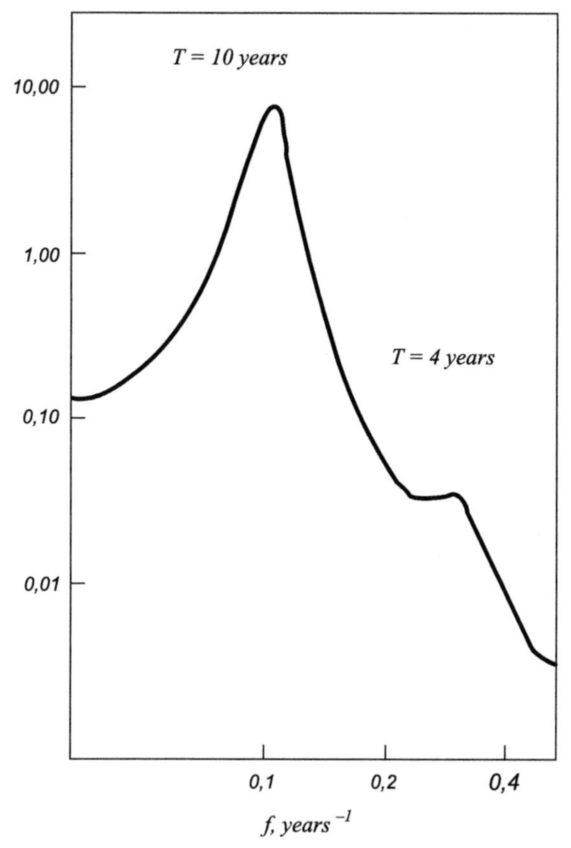

FIGURE 3.3 Mutual power spectra of the ice cover of the NAO and solar activity for the years 1950 to 2005. *Source: Pérez-Peraza et al., 2005.*

Lately, a lot of publications have appeared that talk about the fact that solar radiation variations are not the energetic source of the observed atmospheric disturbance. In connection with that the results obtained by Kondratyev and Nikolsky (1982, 1983, 1995a,b, 2005) and Kondratyev (2004a,b) the authors support a noticeable (about 6%) change in the 11-year solar cycle of the atmosphere transparency, results that are of great interest.

3.2.3.1. The Atmosphere Transparency Variations

A relatively dense atmosphere protects the Earth's surface from fatal short-wave radiation with $\lambda < 3000$ Å. A wide window in the atmospheric screen is observed on wavelengths $\lambda = 3000 - 10{,}000$ Å or in the intensity maximum region in the solar radiation spectrum, which provides penetration of a larger part of solar energy to the lower atmosphere and the Earth's surface.

The existence of a second window on wavelengths $\lambda = 7000$–$15{,}000$ Å is also very important. This wavelength range corresponds to the maximum radiation of a black body heated to the temperature $T \approx 300$ K, which is close to an average temperature of the Earth's surface.

However, optical characteristics of the atmosphere are not given quantities forever. Absorption of solar radiation in the atmosphere depends on ozone contents, water vapor, carbon monoxide, and other small components which have concentrations that can be changed.

As a result, the thermodynamic balance in the atmosphere is very fragile and may be broken easily. A constantly increasing inflow of carbon dioxide formed as a result of human technical activity to the atmosphere leads to a decrease of velocity of warmth withdrawal from the atmosphere (a greenhouse effect) and, consequently, to the rising of the Earth temperature. A noticeable change of chemical composition and contents of small components, as well as the atmosphere transparency is induced, in particular, with ionized radiation flux variations in the atmosphere during magnetospheric disturbances.

A decrease of cosmic particle flux produces an increase of the atmosphere transparency, while an increase of such particle flux should induce a decrease of the atmosphere transparency. Basic calculations show that a total solar energy flux on the latitude zone 55° to 80° increases or decreases by ~ 3×102^6 erg/day which is commensurable with the power of observed atmospheric processes.

Detailed research of possible changes of the atmosphere chemical composition, its optical characteristics, and a high-altitude profile of air temperature in the lower atmosphere was made by Brasseur et al. (1999) and Seinfeld and Pandis (2006). According to the model suggested by them, penetration of energy particles to the atmosphere induces ionization and N_2 and O_2 molecule dissociation.

Ions appearing here participate in a complex of photochemical reactions; one of the products of these reactions is nitric oxide (NO) which interacts actively with ozone molecules

$$NO + O_3 \rightarrow NO_2 + O_2$$

Ozone is destroyed in its interaction with atomic oxygen

$$O_3 + O \rightarrow 2O_2$$

Thus, penetration of energetic particles to the atmosphere induces ozone destruction O_3 and formation of nitrogen dioxide NO_2. In its turn it produces significant changes in the radiation balance of the atmosphere. In the lower atmosphere and on the Earth's surface the solar ultraviolet radiation flux with $\lambda < 3250$ Å increases as a result of a decrease of its absorption by ozone.

In the framework of ecological programs connected with atmospheric pollution a long-term experiment was made (data for 50 years from 1958 to 2008 were used) on the basis of solar radiation measures on some critical points of the Earth which caused serious concern: Mexico, Moscow, St Petersburg, and Vilnius. Together with estimations of contribution of human industrial activity (emissions of dust, combustion products, and exhaust gases to the atmosphere, aerosols, etc.) an attempt to estimate the possible modulation of solar radiation observed on the Earth with solar activity was made.

Analysis of connection of 11-year cosmic ray intensity variation with different solar activity indexes (Wolf numbers, intensity of a green coronal line in the wavelength 5303 Å, number of sunspot groups, solar radio-frequency radiation, sunspot area) has shown that changes of solar activity characteristics under study are connected with changes of the cosmic ray flux.

The existence of well-expressed correlation dependence between long-period changes of cosmic ray intensity and indexes mentioned above is not casual, as each of them reflects common characteristics of solar activity cyclicity. However, direct use of solar activity indexes is possible only with analysis of atmospheric processes (but not always). Such use of indexes for cosmic rays is not correct, as solar wind plasma fluxes from different helio-latitudes possess different effectiveness of modulating influence (Guschina, 1983; Libin et al., 1992).

It is also possible to make an analogical conclusion for solar radiation observed on the Earth, because one of the possible mechanisms of solar activity influence on solar radiation is the modulation of galactic cosmic rays by a solar wind. Moreover, for observations from the Earth of the modulating characteristics of the solar active region, not only its activity and helio-coordinates, but also the angular width of the solar wind flux must be considered.

In the work of Guschina and Dorman (1970) an index was suggested accounting for heliolatitude of the Earth in the moment of cosmic ray intensity registration, nonsimilar activity of the Northern and Southern Hemispheres of the Sun, and changing of sunspot heliolatitude during a solar activity cycle.

$$HL(\Theta_e, \Theta_o, t) = \alpha \int_{\Pi/2}^{\Pi/2} K_i(\Theta, t) \exp\left(- \frac{\Theta - \Theta_e}{\Theta_0} \right)$$

where K_i is a parameter characterizing solar activity in the moment t on the heliolatitude, Θ,Θ_o is a parameter characterizing the angular half-width of solar wind fluxes, Θ_e is the heliolatitude of entrance to the Earth of cosmic rays and solar radiation, except effects appearing because of the Earth orbit inclination to the equator plane, α is a standardized multiplier defined from the condition

$$\alpha \int_{-\Pi/2}^{\Pi/2} \exp\left(- \frac{\Theta - \Theta_e}{\Theta_0} \right) d\Theta = 1 \tag{3.7}$$

where

$$\alpha = \{2\Theta_o[1 - \cosh(\Theta_e/\Theta_o) \exp(-\pi/2\Theta_o)]\}^{-1} \tag{3.8}$$

In the work Dorman et al. (1978) HL-indexes were calculated for 18 to 20 solar activity cycles, where the parameter of activity was sunspot number and area.

Continuous series of monthly average values of a heliolatitude index for more than two solar activity cycles give the possibility to make correlation and ARMA- analysis between total sunspot area, solar radiation, cosmic rays, and heliolatitude index values.

In spite of the fact that all mentioned processes are not stationary, quasi-stationarity does not add significant distortions to spectral estimations received with the help of ARMA-analysis nor significant changes to observed retardation of analyzed processes relative to solar activity (Libin et al., 1996a).

The authors made joint two-dimensional autoregressive spectral analysis of total sunspot area, HL-index, solar radiation, and cosmic rays on the basis of monthly average observation data for the period 1952 to 2000 in Mexico, Russia, and Lithuania. It was found that in a wide frequency range of solar radiation observation data on the Earth, one can observe oscillations with periods 3 to 4 months, 2 years, 4 years, and 11 years tightly connected with solar activity. Comparison of obtained results with analogical research of solar activity influence on storminess shows not only good qualitative, but also quantitative (within retardation) correlation.

The result of behavior of the coherence coefficient and residual dispersion of analyzed processes and solar activity when using an HL-index or total sunspot areas is of a great importance. (Calculations of coherence coefficients between solar activity and solar radiation for quasi-biennial and 11-year variations show that when using sunspot area S, coherence coefficients are 0.7 (for Russia) and 0.8 (for Mexico) and while using an HL-index coherence coefficients are 0.85 (for Russia) and 0.92 (for Mexico). In this case calculations of solar radiation residual dispersion make up a quantity not more than 10 to 20% when describing it with an ARMA-model.)

Calculation of results show that during research of connections between solar radiation on the Earth and solar activity, and building up prediction models on the basis of that research it becomes very important to choose a solar activity index and to take contribution of cosmic ray intensity into account.

3.2.4. Solar Activity and Levels of Closed Water Systems

3.2.4.1. Heterochronism of Closed Water System Level Variations

The conclusion about the existence of a connection between solar activity and lake level variation is not new. In 1917, while processing the observations for 1815 to 1910, Fridtjof Nansen and B. Helland-Hansen showed an inverse relation between air temperature in a tropical zone and a number of sunspots. They found out that the larger a sunspot number was the lower the temperature was and vice versa.

Nowadays trade—antitrade atmosphere circulation is well known according to Brooks' works (e.g., Brooks and Mirrlees, 1932). Air ascending motions are increased in the equatorial zone during years with a great deal of precipitation. In connection with it, the quantity of the air that a countertrade wind (from the equator to the poles) carries away from the equatorial zone is also increased during a year.

The extra-tropical barometrical maximum that appears everywhere in winter and only over the ocean in summer feeds substantially on account of air carried by an countertrade wind. That is why the years with increased air transmission by an countertrade wind have above normal pressure in subtropics.

In (Brooks and Mirrlees, 1932) came to the conclusion that quantity of precipitation in the tropical zone and levels of African lakes are changed parallel to the sunspot number; the highest level of lakes corresponds to sunspot maximum and vice versa. Vize (1925) showed that there is a correlation between the degree of Arctic glaciations and the water level in Lake Victoria located at the equator. Thus we can consider that connection of lake levels and solar activity is proven. But the mechanism of atmosphere circulation that makes this remote connection real is not quite clear. It seems that it is impossible to explain it with any process, but we can find an explanation in a series of basic air-mass transport processes looking at the Earth's atmosphere as a whole.

Berg (1938) noted heterochronism of water level variations in the Aral Sea and Lake Sevan (Gokhcha), on one side, and the Caspian Sea on the other. Shnitnikov (1949) noticed synchronism of level oscillations in the Aral Sea, Lakes Balkhash, Alakul, and other closed lakes of semi-arid and arid zones. Thus, the level of the Caspian Sea, which feeds in general only on account of precipitation in the Volga basin in a damp zone, changes heterochronically with levels of arid zone lakes.

In his report "Heterochronism of Increased Moistening Periods of Damp and Arid Zones" Abrosov (1973) described the results that showed that analysis of meteorological elements (precipitation, pressure, temperature, etc.) pointed to the existence of direct connection between solar activity and frequency and intensity of air mass changing over any territory.

We know that mutual antagonistic west–east transfer and meridional circulation are the main mechanism of the Earth's atmosphere circulation. Frequency and intensity of air mass changing rises together with solar activity (SA) increase and it lowers with SA decrease. The basic transfers are increased or decreased concurrently as well.

In his report, Abrosov (1973) described the results of observation of Lake Balkhash level since 1900 to 1954, and there are hydrometric station observation data since 1934. Analysis of their work shows that a rise in level was of an oscillating character on the general background of level lowering. Rise of yearly average levels (in 1935, 1937, 1942, 1943, 1947, 1949, and 1950) alternated more than once with lowering of them (in 1936, 1938, 1939, 1940, 1944, 1945, 1946, and 1951). In 1952 to 1954 a rise of Lake Balkhash level was observed. By 1931 observations had been only visually described (see Table 3.1).

TABLE 3.1 Observational Visual Level of Lake Balkhash

1900–1910	Rise of level
1911–1920	Lowering of level
1921	V.S. Titov notes rise of level
1922–1930	Lowering of level
1931	P.F. Domrachev notes rise of level

Based on data described above we can make the following conclusions.

1. When solar activity in the first cycle was weak, rise of Lake Balkhash level was observed.
2. During subsequent years, the rise of Lake Balkhash levels were observed only for a short time when solar activity showed that one cycle was changed by another that was weak (1926, 1931, 1942–1943, 1952–1954) (Selevin, 1933; Pokrovsky, 2005; Libin and Pérez-Peraza, 2009).
3. Since 1911 Lake Balkhash level has lowered, which correlates well with general increase of solar activity with each new 11-year cycle. All three conclusions can be combined into one general conclusion: the rise of Lake Balkhash level is observed during years when solar activity is relatively weak.

In connection with this we should note that the maximum level of the Aral Sea was in 1911, that of Lake Sevan in 1912. The Balkhash level had been raised until 1911, and the most sudden rise was observed from 1908 to 1909.

The lowering of the Caspian Sea level had begun in 1930 and by 1946 it was 2 meters lower. Such sudden level lowering had not been observed for 100 to 125 years. The largest timely rises of the Caspian Sea level were in 1931 to 1932, 1942, and 1944, and they corresponded to the moments of solar activity when one 11-year cycle was changed by the new one.

Thus we can conclude that oscillations of the Caspian Sea and Lake Balkhash levels change with the same regularity, but changes of their levels usually happen heterochronically and only sometimes synchronically.

According to the results of Vize (1925), the periods with low solar activity and slow atmosphere circulation are years with relatively high atmospheric pressure in the northern and southern polar regions and high glaciations of them. In periods of high solar activity in warm seasons over the Atlantic Ocean the axis of the barometrical hollow situated from Iceland to Eurasian coast moves to the north. As a result, cyclones move about Eurasia more to the north, then over latitudes of Lake Sevan, the Aral Sea, and Lake Balkhash in a damp zone.

Solar activity lowering moves the axis of the barometrical hollow over the Atlantic Ocean to the south in summer which results in some movement of cyclone route to the south, and it seems that in this case their movement along the Iranian branch of the polar front becomes more frequent. During the years when the atmosphere circulation is intensive and an eastern front of the Atlantic–Arctic barometric hollow moves far to the east through the utmost north of Eurasia, there may be little precipitation not only in closed basins of an arid zone, but in basins of the rivers Don and Volga, situated in the north, and rivers of Ural and Western Siberia.

Spectral Analysis of Lake Level

To reveal retardations, to find out values of them, and study general cyclicities in given water content of closed ecosystems (lakes) and solar activity, we have used traditional methods of spectral analysis (developed in supposition of quasi-stationarity of the processes under study) and ARMA methods of autoregressive spectral analysis (useful for nonstationary processes).

Analysis of solar activity parameters (Wolf numbers, sunspot areas, intensity of a coronal line with a wavelength 5303 Å, HL-index, solar radio-frequency radiation on 10.7 cm frequency) and levels of lakes was made with a help of Tukey's correlation and spectral analysis (Andersen 1976; Bendat and Pirsol, 2008; Key and Marple, 1981) and methods of autoregressive spectral analysis (Dorman et al., 1987a,b; Libin et al. 1992; Libin, 2005; Prilutsky, 1988). The analysis was based on monthly average measuring values of solar activity (sunspot areas, HL-index, radio-frequency radiation on 10.7 cm frequency), cosmic rays and surface temperature (Friis-Christensen and Lassen, 1992) and levels of isolated lakes in Mexico (Lake Pátzcuaro), Estonia-Russia (Lake Chudskoye), and Russia (the Caspian Sea and Lake Baikal) for 1880 to 2008.

After choosing relevant intervals for the analysis (we took solar activity cycles and then went through the whole data set moving each time 5 years further, so only two neighboring results appeared to be partially dependent), we began calculation based on a standard procedure described in Libin et al. (1994).

In spite of the variety of applied methods, it is necessary to understand that mutual power spectra give rather reliable quantitative estimations of connections between observed processes and allow us to estimate drifts between them, but the veracity of the whole series of received correlations is dependent on the accuracy of these results. It is also necessary to understand that the veracity of the obtained results of mutual power spectrum calculations is in a high degree determined by the researcher's capability (choice of methods of analysis and approach to estimation of result validity). That is why the autoregressive spectral analysis (ARMA) firstly described in works of Dragan et al. (1984), Prilutsky (1988), Yudakhin et al., (1991), Libin et al. (1992), Danilov and Solntsev (1997), Braulov et al. (1999), Gazina and Klimenko (2008), and Libin and Pérez-Peraza (2009) was used additionally as a criterion of validity.

The autoregressive analysis differs from standard methods, as it allows us to estimate correlations between analyzed data series with 100% veracity in the frequency region, and, more importantly, it is workable for quasi-stationary (and sometimes even nonstationary) processes, which is the case of the analyzed data series of water content and solar activity. However, it is always necessary to remember that all amplitude estimations received during the autoregressive analysis are relative and cannot be absolutely correlated to the initial series, although the behavior of amplitude estimations in time is quite comparable. This means that although we cannot attach results of amplitude spectrum calculations to the initiate data exactly, their dynamics in time is observed.

3.2.4.2. Description of the Researching Objects

Lake Chudskoye (Shpindler and Zengbush, 1896) is one of the largest lakes in Russia and Estonia (and Europe as well). Its water-surface area (3.6 thousand km^3) is the fifth largest in Europe. The common water-collecting area (including the area of the lake itself) is 47,800 km^3. The catchment area is elongated 370 km in the meridional direction from 56° 10′ to 59°30′ degrees of latitude north and has an average width of 160 km (Lake Pskovsko-Chudskoye based on the data of 1983). The lake itself is also elongated almost 140 km in the meridional direction and situated between 57°51′ and 59°01′ degrees of latitude north and 26°57′ and 28°10′ degrees of longitude east. It consists of three parts differing from each other in morphometric and regime characteristics, but make up the whole water body (Jaani, 1987).

The average many-year level is 30 meters (taken by us) and the volume of water mass is 25.07 km^3. The lake is shallow; its average depth is 7 meters. About 240 rivers and streams flow into Lake Chudskoye. Among them the largest rivers are Velikaya (catchment area is 25,200 km^3), Amaiygi (9960 km^3), Vykhandu (1410 km^3), and Zhelcha (1220 km^3).

Based on the results of observations of the lake level from 1840 to 1852 years, academician K. Barr, who worked as the head of the 'Government Commission on the Tchudskoye lake' in 1851 to 1852, came to a conclusion that there was a gradual rise of the lake level.

3.2.4.3. Level Behavior of Lakes Chudskoye and Pátzcuaro (see Figure 3.4)

In 1864 academician G. Helmersen noted the beginning of the lake level lowering. Analyzing possibilities for building a Chudsko-Baltiysky road (Baer and Helmersen, 1886), Spindler and Zengbush (1896) did not find confirmation to the generally accepted opinion about lowering of the lake level, as it was discovered for the first time that the level of the lake changed from year to year.

The latest discovery led to the decision to open the first hydrological stations on Lake Chudskoye in 1902, one of which in the village of Vasknarva is still in operation today.

In 1921 a lake hydrological station was built in Mustvee, which is the main one today and has a continuous and qualified series of observations. So, using data from all the stations we have managed to get a reliable 102-year series of

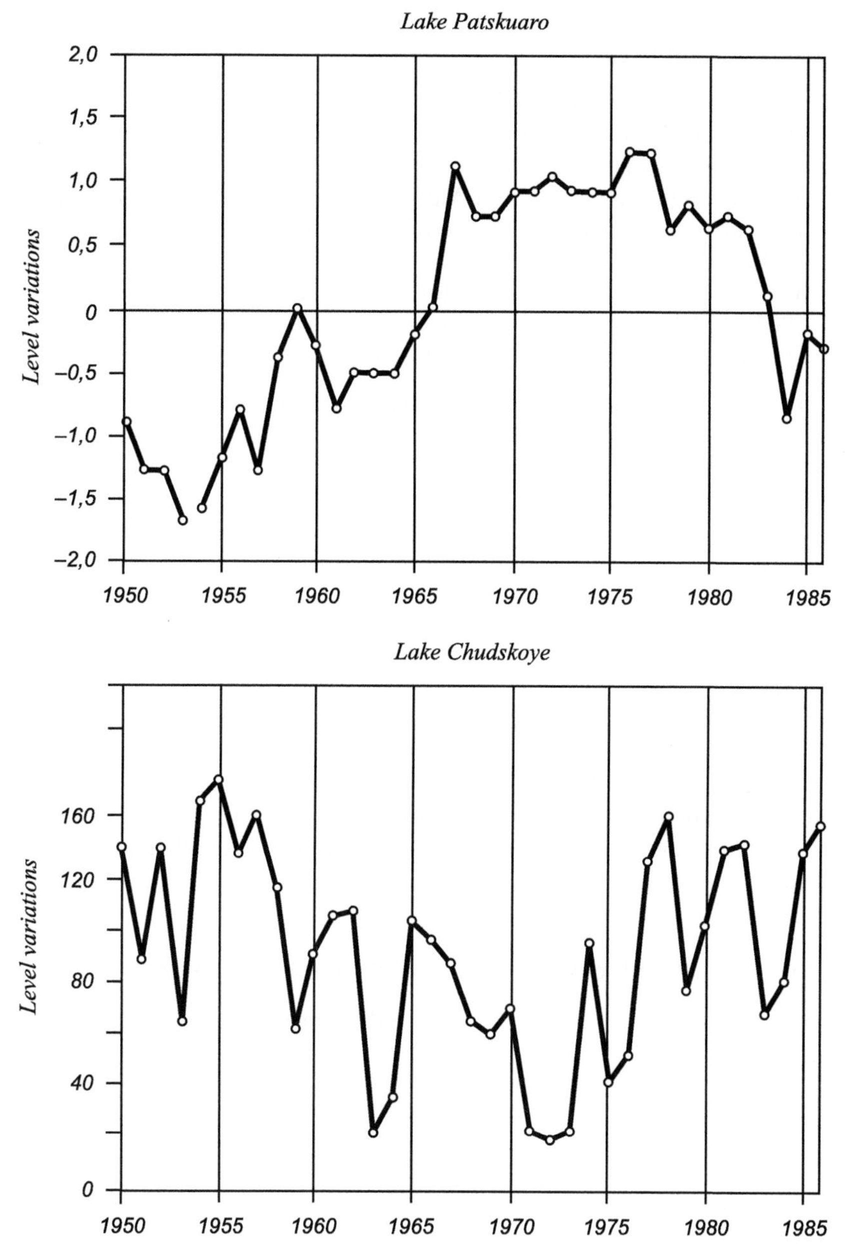

FIGURE 3.4 Behavior of the levels of lakes Peipsi and Pátzcuaro. *Source: Pérez-Peraza et al., 2005.*

observations of monthly and yearly average levels of Lake Chudskoye. The series consists of three parts.

1. Daily natural observations in Mustvee since March 1921 (separate breaks in 1937 and in 1941–1944 are exactly restored on the basis of observations from the other stations)
2. Monthly average observations in Mustvee for separate periods from 1903 to 1917 are restored on the basis of correlation with the levels from the station Vasknarva
3. Yearly average levels in Mustvee for separate years in the periods 1885 to 1902 and 1918 to 1920 are restored on the basis of correlation with data of water level in the river Amaiygi (correlation coefficient is 0.92)

As mentioned before, years 1840 to 1844 were flood ones; a high level was also observed in 1867 (almost 22 years later) and from 1879 to 1884, as it seemed to be. In the period of instrumental observations the highest levels were observed from 1924 to 1928, 1957, and 1987.

As reported by Jaani (1987), in 1940 A. Wollner and in 1971 T. Apere noted the existence of cyclicity in observations of the lake level (Babkin, 2009). Jaani (1973) showed that maximal levels were observed during solar minimums; he also discovered intercentury cycles of water content with duration 19 to 34 years (close to the so-called Brickner cycle) and short-time cycles with duration 4 to 5 years.

Reap (1981) separated cycles with duration 6.1 to 6.4, 10 to 11, and 80 to 90 years. He supposed that in the flow of northwestern rivers minimum of 11-year cycle is observed 1 to 3 years after solar maximum, and maximum of the flow was observed 2 to 4 years before solar minimum.

Results of several authors on the cycles of water level oscillations of Lake Chudskoye are summarized in Table 3.2.

TABLE 3.2 Periods of Water Level Oscillations (Lake Chudskoye)

Author	Periods (number of years)					
Jaani (1973)		5.1–8.0		19–34	80–90	
Reap (1981, 1986)		5.1–6.4	10–11		80–90	
Doganovsky (1990)				26, 33		
Libin, Jaani (1987)	2.6		11.2	22	80–90	
Libin, Jaani (1989, 1990)	2–4		9–11	22		
Current work on three lakes (2009), Libin and Perez-Peraza (2009)	2.6	5.4–7.0	11.0–12.8	22.1–23.0 28.0–35.0	84–91	300–380

Lake Baikal is one of the oldest lakes on the planet. According to scientists it is about 25 million years old. The majority of lakes, especially of glacier origin, live 10 to 15 millions years, and then they are filled with precipitation and disappear. Different from many lakes in the world, Lake Baikal does not have any signs of aging. On the contrary, researches of the last few years let geophysicists express the hypothesis that Lake Baikal is an incipient ocean.

When someone mentions Lake Baikal everybody recalls the fact known from the childhood that the lake contains one-fifth of the available drinking water of our planet. It is true, if the water surface area of Lake Baikal is less than that of the great American lakes Lake Superior and Lake Huron and the great African lakes Lake Victoria and Lake Tanganyika, the volume and depth of it make it the champion among lakes.

The size of the Lake Baikal really make a great impression. The water surface area is 31,500 km^3, more than the size of Belgium or Israel, and the area of the largest of the 22 islands of Lake Baikal—Olkhon—is bigger than the European countries of Andorra, Liechtenstein, San-Marino, Monaco, and Vatican taken together. Lake Baikal is situated in a mountain kettle; its length is 1500 km, its width is 500 km, and the lake contains 23,000 km^3 of the purest water. The enormous water volume of the lake serves as a stabilizer of the climate in the suburbs. It is colder in Lake Baikal region in summer and warmer in winter than in other Western Siberia regions.

The area of the catchment basin of the lake (the territory from which water flows down to Lake Baikal) is more than 500,000 km^2. Water coming to the lake with the rivers Selenga, Bargusin, and Upper Angara stays there for more than 400 years, and then leaves through the head of the Angara. This explains the unique purity of the lake water.

An important peculiarity that makes Lake Baikal different from other deep lakes is its high concentration of oxygen in the water spreading to its depth. Maybe that is why Lake Baikal is covered with ice quite late (in January) and only gets free from the ice in June and July.

The first reliable (from scientific point of view) information about Lake Baikal refers to the eighteenth century. In the middle of the seventeenth century the first scheme of Lake Baikal was made. In the second half of the nineteenth century regular scientific study of Lake Baikal was connected, first of all, with the names Benedict Dybovsky and Victor Godlevsky (exile participants of the Polish uprising of 1863–1864) who laid the basis of studying bottom topography of the lake, bottom precipitation, temperature and ice regime, and Baikal winds [*www.rgo.ru/2010/09/ekspediciya-benedikta-ivanovicha-dybovskogo-i-viktora-aleksandrovicha-godlevskogo-na-ozero-bajkal/*].

Unfortunately, scientists (of Irkutsk State University) only started regular reliable observations of water level in Lake Baikal from the 1920s; that is why for our purposes data series since 1921 are used. There is the possibility to restore (but not with a high veracity) data of Lake Baikal level up to 1880. Such data are very important not only because the lake is the largest source of drinking water, but because a substantial sudden lowering (or rising) of water level puts the unique ecology of this region in hazard. Lake Baikal is a unique system and regulates itself when there should be more or less water, and we must not interfere in this process.

After building Irkutsk hydropower station, the average level of Lake Baikal became 1 m higher. But amplitude of level oscillations and the highest notes of the level are kept in the former range. Over the last decade, the water level in Lake Baikal has been much lower, and its minimal values have come back to values they had been before the building the dam.

The state of Lake Baikal, the largest drink-water object of our planet, which contains 10-year inflow of the Volga, Ob, Yenisei, Lena, and Amur, taken together, has not undergone any important changes since 2000, although in some years there were several unusual situations. The first situation in the middle of 2003 was connected with the level of Lake Baikal. The reason was, a low water content of basic inflows since 1996. For seven years, only 382 km^3 of water had come to the Baikal, representing 90% of the norm. The side inflow to the Bratskoye reservoir in the first half of 2003 was 1.5 times below normal. In 2002 and 2003 the level of Lake Baikal was close to a minimal mark of 456 m. On the May 8 and 9, 2003 the minimal daily average value of a level was marked (456.02 m). However by August 2003, the water level had risen and finally reached 456.71 m in October 2003. The situation was normalized, but it could be repeated again.

This is why the problem of regulating the water level in Lake Baikal became the subject of discussion in Irkutsk at the Chairman of Government of the Russian Federation's meeting on June 25, 2003. The meeting was devoted to the question of measures to protect the unique ecological system of Lake Baikal.

Then, because of water evacuation by workers of the Angara cascade of the hydropower station, which lasted from October 2007 to January 29, 2008, the level of the Baikal was lowered to 456 meters again. In the opinion of scientists, in April 2008, as a result, the lake level could have fallen down to the mark comparable to the most water-short dry years. Fortunately, the level had restored again by October 2008. That is why it is important to understand where natural processes and where anthropogenic influence work.

Lake Pátzcuaro (coordinates are 19°35′ latitude north and 101°35′ longitude west) is one of the highest in the world (its altitude above sea level is 2220 m). The size of the lake is not large, 20 × 14 km, and the average depth is 50 meters. The main wealth of the lake is the purest fresh water, and it is one of the basic resources of drinking water in the state of Michoacán. In summer one can feel a sweet melancholy from the background of constant low clouds around the lake. It is usually cool in Pátzcuaro (in comparison with weather in Mexico); the maximum temperature is 24 °C in the daytime and 11 °C at night.

The earliest mention of Lake Pátzcuaro was from 1526, after the conquest of the lake region by the Spaniards and the building of the city of Pátzcuaro. This historical city on the southern shore of the lake was a large and religious center of the Indian tribe Tarasque. During the colonial period, the city turned to an administrative, religious, cultural, and scientific center of the state for some time due to Vasco de Kiroga, the first bishop and scientist.

Among well-known cities on the shore of the lake there is an ancient Tarasque capital Tzintzuntsan, which is famous for its unusual Tarasque pyramids—yakatas—one of which was devoted to a pre-Spanish humming-bird-idol; due to its rhythmical flapping this city got such a descriptive name. For a short time Tzintzuntsan, the capital of colonial Michoacán, was a refuge of the first Franciscan mission in this region and the first mission studying culture of the Tarasque and Lake Pátzcuaro (Perry, 2008).

Regular observations of the lake level began in 1921 on the hydrological station situated on the shore of the lake, 3 km from city of Pátzcuaro and on the hydrological station on Janitsio island. The second and the third hydrological stations were organized on the western shore of the lake and on Jaracuaro island.

On the western shore of Lake Pátzcuaro there are villages, each of them having its own colonial monument to draw visitors' attention. A thin cross made of stone is situated in front of St. Pedro Pareo colonial church, the façade of which is decorated with various animals and figures made of stone, such as the Sun and the Moon.

In ancient times, the top of Jaracuaro island unprotected from wind was a temple for Ksaratanga, a Tarasque goddess of the Moon. Nowadays there is St. Pedro church and a hydrological station here.

It is rather easy to reach hydrological stations situated on islands—from piers San Pedrito (to Haracuaro) and General (to Janitsio). You cross the lake by ship of a certain color depending on the pier from which you start and come back by any ship of the same color at any convenient time (till 6 P.M.). Since the end of the 1930s the Mexican government assumed some measures to increase monitoring and controlling of the lake and made some capital investments into observing a network of water monitoring and complex monitoring of surface and underground waters in the region of Lake Pátzcuaro. Since then, observation data of water level of the lake have been quite reliable.

Spectral and Autoregressive Analysis of the State of Water Resources (Lake Levels)

In Libin and Jaani (1989), the results of preliminary calculations of spectral water content (level) characteristics of Lake Chudskoye and solar activity were given. The existence of statistically important water content variations with periods 4 to 5 years, 11, 22, and 80 to 90 years were shown. The authors discovered that retardation of water content relative to solar activity oscillates from 1.5 to 3 to 4 years and depends on the solar activity cycle. (A water content maximum is 2 years late relative to solar minimum for uneven cycles and about

3 years for even ones.) Herewith a structure of water content histogram for even and uneven cycles is different, which tells us that 22-year cyclicity prevails in hydrological processes (Anisimov and Belolutskaya, 2002; Libin, 2005; Birman, 2007; Anikeev and Chyasnavichyus, 2008; Afonin et al., 2009).

Comparison of results of water content spectral analysis obtained by the authors and analogical spectrums of galactic and solar cosmic rays shows very good frequency and phase correlation (see Attolini et al., 1985; Venkatesan 1990; Libin and Pérez-Peraza, 2009). Based on data of cosmic ray measuring from 1952 to 2006 scientists observed variations of 1-year, 5- to 6-year, and 11-year periodicity well correlated with solar activity and surface temperature, which coincided with variations of solar activity and temperature for the same period.

Calculation of spectral characteristics of Lake Chudskoye level oscillations with the help of the 7th ARMA model for the period 1921 to 2006 revealed noticeable oscillations of water content (level) with periods 1.0 to 1.2, 9.0 to 11.0, 21.5 to 22.8, and 80 to 90 (as it seems) years connected with solar activity.

Similar calculation results for water content spectrums for Lake Pátzcuaro in 1932 to 2004 and for Lake Baikal in 1927 to 2006 show similar pictures: periodicities of 1 to 2, 9 to 11, 22, and 90 years. It is absolutely clear that all discovered periodicities are connected with solar activity and have single-type retardations of water content in relation to solar activity for even and uneven cycles.

From the calculations we see unstable fluctuations in water level with periods ranging from several months to 4 years (on the background of relatively stable annual, 11-year, 22-year and 90-year changes in water content). (The fluctuations with periods ranging from several months to 4 years did not are observed in all cycles of solar activity, although it is well followed the solar rhythm).

To be certain of received estimations, scientists decided to make a full autoregressive analysis of water content oscillations of all three lakes and solar activity in the following succession.

1. ARMA-analysis of water content oscillations (levels) of Lakes Pátzcuaro, Baikal, and Chudskoye separately for 1932 to 2006
2. ARMA-analysis of level oscillations of Lake Chudskoye and solar activity for the same period (based on existed data)
3. ARMA-analysis of level oscillations of Lake Pátzcuaro and solar activity (based on existing data)
4. ARMA-analysis of level oscillations of Lake Baikal and solar activity (based on existing data)
5. Joint multivariate ARMA-analysis of level oscillations of three lakes and other meteorological and hydrological parameters and also of the variations of galactic cosmic ray intensity on Earth (based on existing data for 1951–2006)

The level behavior of Lakes Pátzcuaro, Baikal, and Chudskoye from 1921 to 2006 showed at least two curious peculiarities: 22-year variations are well observed and, primarily, oscillations of Lakes Pátzcuaro and Baikal are out of phase with oscillations of Lake Chudskoye.

Calculations of mutual correlation function Pátzcuaro–Chudskoye give 0.6 anticorrelation with 1 to 2 year retardation. Analysis of Lakes Chudskoye and Baikal behavior does not give such a beautiful picture (anticorrelation is about 0.4), although a 22-year wave is observed in water content behavior for Lake Baikal as well.

Obtained results correspond well with other researchers' results for Lake Chudskoye (see Table 3.2; Doganovsky, 1990; Reap, 1981, 1986). It is rather difficult to identify obtained results of analysis in all researched data sets with results in literature data. It is necessary to know a level of accuracy of estimations obtained by other authors.

As for our results, we can be sure in coincidence of estimations of separated periods including the 28-year period which were obtained with independent methods (the track method (Rozhkov, 1979) and methods of autoregressive analysis.

It should be noted that the study and forecast of lake levels is very important because they allow us to predict the yield of agricultural products. Figure 3.5 shows the results of spectral analysis of solar activity, the level of Lake Peipsi, and productivity in Estonia in years 1950 to 2000.

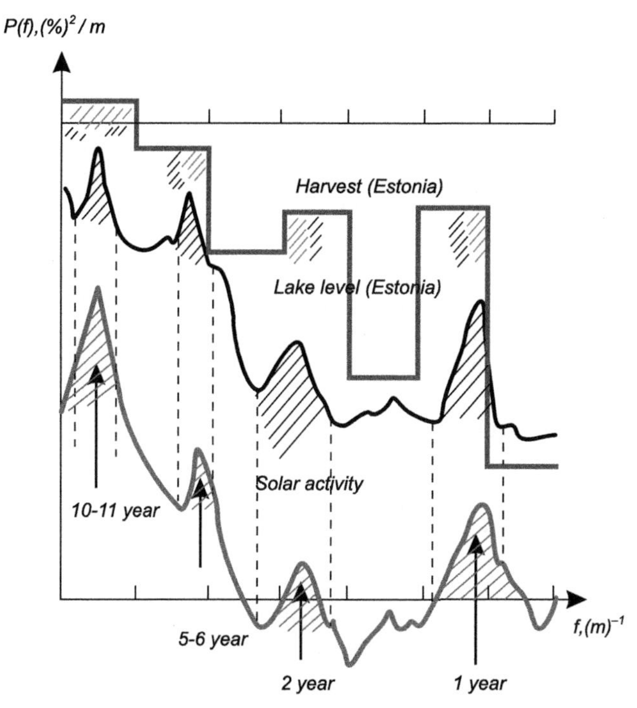

FIGURE 3.5 Spectral estimates of solar activity, the level of Lake Peipsi, and productivity in Estonia from 1950 to 2000. *Source: Libin and Pérez-Peraza, 2009.*

Autoregressive analysis between oscillations of the given lakes (based on yearly average data) has confirmed the existence of well-expressed oscillations with periods 11, 22, 35, 90, and 380 years (and even 720 years with an accuracy of 90%). In this case, in general, coherence of both processes is very high, and for 22-year and 90-year oscillations a coefficient of coherence (square coefficient of process correlation on the given frequency) is 0.75 to 0.85 (0.6 for 11-year and 35-year oscillations and, unfortunately, only 0.4 for 380-year ones).

Comparison of process amplitude show some relative exceeding of oscillations of Lake Chudskoye level (that is connected with solar activity) in comparison with the Caspian Sea by 40 to 60%, by 80 to 90% with Lake Baikal, and by 100 to 120% with Lake Pátzcuaro.

ARMA-analysis of oscillations of the lake level and solar activity has enabled us to draw the following conclusions, important for future forecast of water content:

- Behavior of water content (level) oscillations of isolated lakes reflects dynamics of solar oscillations
- Retardations of water content oscillations relative to solar activity coincide with retardations of other hydrometeorological processes and reflect a joint mechanism of solar activity influence on the Earth climate

Estimation of Water Resources in Russia, Mexico, and Estonia and the Forecast of them for the Nearest Decade

So, the use of the whole existing spectral apparatus and comparison of results of different spectral calculations show that solar cyclical activity and its influence on the Earth's atmosphere is a source of mechanism of influence on closed lake water content changing (Libin, 2005).

Results received in a whole series of works allow the use of an autoregressive prognostic model applied by the authors (Libin 2005; Libin and Pérez-Peraza, 2009) for forecasting water content of closed lakes with the purpose of raising exactness of the model on summarizing the series of used predictors (today's estimations give quantitative forecasts with a 35–40% error).

In a result of calculations based on a large volume of measuring data, a probable interconnection of processes on the Sun and in the Earth's atmosphere is shown. In this case analysis of behavior of retardation between atmospheric processes and solar activity shows the existence of stable drifts from 12 to 42 months between processes, and it correlates well with the results of calculations based on other techniques (Reap 1986; Rozhkov 1988; Kondratyev, 2004a). Moreover, it is found during joint analysis of water content in different points of the Earth and solar activity (and analysis of temperature behavior as well) the choice of solar activity indexes does not play a decisive role, so that the sunspot area in the near-equatorial zone of the Sun, as it was used by the authors earlier (Libin et al., 1994), is the most admissible index for calculations. That is why when solving problems of revealing large-scale processes in the atmosphere or trying to create prognostic models of climatological or hydrological processes it is necessary to take solar activity variations, processes in the interplanetary medium, and cosmic radiation variations observed on Earth into account.

As a result of application of an autoregressive prognostic model the authors predict insignificant rise of water content of all three lakes connected with solar activity by 2020. Calculations show 1.0% increase of water content for Lake Baikal, 2.5% increase for Lake Chudskoye, 3.0% increase for Lake Pátzcuaro (only if no anthropologic disasters leading to extreme water consumption from lakes occur like happened with Lake Baikal at the end of 2007).

Analogical researches were made in the work by Goliyandina et al. (1997), where the author studied the behavior of three series of yearly average hydrological characteristics.

1. The level of water of Lake Chudskoye since 1885 to 1993: the leading period in a series of observations is a period approximately equal to 23 to 28 years.
2. Water inflow to Lakes Pesvo and Udomlya, lakes-coolers of the Kalininskaya nuclear power plant (NPP) since 1882 to 1992: the leading period in a series of observations is a period approximately equal to 24 to 28 years.
3. Maximal ice areas of the Baltic Sea since 1720 to 1992: the leading period in a series of observations is a period approximately equal to 23 to 28 years.

Obtained results are of special interest to authors in connection with the fact that researchers (Goliyandina et al., 1997, 2001) apply independent methods of analysis (multivariate statistics): the track method and methods of statistic modeling for researching temperature regimes of water objects, a possibility to rule a water balance of lakes-coolers, and ecological consequences of hydro-economic activity.

Multivariate Statistics

Multivariate statistics works (Solntsev, 1997) were directed to the research of connections between different methods (analysis) of multivariate statistics and to development of a single approach to the system of methods of multivariate statistics which does not rest on traditional use of nondegenerate multivariate normal distribution.

The works are directed to studying the hydrological and ecological conditions of water objects of the northwest including Lakes Chudskoye and Pskovskoye, Pesvo, and Udomlya, used nowadays as basin-coolers of the Kalininskaya NPP, the Baltic Sea, and others. Using methods of multivariate statistics and the track method, scientists try to make long-term (up to 15 years) hydrological prognoses with the aim of estimating possibilities of development of economic objects, in particular, extension of NPP.

Analysis of Time Series

Works on analysis of time series gave special consideration to working out theory, development, program realization, and practicing technique of applying the method Track-SSA, the idea of which was firstly formulated by O.M. Kalinin in 1971 (Goliyandina et al., 2001; Russian Academy of Sciences, 2006). Modes of time series forecast, finding out disorder moments, and analysis of multivariate time series and point images were suggested on the basis of the main algorithm of the method (Solntsev, 1997).

A.V. Shnitnikov, who researched level changes of steppe lakes between Ural and the Ob for more than last 200 years, got the same results. It appeared that the water level in the lakes oscillated constantly; the lakes were full of water or dried up.

Zverinsky (1871) wrote in the last century, "Grass grew up on the bottoms of many lakes, and the lakes turned to grasslands which people used to cock hay and cultivated to sow wheat and flax." Since 1854 all dried lakes began to fill with water and in 1859 they became real lakes. As for large lakes, such as Lake Baikal, the water level in them can change by 5 m. These changes of shallow lakes are less (about 3 m).

Cyclic changes in a range of a century cycle of solar activity can be observed in various natural processes, for example, in precipitation level, droughts, changes of water in the rivers and water level in the lakes. It is important to take into account that lakes, especially isolated ones, are very sensitive to precipitation amount and air temperature changes. The more precipitation there is, the higher the water level in the lake should be. On the other hand, a rise of air temperature leads to faster expulsion of water from the lake surface. As a result, the water level lowers. Thus both factors work simultaneously.

It is clear that the water level changes while precipitation amount changes, but not quickly and not at once. There is certain retardation. Observations show that after 2 to 4 years of the most active precipitation, the water level in the lake reached maximal value. As the lake is not a reservoir with impermeable walls, a part of water goes to soil, and it becomes gradually replete.

Solar activity change induces atmospheric circulation change, as a result of which precipitation amount changes too. Changes of precipitation amount and air temperature lead to oscillations of water level in the lakes relative to the rate. Years of low water (very low, low and middle) change with years of high water (middle, high, very high).

3.2.5. Solar Activity Influence on Cyclic Changes of Precipitations

As a result of temperature and hydrological parameter research a prognostic model was built which gave a 10 to 30% error in forecasting levels of Lake Chudskoye and Lake Baikal for the following year.

Analysis of possible mechanisms of lake level oscillations results in the need for different meteorological parameters for a prognostic model, in particular, data of precipitation amount measures in the region of Lakes Chudskoye and Pátzcuaro, and, consequently, a question about a character of these changes has appeared.

Observation data of precipitation in Mexico, Estonia, Lithuania, and Russia for 1910 to 2008 and solar activity (sunspot area *S*) were used for the analysis. Figure 3.6 shows the behavior of the precipitation series and solar activity index *S* series: (a) for January 1910 to August 1950, (b) for September 1950 to April 1993. The results of calculations of spectral estimates are shown in Figure 3.7.

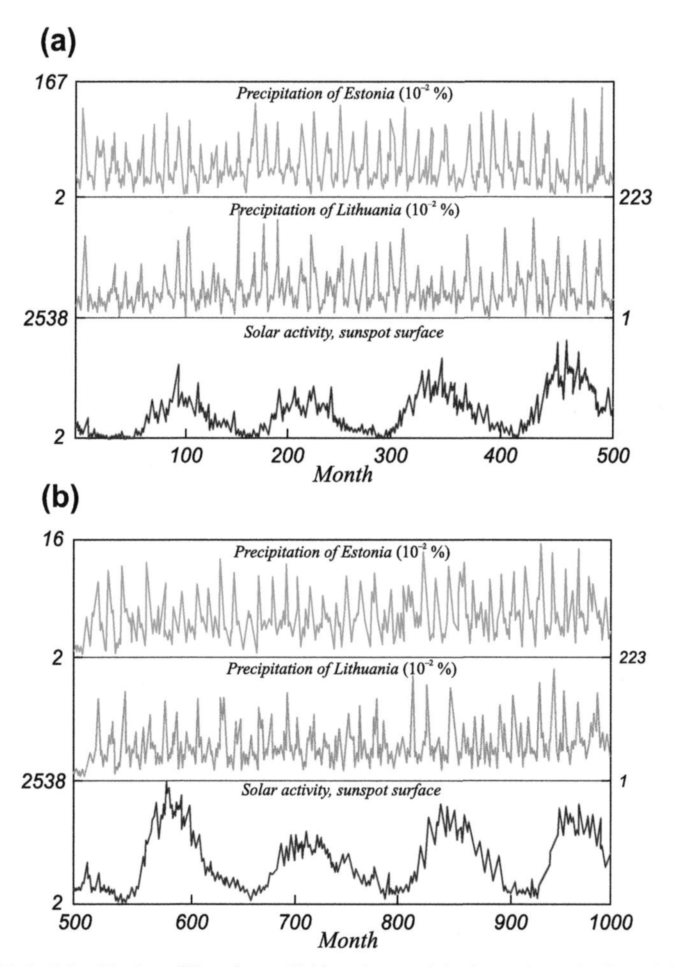

FIGURE 3.6 Monthly data of Estonian and Lithuanian precipitation series and solar activity index S series: (a) for January 1910 to August 1950; (b) for September 1950 to April 1993. *Source: Pérez-Peraza et al., 1997.*

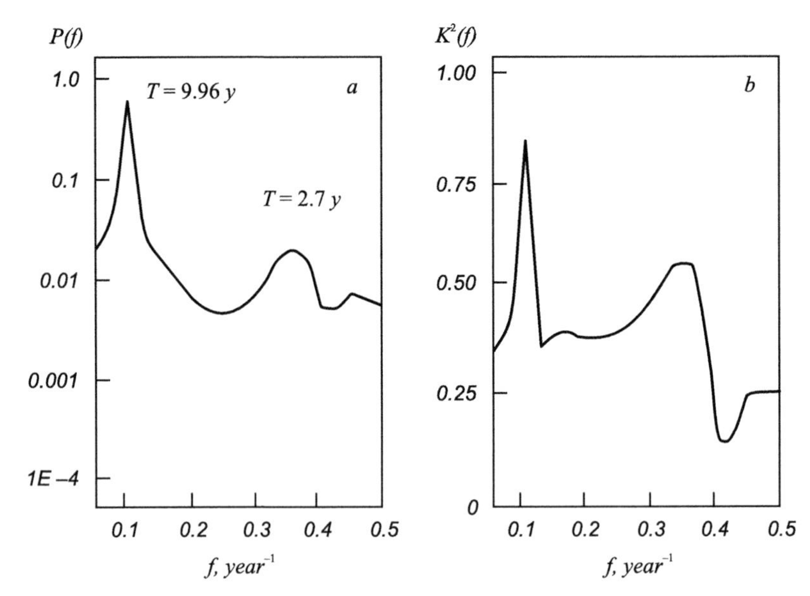

FIGURE 3.7 Amplitude and (b) coherence co-spectra of Estonian precipitation and S annual series for 1915 to 1934; AR (1.5) and AR (2.5), respectively. *Source: Pérez-Peraza et al., 2005.*

Deviation of the mean annual precipitation (averaging 10-year) from present-day values for the Russian territory are shown in Figure 3.8. Correlation and mutual correlation functions demonstrate good coincidence of process behavior.

The analysis was made with the help of autoregressive spectral methods (Libin and Pérez-Peraza, 2007), applied to the connections between solar

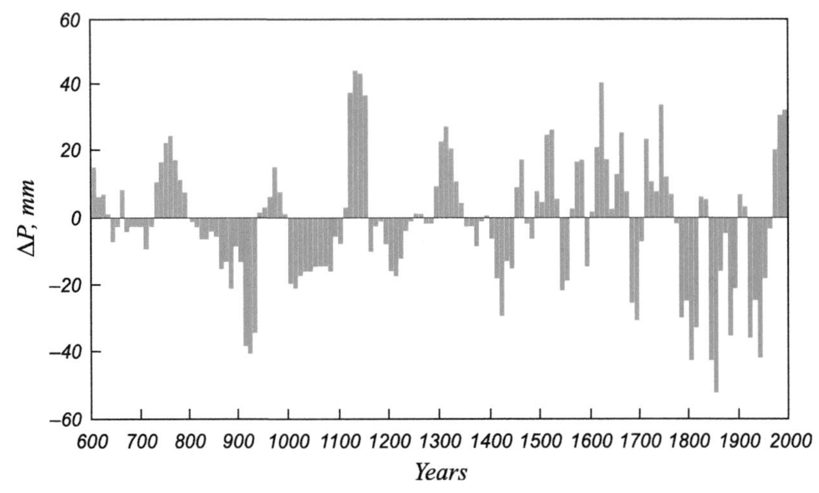

FIGURE 3.8 Deviation of the mean annual precipitation (averaging 10-letiyam) from present-day values for the territory of Russia. *Source: Klimenko, 2009a,b.*

activity and oscillations of precipitation amount in each region and connections between precipitation oscillations in all given regions studied.

Calculations of spectral characteristics were based on monthly and yearly average values of the analyzed series which allowed us to estimate precipitation amount and interconnection in a large frequency range. Correlation and mutual correlation functions of both processes (precipitations) demonstrate good coincidence in the behavior of these processes.

Spectrums for each analyzed series and mutual spectrums are also alike; they show the existence of the same separated peaks and confirm assumptions about identity of both processes.

In this case, 2-year and, maybe, yearly periodicities in all data are connected with solar activity which is proved with results of ARMA-spectral analysis of monthly average values of solar activity and oscillations of precipitation amount in Mexico, Russia, and Lithuania.

Autoregressive analysis for yearly average values of precipitation amount oscillations was analogously made. Calculations of amplitude spectrums and spectrums of coherence showed the existence in the analyzed data of a 11-year component and a quasi-biennial wave.

Obtained results correlate well with the data of analogous calculations for temperature and lake levels (Libin and Pérez-Peraza, 2009). Obtained results correlate well with results of analysis of average wind velocity in energetically active zones (Dorman, 2005) and, consequently, fit themselves well in the common picture of connections of atmospheric processes with solar activity.

Calculations based on yearly average data of solar activity and temperature in the same regions during the same periods show similar results: temperature oscillations with periods 2 to 4 years and 9 to 11 years connected with analogous solar activity oscillations are found.

Dynamics of oscillation behavior coincides too; if 9- to 11-year oscillations exist always, 2- to 4-year oscillations are more casual that also correlate well with the behavior of similar solar activity oscillations. In this case phase spectrums show retardations of temperature changes which correlate with results of works devoted to studying solar activity influence on geographical and hydrological processes (Pudovkin and Raspopov, 1992a,b).

Comparison of spectral characteristics of atmospheric parameters with analogical spectrums of galactic and solar cosmic rays shows a good frequency and phase correlation: from 1952 to 2000 one could observe 3- to 5-month, 1-year, 2- to 4-year, and 11-year variations of well-correlated cosmic rays with solar activity and temperature that coincided with variations of solar activity and precipitation for the same period.

Based on received regularities an autoregressive model was built

$$Pr(\mathrm{t}) = \sum_{\mathrm{i}=1}^{q} a_i\, Pr(t-i) + \sum_{\mathrm{i}=1}^{s} b_i\, W(t-i-\mathrm{w}) + \sum_{\mathrm{i}=1}^{r} c_i\, I(t-i-i) + \xi_t \tag{3.9}$$

where $Pr(t-i)$ is current data of registration of precipitation amount, $W(t-i-w)$ is solar activity, and $I(t-i-i)$ cosmic ray intensity; $Pr(t)$ is a predicted parameter; a_i, b_i, and c_i are autoregressive coefficients; and ξ_t is a residential noise value minimized during calculations.

Involvement of registration data of cosmic ray intensity (parameter) improves exactness of the model a great deal; application of CRI decreases a definition error from 40 to 20%.

Obtained results correlate well with similar calculation data of temperature and lake levels as well as results by Australian scientists (Robert Baker, *http://www.science.mcmaster.ca/~igu-cmgs/publications/CMGC00-18Rep2004-2008Fin.pdf*, 2008; IGU Commission C00-18 Modelling Geographical Systems, 2004–2008 report). Obtained results also correlate well with results of analysis of average wind velocity in energetically active zones and, consequently, fit themselves in the common picture of connection of atmospheric processes with solar activity. Robert Baker from the University of New England (Australia) put forward a theory about interconnection of the Earth's climatic system and cycles of the solar magnetic field which change poles each 11 years. In his work the research is based on data about the precipitation level in Australia.

After estimations of Baker et al., (2005) by the end of 2007 and in 2008 there should have been highly intensive precipitation in Western Australia. The sun is now in a similar position in terms of its magnetic field as it was in the 1920s, says **The authors claimed that**. Eastern Australia in 2007 and 2008 should follow a similar path to the particularly wet years in the decade of the 1920s, as, for example, it used to be the dampest in 1924 and 1925. According to the model of Baker' (2008) the following drought in the continent was to occur after 2009. Forecast Bikers confirmed all those predictions, but not in its entirety.

In general, there is a tendency to increase precipitation amount on the whole territory of the smallest continent. Baker based his theory on a physical sunspot model that he made. As a result of working on it, Baker found convergence between sunspot minimum and droughts in Eastern Australia over the last 100 years. In recent years in Australia were observed heavy rainfall and flooding. However, heck-Baker model for 2010, made using autoregressive model (in 2011 did forecast for 2010, based on measurement data from previous years since 1970), showed that the Baker-model can be used in the long run. Also, the long-term results of calculations by Baker-model coincide with the predictions of the authors: in 2030 the average temperature in Australia could rise by 1.5 degrees Celsius. After the present cycle of increased sunspot activity, the following cycle will be dominated by the lowest activity from sunspots and magnetic activity in 100 years. This raised the possibility of extensive drought in the 2020s. The theory will help in better predictions of droughts and floods. The model could also be used as a significant decision-making tool in agriculture and natural resource management.

The Earth's weather changes depend on the influence of solar activity on the atmosphere of our planet because the subsequent to intensification of cloud formation.

3.2.6. Solar Activity Changes and Possible Influence of Them on Long-Period Variations of Surface Temperature

This analysis was based on monthly average values of surface temperature measured in Mexico (Tacubaya, Sonora, Sinaloa, Baja-California, and other Mexican states), Estonia (Tartu), Sweden (Stockholm), Lithuania (Kaunas), and Moscow from 1910 to 2008. Temperature behavior in Mexico and Estonia against the background of solar activity based on measurements in the years 1950 to 2000 are shown in Figure 3.9.

Calculation results of solar activity (SA) spectra and similar results of mutual SA and temperature spectra demonstrate precise coincidence of separated frequencies (correlating to 2–4 year and 9–11 year periodicities) and good coincidence of 2- to 3-year retardations between processes. In this case, dynamics of the oscillation behavior also coincides; if 9- to 11-year oscillations exist always, 2- to 4-year oscillations are more casual that also correlate well with behavior of analogous solar activity oscillations.

Comparison of spectral characteristics of atmospheric parameters with analogous spectra of galactic and solar cosmic rays shows a good frequency and phase correlation. In a comparison of calculation data of amplitude and phase temperature spectra for 1937 to 2004 in Estonia and Sweden, cosmic rays for the same solar activity period gave the most impressive result. Not only 10.5, 2 to 3.7, and 1.3- to 1.7-year oscillations common for all data sets, but simultaneous phase change of all given oscillations in 1958 to 2002 were found. These results agree well with those of Jones (1986).

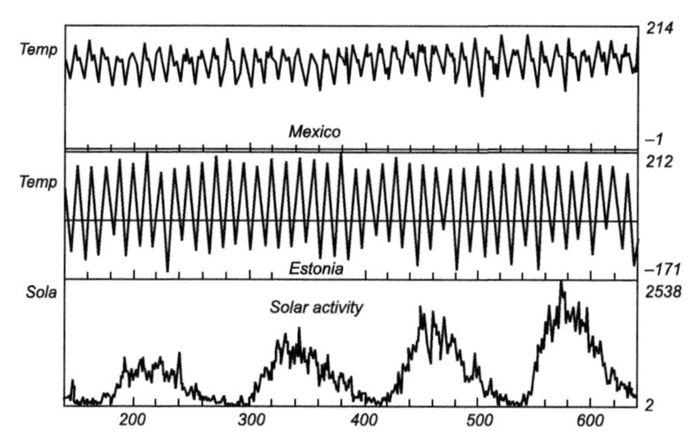

FIGURE 3.9 Temperature behavior in Mexico and Estonia against the backdrop of solar activity based on measurements in the years 1950 to 2000. *Source: Pérez-Peraza et al., 2005.*

On the other hand, if one takes away all separated within-year and climatic variations from temperature observation data of all researched regions, and analyzes the data, free from variations, it is possible to reveal a 700- to 800-year temperature trend which is called *global warming*.

It is clear that the veracity of the spectral analysis of the given temperature data set is not high (practically we have a series that is just a bit larger than a half period of the studied phenomenon), nevertheless, the existence of global warming is not in doubt (Pérez-Peraza and Libin, 2009), as was stated in Chapter 1. On the other hand, the situation is not catastrophic (Klimenko, 2009a). Figures 3.10 through 3.12 show the average (10 years) temperature for the territory of Eastern Europe, world temperature change, and global average-annual surface temperature. Figure 3.13 shows measurements, model calculations, and forecast of the behavior of temperature (Klimenko, 2009a,b) and the average (10 years) temperature for the territory of Eastern Europe. It is seen that the temperature was raised only in the last 30 to 50 years. It is important that the mark of the medieval optimum is not achieved until now.

Based on the results of joint analysis of solar activity and temperature, we think that a tendency of temperature rise on the Earth will continue, at least until 2050 (we should take into account that the exactness of our estimation is, at least, ±15 years), and then a process of global cooling will begin.

K. Willet, an American meteorologist, on the basis of studying sunspot cycles predicts that the Earth temperature will begin to fall in the nearest 25 years, and in a result, it will fall much more than it has risen for the last decades [*www.bcetyt.ru/science/researches/soln-aktivnost-klimat.html*] (Willet, 1957, 1974). He confirms that there will be less long-term droughts

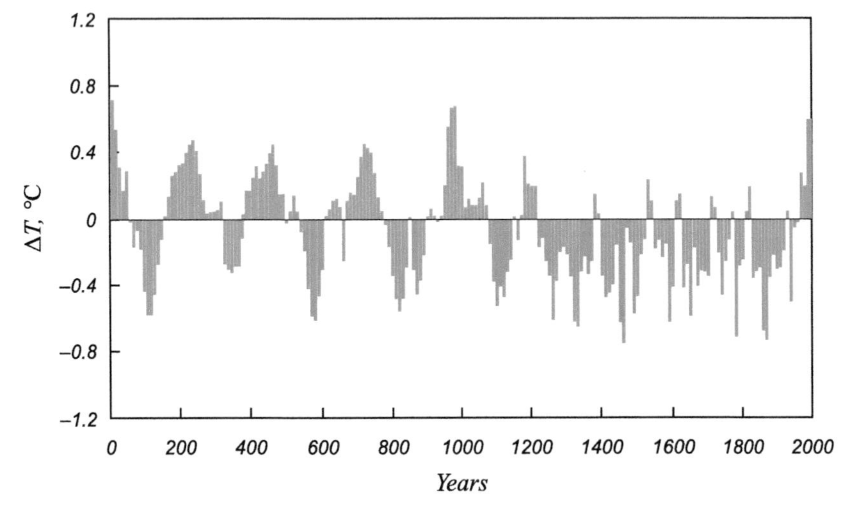

FIGURE 3.10 The mean annual temperature (averaging 10 years) of modern values (for Russia). *Source: Klimenko, 2009a,b.*

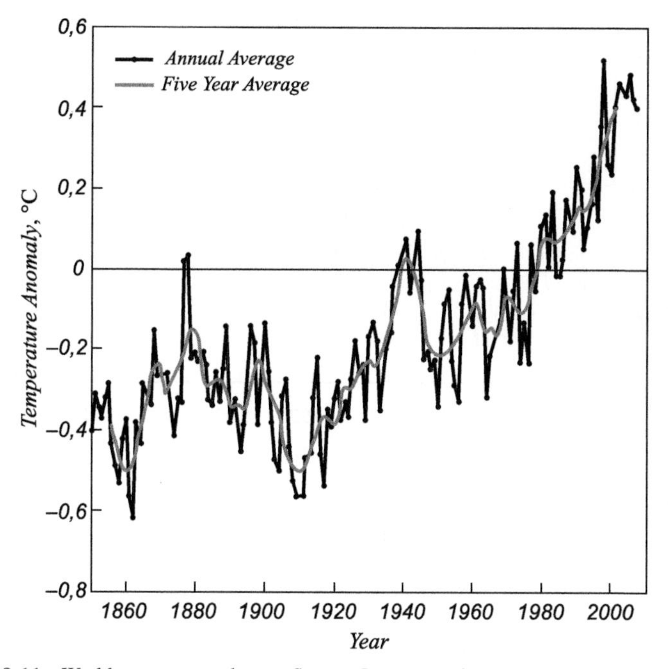

FIGURE 3.11 World temperature change. *Source: Instrumental Temperature Record (NASA).svg.*

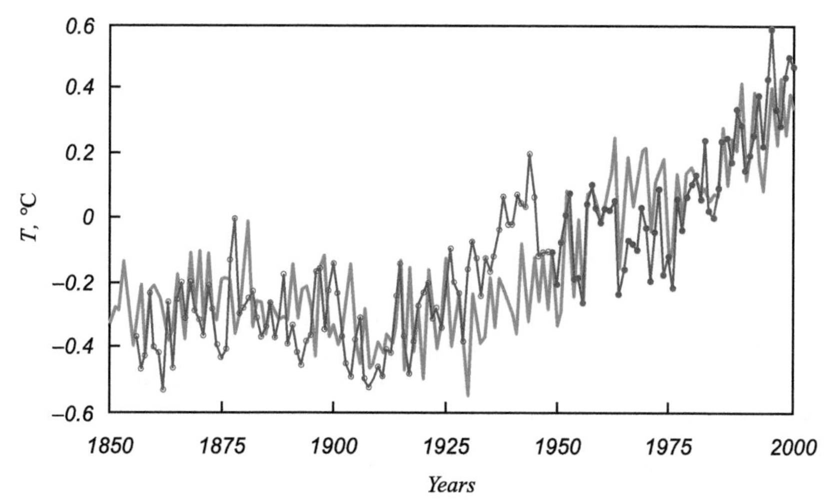

FIGURE 3.12 Global average—annual surface temperature. *Source:* http://data.giss.nasa.gov/gistemp/.

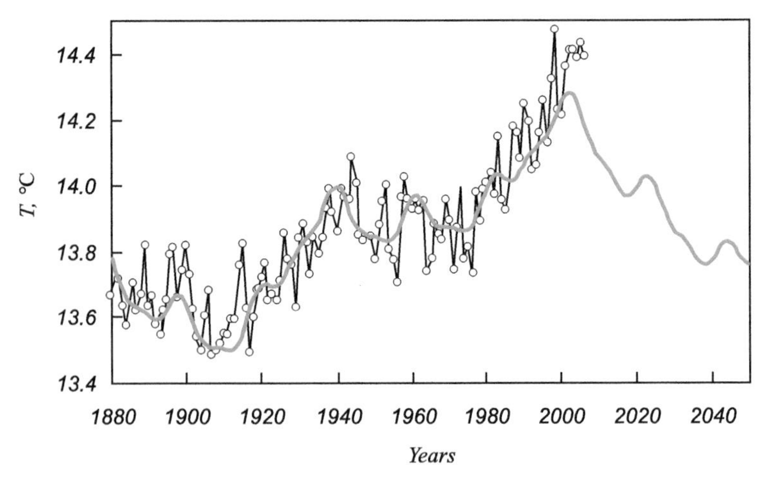

FIGURE 3.13 Measurements, model calculations, and forecast of the behavior of temperature. *Source: Klimenko, 2009a.*

in the middle latitudes and longer periods of insufficient precipitation amount in northern latitudes. This mainly concerns Canada and Northern Europe. In Africa and Asia there will be a 10-year period of dry weather. According to Willet's hypothesis, today's temperature is rising on Earth (from 2000 to 2030). In the subsequent years temperature will fall substantially, and in 2150 (during the peak of the 720-year solar activity cycle) a short glacier period will come. Meanwhile (Libin and Pérez-Peraza, 2009) show that in the decade of 2040 to 2050 the global temperature will peak and after that begin to decline.

3.2.7. Solar Activity Changes and Possible Influence of Them on Processes in the World Ocean

Weather in Europe is much influenced by the so-called *North Atlantic Oscillation*, or NAO, which describes changes of atmosphere pressure on sea level measured over Iceland and the Azores, where, as we know, centers of atmospheric action are situated. Iceland is the minimum and the Azores have the maximum pressure (Reid, 1987; Monin and Shishkov, 2000; Birman, 2007; Bochkareva and Romanov, 2007; Kondratyev, 2004a,b; Pokrovsky, 2005; Libin and Pérez-Peraza, 2007, 2009).

Over the last 50 years in winter months we could observe a tendency of atmospheric pressure fall over Iceland and rise over the Azores. But in recent years this trend has had a slight increase; however, influence of this climatic system spreads from the upper troposphere to the ocean floor. Scientists discovered that for the last 50 years this tendency has led to frequent droughts in Southern Europe in winter and more precipitation on the north of the continent. As for other seasons, this effect does not appear as clearly as in winter, although today's summer heat in Southern Europe and a great deal of rain in its northern regions illustrate well the influence of this climatic system on European weather. It is not generally accepted by scientists that anthropogenic influence on the environment is the reason for changes in North Atlantic Oscillation. There are probably some mechanisms unknown to scientists, but to establish them a long series of meteorological observations is necessary.

Over the same time period the models show that today's tendencies in North Atlantic Oscillation will be kept in future that will make the climate in Southern Europe drier, when Northern Europe will subject to heavy storms. In this case the Earth climate will become warmer and warmer which will lead to decreases of pure water stocks on the planet.

On the basis of the difference of North Atlantic oscillation index (NAOI) from its rate value it is possible to estimate what kind of winter will be in the near future in Europe, cold and frosty or warm and damp. Such calculation models have not been worked out yet, and it is very difficult to make reliable forecasts.

A great deal of research work is in store for scientists; they already understand the most important components of this weather mix in the Atlantic Ocean and can understand some consequences of it. Gulf Stream plays one of the decisive roles in the game between the Ocean and atmosphere. Today it is responsible for warm, mild weather in Europe, without it climate in Europe would be more severe. If a warm flow of Gulf Stream is strong its influence increases atmospheric pressure difference between the Azores and Iceland. In this situation a high pressure zone near the Azores and a low pressure zone near Iceland arouse leeway of a western wind. A result of it is a mild and damp winter in Europe.

If Gulf Stream is cool there is an opposite situation: pressure difference between the Azores and Iceland is much lower, or NAOI has a negative value. The result is a weak western wind, and a cool air from Siberia can penetrate unobstructed on the territory of Europe. In this case, a frosty winter would come.

NAO which creates a pressure difference value between the Azores and Iceland lets us understand what kind of winter can we expect. Whether or not it is possible to predict summer weather in Europe on the basis of this method is not yet clear. Some scientists, among which Latif (2009), a meteorologist from Hamburg, and Mikalajunas (2010), a meteorologist from Lithuania, predict high probability of heavy storms and precipitations in Europe. According to Latif (2009) in future, if a high pressure zone near the Azores becomes weaker usual storms in the Atlantic will reach southwestern Europe. He also supposes that in the given phenomenon, like in El Niño, circulation of warm and cold ocean flow during uneven periods of time plays a big role. There is still much not yet studied in this phenomenon.

Hoerling et al. (2006) describes the comparison of NAOI indicators with the real temperature in Europe over many years. According to James Hurrell, a climatologist from the National Center for Atmospheric Research in Boulder, Colorado, the result was satisfactory: an undoubted interconnection was revealed. Thus, for example, a severe winter during the Second World War, a short warm period at the beginning of the 1950s, and a cold period in the

1960s correlate with NAOI indicators and solar activity cycles. The authors researched storminess in NAO over many years. The results of solar activity influence on NAO are given in this chapter.

Simultaneously with researches of atmosphere circulation processes in the Atlantic Ocean (study of storminess in the region of North Atlantic energetic active zone—North Atlantic Oscillation), the phenomenon El Niño (ENSO) attracted researchers' attention (Bucha, 1988; Philander, 1990; Anderson, 1992; Mendoza and Pérez-Enriquez, 1993; Friis-Christensen and Lassen, 1992). South oscillation or El Niño is a global ocean–atmospheric phenomenon. This is the name of abnormal warming of surface water of the Pacific Ocean near Ecuador and Peru which happens once every several years. This tender name reflects the fact that the beginning of El Niño often occurs at Christmas, and fishermen off the west coast of South America connected it with the name of baby Jesus. Being a characteristic feature of the Pacific Ocean, El Niño and La Niña are temperature fluctuations of surface water in tropics of an eastern part of the Pacific Ocean. The names of these phenomena were taken from local people's Spanish language and first introduced for scientific use in 1923 by Gilbert Thomas, meaning baby boy and baby girl (e.g., Libin and Pérez-Peraza, 2009).

Circulation is a substantial aspect of Pacific phenomenon ENSO (El Niño Southern Oscillation). ENSO is many interacting parts of one global system of ocean–atmospheric climatic fluctuations that are a succession of ocean and atmospheric circulations. ENSO is the most famous resource of interannual variability of weather and climate in the world (from 3 to 8 years).

During normal years temperature of the ocean surface oscillates in a narrow seasonal range from 15 °C to 19 °C along the whole Pacific coast of South America because of coastal rise of depth water induced by a Peruvian cold flow. During the El Niño period temperature of the ocean surface in a coastal zone rises by 6 to 10 °C. According to geological and paleoclimatic research, the phenomenon has been existed no less than 100,000 years. Temperature oscillations of the ocean surface from extremely warm to neutral or cold happen once every 2 to 10 years.

Michalenco and Leonova (2007), the leading weather forecasters of the Forecast Department of the Far Eastern Scientific Research Hydrometeorologic Institute (FESRHMI), write in their report,

There is a constant warm flow springing from the Peruvian coast and spreading to the archipelago which is situated to southeast from the Asian continent. It looks like an elongated tongue of heated water and its area is equal to the territory of the U.S.A. Heated water evaporates intensively and fills the atmosphere with energy.

During El Niño in the equatorial region this flow warms up more than usually, that is why trade winds become weaker or stop blowing. Heated water spreads over, and flows back to the American coast. An abnormal convection zone appears. Rains and tornados fall on Central and South America.

The La Niña phenomenon is opposite to El Niño and manifests itself as a fall of surface water temperature lower than the climatic rate in the east of the tropical zone of the Pacific Ocean. During La Niña formation trade (eastern) winds from the western coast of both Americas become stronger and move the warm water zone, and a cold water tongue 5000 km elongates in the same place where during El Niño a warm water zone would be. In this case Caribbean countries, Mexico, and the U.S.A. suffer from storms and droughts.

La Niña like El Niño appears more often between December and March. The difference between them is the following: El Niño appears on average once every 3 to 4 years, and La Niña once every 6 to 7 years. Both phenomena bring a larger quantity of storms, but during La Niña they can be 3 to 4 times more than during El Niño.

In the Pacific Ocean during substantial warm events El Niño warming becomes wider by a large part of the Pacific tropics and finds itself in a direct connection with intensity SOI (South Oscillation Index). When ENSO events are mainly between the Pacific and Indian Oceans, ENSO events in the Atlantic Ocean are 12 to 18 months later than the first. As ENSO is a global and natural part of the Earth climate it is important to find out if intensity and frequency change can be a result of global warming.

Low-frequency changes were found by Bucha (1988), who used the autoregressive spectral method to uncover 22-year and 400-year oscillations. Ninety-year oscillations of El Niño appearance were discovered by Libin et al. (1987). It should be mentioned that one of the authors I. Ya. Libin was directly involved in research on the Niño and the Niña events onboard the Soviet deep-diving vehicles *Pices* and *Mir* (Figures 3.14 and 3.15).

FIGURE 3.14 Underwater vehicle *Pices* explores the Pacific Ocean (Institute of Oceanology). *Source: Courtesy of Igor Libin.*

FIGURE 3.15 Underwater vehicle *Mir* explores the CAO. *Source: Institute of Oceanology. Courtesy of Igor Libin.*

Inverse correlation between ENSO number and quantity *W* and sunspot areas *S* was also discovered. We can observe well-expressed correlation of ENSO with atmospheric quasi-biennial oscillations (AQBO) (Libin et al., 1999; Anderson, 1992).

Research of the last 50 years has allowed us to find that El Niño means something more than just correlated oscillations of surface pressure and temperature of ocean water. El Niño and La Niña are the most brightly expressed manifestations of within-year climate change on a global scale. These phenomena are large-scale changes of ocean temperature, precipitation, atmospheric circulation, and vertical air movements over a tropical part of the Pacific Ocean.

El Niño phenomena are also responsible for large-scale abnormalities of air temperature across the whole world. There can be important rises of temperature. Warmer conditions than normal in January and February occur over southwestern Asia, the Far East, Japan, the Japan Sea, southwestern Africa and Brazil, and southwestern Australia. Temperatures warmer than normal are also noted in June through August along the western coast of South America and over southeastern Brazil. Colder winters (December–January) occur along the southwestern coast of the U.S.A.

During La Niña periods drier conditions than normal are observed over the coast of Ecuador, northwestern Peru, and an equatorial part of eastern Africa from December to February and over south Brazil and central Argentina from June through August. All over the world scientists note large-scale abnormalities with the largest quantity of regions suffering from abnormally cold conditions: cold winters in Japan and the Far East, in southern Alaska, and western and central Canada; and cool summers in southwestern Africa, India, and southeastern Asia. Warmer winters can be observed in the southwest of the U.S.A.

During El Niño more energy is carried to the troposphere of tropical and middle latitudes. It is manifested in an increase of thermal contrasts between tropical and polar latitudes, and activation of cyclonic and anticyclone activity in the middle latitudes (Velasco and Mendoza, 2008). A scheme of La Niña and El Niño are shown in Figures 3.16 and 3.17. Comparison of historic El Niño events and the current event (multivariate hazard index of El Niño) are shown in Figure 3.18.

(In the Far Eastern Scientific Research Hydrometeorologic Institute calculations of cyclone and anticyclone repetition along the northern part of the

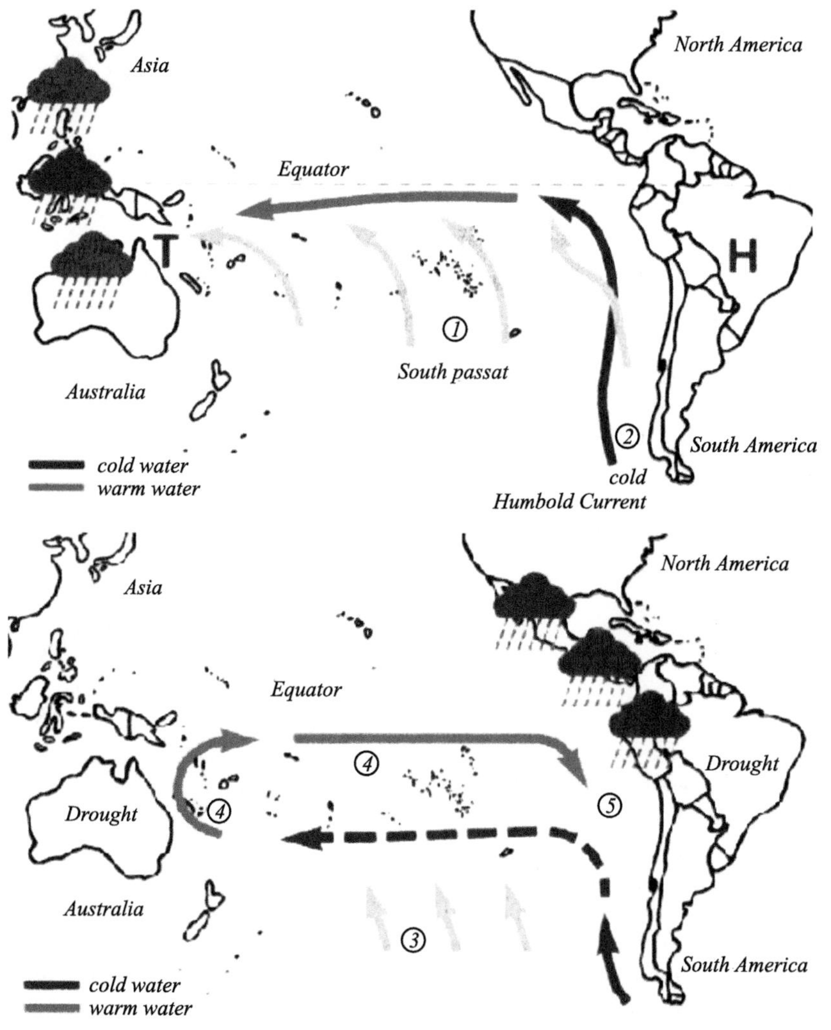

FIGURE 3.16 La Niña and El Niño. *Source: Queensland Climate Change Centre of Excellence; http://www.longpaddock.qld.gov.au/images/seasonalclimateoutlook/WalkerCirculation.gif.*

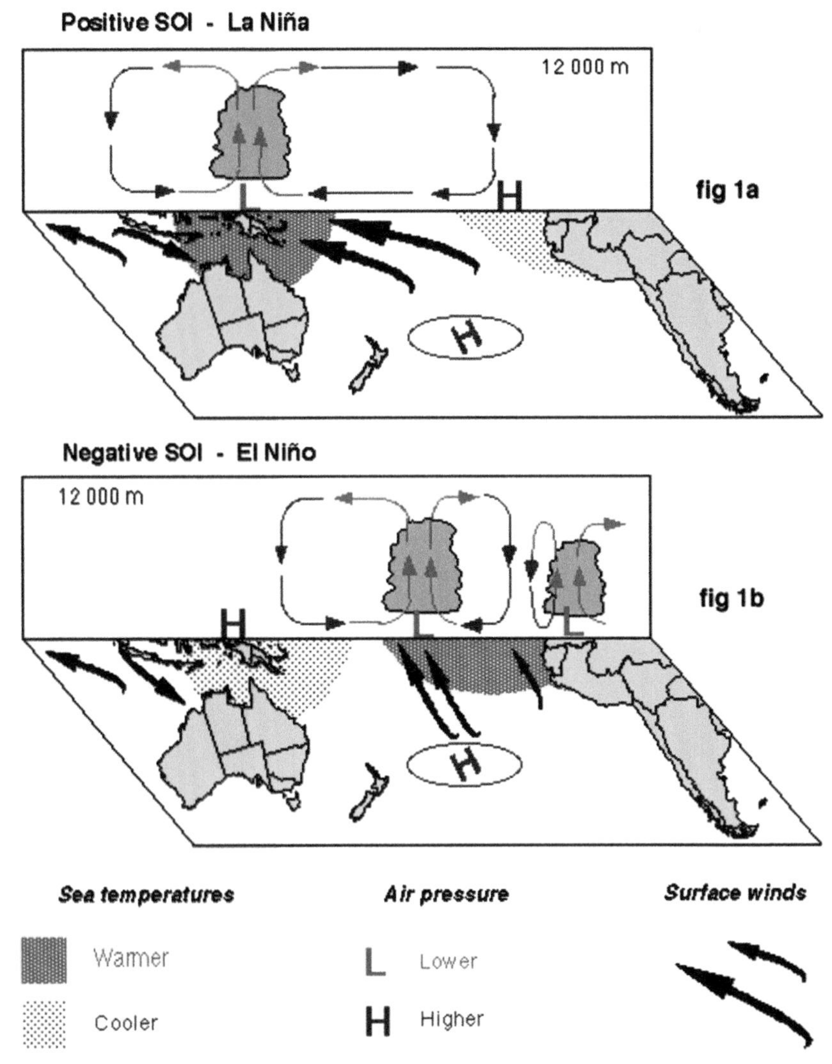

FIGURE 3.17 La Niña and El Niño. *Source: www.longpaddock.qld.gov.au/images/seasonalclimateoutlook/WalkerCirculation.gif.*

Pacific Ocean from 120° longitude east to 120° longitude west were made. It appeared that more cyclones in the 40° to 60° latitude north zone and anticyclones in the 25° to 40° latitude north zone appear in winters following after El Niño than in preceding ones. So, processes in winter months after El Niño are more active than before the period.)

Voyskovsky et al. (2006) estimate quantitatively the influence of solar and geomagnetic activity and El Niño phenomenon on the origin and evolution of

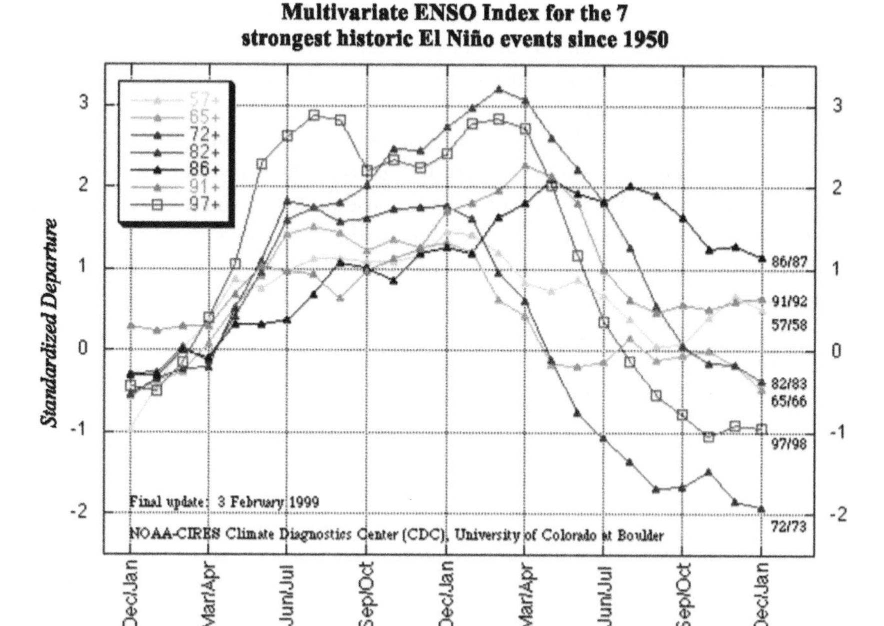

FIGURE 3.18 Comparison of historic El Niño events and the current event (multivariate hazard index of El Niño). *Source: http://www.esrl.noaa.gov/psd/enso/mei/; http://www.esrl.noaa.gov/psd/enso/mei/comp.png.*

tropical cyclones (TC) by using the correlation depth between solar and geomagnetic activity and basic characteristics of TC in a northwestern part of the Pacific Ocean (N-WPO). It is shown that from 1945 to 2005 there were periods in some regions when high (up to 75%) coefficients of TC parameter correlations with solar and geomagnetic activity indexes were observed. In other regions there was no such correlation, but there was TC parameter correlation with the index SOI characterizing El Niño phenomenon.

Coefficients of correlation between a tropical cyclogenesis (TC) phenomenon and different indicators describing solar (Wolf numbers) and geomagnetic activity (Aa and Ap) and El-Niño influence (SOI index) in N-WPO were defined. Obtained correlation coefficients show that there can be a connection between solar-magnetospheric activity and TC: particle precipitation leads to a decrease of transparency of the upper atmosphere which results in a decrease of solar energy penetration to the atmosphere and reduce the probability of TC appearance.

In the report "Solar Activity Oscillations are the Main Climate Formation Factor on a Millennium Scale" (Khorozov et al., 2006) it is suggested a three-layered warm-balance physical-statistical model of the oceane atmosphere system built on the basis that solar activity influences El Niño formation and accounting circulation factor (imitating satisfactorily a yearly average state of global and regional climatic systems). According to the authors, the model can be used to explain other famous and large-scale events in the past and also to predict them in future. Their estimations place a well-expressed phenomenon El Niño from 2010 to 2012.

The authors connect this phenomenon in particular, with droughts (in Indonesia and Australia), heavy showers and floods (Peru), appearance of warm water near the central American Coast, and, consequently, a sudden decrease of fish resources in this region expected during these years. The minimum of yearly average temperature in 2011 in St. Petersburg, Kaliningrad, and Ekaterinburg also correspond to it. We should also expect an increase of tropical storms in the Atlantic Ocean and, consequently, increasing risk of their coming out on the American continent from 2013 to 2015.

The climatic phenomenon El Niño with all its manifestations in different parts of the world is a complex operative mechanism. We should especially underline that interaction between the ocean and the atmosphere induces a series of processes that are further responsible for El Niño appearance.

The conditions of El Niño phenomenon appearance are not fully studied yet. We may say that El Niño is a globally influencing climatic phenomenon, not only in the scientific sense of this word, but it also influences significantly the world economy. El Niño substantially influences people's everyday life in the Pacific Ocean basin—many people can suffer from sudden rains or long-lasting droughts.

El Niño influences not only people, but animals as well. Thus, during El Niño there are problems with anchovy fishing on the Peruvian coast. It happens because anchovies were caught earlier by many fishing flotillas, and a small negative impulse is enough to make this wavering system out of balance. Such El Niño influence has the most damaging influence on a food chain including all animals.

Decrease of storm number in North America is a positive effect of El Niño. In contrast, in other regions of the planet the number of hurricanes increases during the El Niño.

3.2.8. Solar Activity Changes and Their Possible Influence on Biological Processes

One of the fundamental problems of modern solar-terrestrial physics is revealing connection mechanisms between solar activity and operation of different objects of biosphere including human beings (Takata, 1941; Piccardi, 1962; Vladimirsky et al., 1995; Oraevsky et al., 1998; Dmitrieva et al., 1999; Dmitrieva et al., 2000; Halberg et al., 2000; Cherry, 2002; McMichael et al., 2003; Palmer et al., 2006; Stoilova and Dimitrova, 2006; Babayev and Allahverdiyeva, 2007; Stoilova et al., 2008). Many data in literature resources about influence of external fields on biological objects are spread across different scientific specialties and interpreted in different ways, without a satisfactory theoretical explanation.

In the 1920s, Chijzhevsky (1976) was the first who found solar activity influence on diseases. He is usually considered to be the founder of heliobiology. Since that time research has been conducted and scientific data proving influence of solar activity and magnetic field changes on human health have been collected. That is why participants of the interdisciplinary seminar Biological Effects of Solar Activity (April 6–9, 2004 in Pushchino-on-Oka) made a decision in which they emphasized the importance and currency of studying the influence of solar and magnetosphere process dynamics on the Earth's biosphere and the importance of such influence forecasting.

The fact that such influence exists is not in doubt, however it will take a lot of time and effort to work out the techniques of making experiments and processing the obtained solar bio-effect data that would provide regularity of such effects in different periods of time and on different places on the Earth.

For example, biological experiments revealing weak magnetic fields tell us about the importance of a variable component in the 1 Hz range and higher, during the time when the majority of geomagnetospheric indexes describing disturbance of the Earth's magnetosphere has a characteristic storage time of about 1 hour. That is why analysis of terrestrial magnetic data with 1 c resolution (for example, Pc-pulsation) is of great interest. It is necessary to continue research to reveal mechanisms of solar disturbance influence on circumterrestrial processes, magnetosphere, ionosphere, atmosphere, and biosphere, including human organisms.

The life on our planet is connected with the Earth's rotation round its axis determining the daily rhythm and the Sun's rotation, on which the Earth's seasons depend. The majority of living organisms perceive seasonal rhythms as season changes. That determines growth, development, and death of plants. The Earth's rotation around its axis determines the rhythmical change of environment factors: temperature, light, relative air moisture, barometrical pressure, atmosphere electric potential, cosmic radiation, and gravitation. All these environmental factors influence the life processes of living organisms. Among them alteration of light and darkness is of great importance. Plant metabolism—carbon dioxide absorption in the day time and oxygen release at night—depends on diurnal rhythm.

The daily rhythms of animals are revealed as alteration of wakeful and active periods with sleeping and quiescent periods. (Some seasonal rhythms of animals also have alteration of activity and rest.) An example of the dependence on solar activity on catch crabs is shown in Figure 3.19. The presence of similar periodicities in cosmic and biological processes are well illustrated by the results of spectral analysis in the thickness of annual rings of trees of the Tian Shan Mountains (see Figure 3.20).

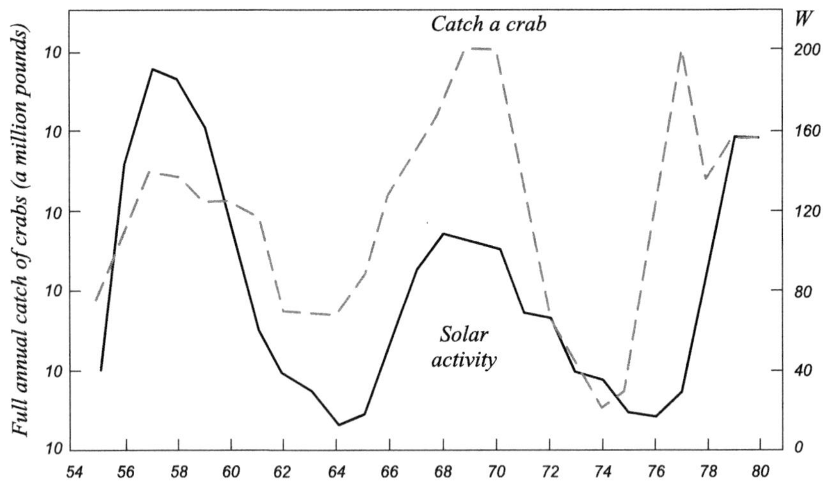

FIGURE 3.19 Solar activity and catch crab. *Source: Libin and Pérez-Peraza, 2009.*

All living organisms on Earth were developed under the influence of daily or seasonal rhythms. But have they always had the same duration as now? Many scientists think that millions of years ago the Earth rotated quicker and the day was shorter. Friction of substance in ocean tidal waves and in the solid body of the Earth was the reason for deceleration of the Earth's rotation. The ocean tidal waves have already stopped the Moon's rotation as it is lighter than the Earth.

It is under the influence of solar cyclic activity and the Earth's rotation round its axis and round the Sun that periodicity of the phenomena occurring in nature appeared. It is revealed in weather change, volcanic explosion, earthquakes, floods, and so on. This periodicity has created the rhythm in living organisms that makes up the essence of their life.

It is noted according to the results of much research in different countries that patients' health is aggravated, firstly, after a solar flare, secondly, when a magnetic storm begins. We can explain this by the fact that several minutes after the start of a solar flare, solar particles reach the Earth's atmosphere and induce processes influencing the organism's operation.

Available data (unfortunately, often fragmentary) show that among all the diseases that are subject to influence of magnetospheric storms, cardiovascular diseases take the first place, as their connection with solar and magnetic activity is more evident. Scientific groups compared dependence of quantity and severity of cardiovascular diseases on many factors (atmosphere pressure, air temperature, precipitation, cloudiness, ionization, radiation regime, etc.), but true and stable connection of cardiovascular diseases with chromospheric flares and geomagnetic storms is revealed specifically. During magnetic storms subjective symptoms of patients' health aggravation were revealed, blood pressure rose, and coronary circulation deteriorated more often, accompanied by negative ECG dynamics.

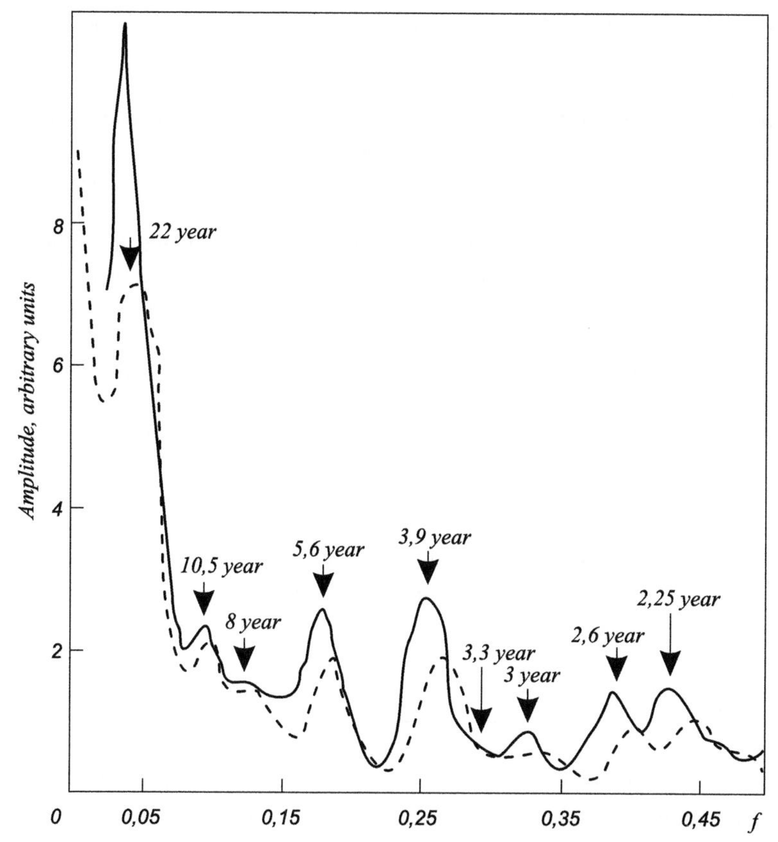

FIGURE 3.20 The power spectrum of fluctuations in the thickness of annual rings of trees in the Tian Shan Mountains. *Source: Lovelius, 1979.*

The research showed that during the day of a solar flare the number of cases of myocardial infarction arises (Halberg et al., 2000). It reaches a maximum the day after the flare (about twice as much in comparison with calm days). A magnetospheric storm induced by a flare begins on the same or the following day.

Research of heart rate during a long period of time among large groups of patients showed absence of heart rate abnormalities during weak disturbances of the Earth magnetic field. But during the days with middle or strong geomagnetic storms heart rate abnormalities happen more often than in the days when there are no magnetic storms. It concerns observations of people in a quiescent state and during physical activity.

Observations of patients with essential hypertension showed that a proportion of patients had reactions a day before a magnetic storm. All the rest felt unwell in the beginning, middle, or end of a geomagnetic storm.

Research in different countries showed that the number of accidents and injuries connected with transport also rise during solar and magnetic storms.

Time of reaction to different visual and sound stimuli increases, quickness of wit comes down; thus the possibility of making an incorrect decision becomes more likely.

In some countries observations of influence of magnetic and solar storms on patients suffering from mental diseases were held. Connection between people's visits to mental institutions and disturbances of the Earth's magnetic field was evident.

It is necessary to note that ill and healthy organisms react differently to changes of cosmic and geophysical conditions. Characteristics of biological energy, immune protection, the state of different organism's physiological systems of depression, tiredness, and people's emotional instability become worse; mental strain appears. But a psychologically and physically healthy organism finds itself in a condition to rebuild inner processes in correlation with changing environmental conditions. In this case an immune system is activated, nervous processes and an endocrine system are reformed, and capacity of work is kept or even rises. Subjectively, a healthy person perceives it as general well-being mend and mood improvement.

Meanwhile, the question about influence of solar activity and a full series of physical agents transmitting such influence is still open. The majority of works devoted to the adaptation of the human organism to natural fields concern two utmost cases: solar activity periodicity manifestations in the results of medical statistics and statistics of disasters, or, external field influence on very healthy people, such as cosmonauts and sportsmen.

The general mass of healthy working people is not covered by this kind of scientific research. Moreover, absolute majority of researches of solar activity influence on human beings have been held without a strict mathematic apparatus. Ragulskaya (2005) proved the existence of statistically nonrandom reaction of comparatively healthy organisms (not to speak of ill ones) to solar activity variations and the leading role of solar-terrestrial interactions in formation of organism rhythms.

The technique of measuring, processing, and an instrumental base for studying influence of different external factors on human organisms, including solar activity variations and environmental factors connected with them, have been worked out. Over seven years of monitoring an experimental base covering more than 100,000 daily measurements of functional parameters of a constant group of people on the backdrop of more than 350 magnetic storms has been formed. Nonrandomness of coincidence of biophysical and geomagnetic events on the level of 0.01 statistic significance has been proved.

This spectral analysis has revealed coincidence of all basic periods of biological parameters of individual and groups on different latitudes that proves the existence of global regulating environmental factors. Ragulskaya (2005) found coincidence of all basic periods of resuscitation cases with periodicity of data of monitoring biophysical experiment and periods of atmospheric pressure variations, geomagnetic activity rise, and increase of sunspot number provided by the IZMIRAN (of the Academy of Sciences of Moscow).

It is found that flare processes on the Sun and following changes in the spectrum of natural very low-frequency electromagnetic fields, cosmic rays, and atmospheric pressure fluctuations induce a human's stable and producible reaction on the level of operation of separate systems (vegetative nervous system, inner organs, changes of cardio cycle parameters) and of the organism as a whole.

The existence of amplitude, latitude, timely, trigger, and cumulative effects of solar activity variation influence on the organism are found.

3.2.8.1. Trigger Effect

A human organism reaction under influence of natural external fields has a trigger character. In this case when geophysical fields change suddenly the amplitude of physiological reactions practically does not depend on the increase of external field amplitude, but it is determined by inner characteristic of the bio-system.

3.2.8.2. Latitude Effect

Experiments made simultaneously on different latitudes showed coincidence (in a 24-hour range) of measured physiological parameter variations with 0.7 correlation coefficients between series of data at the 0.01 significance level. A fixed reaction is a mass all-round one and keeps its characteristics while an object of study changes from a separate human organ to socially organized groups. When geophysical latitude of experiment place increases, it raises the percentage of reacting people (from 50–60% in Odessa and Kiev to 90% in St. Petersburg) and makes amplitude of reaction 1.4 times higher.

3.2.8.3. Timely Effect

During the analysis of long-term (yearly) series of observations a growth trend of monthly average values of the individual rate of the majority of observed physiological indicators of people in the period of solar activity increase, and maximums (1998–1999 and 2000–2002) have been revealed, as well as a decrease trend of monthly average values of an individual rate on the phase of solar activity decay (2003–2005).

This effect reflects integral influence of solar activity variations on human organisms and biosphere and it is conditioned with peculiarities of adaptation processes under a long-term weak external influence, but not qualitative difference of magnetic storms in different phases of solar activity.

3.2.8.4. Cumulative Effect

The experiments showed the existence of a synergetic cumulative effect; simultaneous influence of external factors becomes greater and effective, even if amplitude of each of these separate factors is too small for a stress-reaction to triggered.

It is shown that there are general planetary external factors ruling rhythms of the human organism. Only variations of natural external fields (the magnetosphere, ionosphere, and atmosphere) induced in their turn with solar magnetic field variations can be such factors. According to the results of the experiment, anthropogenic fields, local and unique in each city, do not contribute to formation of a spectrum of general collective reaction.

About 70% of biological parameter emissions fall on solar activity processes (flares and coronal holes) and magnetic storms, 10 to 20% react to sudden changes of atmospheric pressure. Correlation is opposite in dependence of blood pressure on external parameters—the majority of sudden rises of blood pressure is connected with sudden changes of atmospheric pressure while magnetic fields play a role of a modulator of physiological amplitude frequency characteristics.

However, multivariate analysis of bio-effective external fields has shown that not only flares and magnetic fields are biologically effective, but disturbances having reached the Earth that not necessarily induce the development of magnetic storms.

3.2.9. Solar Activity Influence on Economic Processes

There are two things that many people try not to think of: that prosperity periods are inevitably broken by crises and that globalization of world economy will inevitably meet the world crisis. Figure 3.21 shows the dynamics of

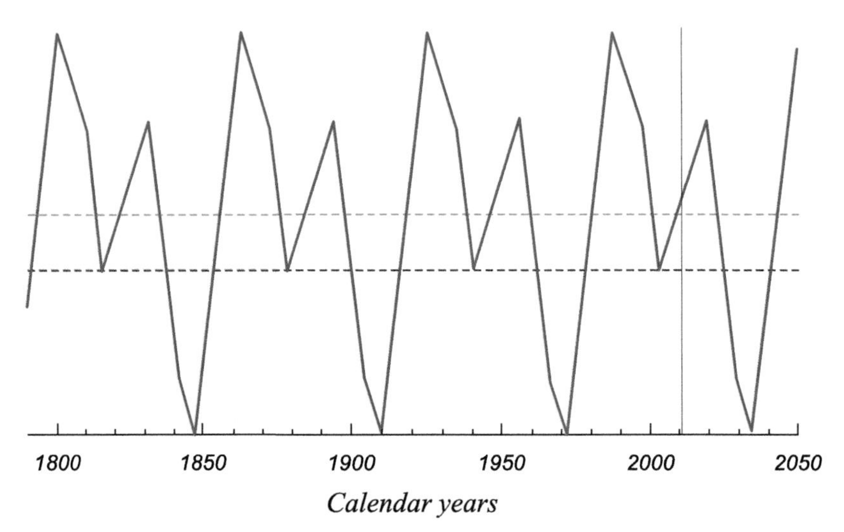

FIGURE 3.21 Dynamics of economical cycles. *Source: www.timingsolution.com/TS/Articles/w_economic/.*

economical cycles. From the figures the presence of similar periodicities in space and economic processes are clear (even without using special techniques). For instance, the so-called Juglar Cycle (9–11 years) agrees with the solar activity cycle.

The Great Depression in the U.S.A., the leader of the World economy, did not only impact on economy of developed capitalistic countries tightly connected with the U.S.A., but touched even the East countries which separate from the rest of the world were trying to built up a self-sufficient economic model (Klinov, 2002). Klinov (2003) suggested a modern conception of large cycles of economic growth in connection with regularities of scientific and technical progress (STP), estimated times of collecting and wasting scientific and technical potential, and made a forecast for the first quarter of the twenty-first century.

Solar activity influence on economic processes reflects in the economical theories of Kondratiev (1935, 1984), Kondratiev et al. (2002), Freeman and Louca (2001), and Ayrapetian (2003).

At the end of the eighteenth century William Herschel, an English scientist, one of the founders of stellar astronomy who built up the first model of the galaxy and discovered Uranus, tried to find a connection between the sunspot number, crop failures, and prices of bread, and determined rather a large correlation between them.

Solar activity influence on economy is not in doubt. In 1989 a magnetic field left Canada's capital Ottawa and Quebec without electricity for 8 hours. In 1997 a solar storm cut off television satellite Telstar 401 of AT&T. The next year a storm destroyed the work of the satellite Galaxy IV which managed automatic cash terminals and aviation tracking systems. In 2000 the Japan satellite Asko was damaged by a solar storm, failed, and sank in the Pacific Ocean.

Magnetic fields affect mobile phones, induce failures in the Internet and automatic systems, an disturb high-frequency aviation radio communication. On Russian railways there have been events connected with failures of automatic devices.

In the second half of the twentieth century in connection with launching spacecrafts and nuclear tests, especially those that were made in the upper atmosphere, an ozone layer protecting the Earth from Sun strokes began to melt quickly.

Coincidence of several cycles of solar activity (11, 22, 80–90, 320–400, and 720–900 year) at the beginning of the twenty-first century has led to more global ecological changes.

At the end of nineteenth century, Jevons, an English scientist, developed a theory connecting forming of economic cycles with solar activity. According to his theory, years of rich crops repeats each 10 or 11 years and "it seems possible that trade crises are connected with periodic weather change touching all parts of the world and appearing, probably, as a result of increased hot waves got from the Sun in general every 10 years or so."

According to Kondratiev et al. (2002), Javons supposed that "Periodic failures are really psychological phenomena depending on changes of low spirit, optimism, pother, disappointment, and panic. But it is probable that mentality of business circles, although forming the main contents of the phenomenon, can be determined by external events, especially circumstances connected with crops."

In economics and sociology some processes were noted (Adler et al., 2005; Ayrapetyan, 2003; Ivaschenko, 2001). Their phase alteration tells us if not about their cyclicity, then about a wavy character as there are no cycles here which would have cyclic parameters.

First of all an industrial 7- to 12-year cycle was found, as reported by Kondratiev et al. (2002). For this cycle K. Marks separated four phases that change each other consequently: crisis, depression, revival, recovery. An industrial cycle got the name after Juglar (1889) who analyzed fluctuations of interest rates and prices in France, Great Britain, and the U.S.A. He found coincidence of them with capital investment cycles which in their turn initiated change of the gross national product (GNP), inflation, and employment. There are 11 Juglar cycles in the period since 1787 to 1932. Insurance market operates cyclically with a period of 7 to 10 years. J. Kitchin's cycles (Kitchin, 1923) are cycles of inventory movement with a period from 2 to 4 years (Voropinova and Kiselev, 2001).

Kuznets' cycle or long swings possessing the largest amplitude in building have a 20-year cycle. Kuznets (1930) discovered interacted fluctuations of indicators of national income, consumer's expenditure, gross investment to industrial equipment, buildings and installations with long-term intervals of fast growth, and deep recessions or stagnancy.

Kondratiev et al. (2002) created an economical theory of long waves and large conjuncture cycles (40–60 years). He pointed to polycyclicity of economical dynamics, "The real process of economical dynamics is a single one. By analyzing and dividing this process to simplest elements and forms we admit the existence of different cycles in this dynamics. Together with it we should admit that these cycles interlace and exert influence on each other." Moreover, Kondratiev found interconnection of economical cycles with cyclic processes in other spheres of societies.

Volume of country production—the gross national product (GNP)—is the fullest indicator of joint economic activity. The British economist Angus Maddison published values of the gross national product of different countries of the world for the period 1870 to 1960 (Maddison, 1962) [*www.ggdc.net/Maddison/*]. For the time interval 1879 to 1954 seven global economical cycles can be distinguished. Thus, an average duration of such a cycle is 11 years (Maddison, 1964, 2007).

The analysis of given data shows that in more than 90% of cases economical indicators have degraded during the years of extreme values of solar activity (minimums and maximums) or during a time period corresponding to its decreasing (descending parts of Schwabe's quasi-eleven-year cycle). There were no economical crises during periods of increasing solar activity (Konstantinovskaya, 2001a,b; Kalashnikov et al., 2002). We should say that similar action of solar minimums and maximums is not something new.

Fast growth of world economy since the second part of the twentieth century has led to the absence of brightly expressed minimums of the GNP values in the given period of time, which, however, does not mean absence of cyclicity. In the given period of time cycles can be separated according to decrease of the GNP growth rate. Extreme characteristic of Solar Activity (SA) value does not have to lead to sudden change of economical growth indicator; its influence can be manifested with some delay, and the existence of economic crises on descending parts of SA cycles can be connected with that.

There are rather reliable data about dynamics of the GNP specific value of the most powerful economical empire of the second part of the twentieth century—the U.S.A. The analysis of given data confirms the supposition that solar maximums are followed by slowdown or fall of American economy growth rate. It is necessary to note that in the second part of the twentieth century a global economic cycle is not of a sinusoidal character—a relatively short-term economic recession (about 2 years) is followed by a much longer period of its growth.

Certainly, there are handmade crises. It is enough to remember oil crises (Bushuev et al., 2002) induced by OPEC policy, the first of which shocked the world economy in 1970s, and the second became one of the reasons for the breakup of the U.S.S.R.

The U.S.S.R. economic development indexes were made with the help of the method of Principal Component Analysis based on data about dynamics of 10 basic economic sectors. The indexes corresponded well to national indicators, but in a larger degree they were correlated with the structure of planned Soviet economy of that time. Then dynamics of these indexes were analyzed with the help of the track method. The result indicates a trend, and two precise cyclic components were revealed: 10 to 11 and 4-year ones.

The trend showed a stable tendency of growth rate slowdown from the middle of the 1950s and 4% of yearly economic recession since 1986. Further, together with a slowdown of economic growth rate in the U.S.S.R., a role of periodic components in economic dynamics became more substantial, as because of growth rate less than 4% a year (1974–1990) about 35 to 50% of economic dynamics fell on cyclic oscillations.

It is curious that by the beginning of 1970s 4-year cyclicity had dominated, but as a result of its decay an increase of 10-year cycle amplitude took place. Further there was more and more promotion of 10-year cycles, the amplitude of which had reached 3.5% points by the end of 1980.

The further analysis of economic dynamics with the help of making forecast and retro forecast with the track method showed that it is possible to separate two trend components—evolutional (system) and transformational (handmade). It was shown that 20 to 50% of recession of the middle of the 1990s fell on the evolutional component. Roots of the system crisis are in the middle of the 1970s when slowdown of Soviet economy growth rate starting in the 1950s accelerated suddenly.

If at the beginning of 1980 there had not been some stabilization it would have turned into a disaster for national economy of the U.S.S.R. Only in 1986 when stabilization potential ran out, crisis phenomena began to dominate the economy again and were over only at the end of the 1990s after Gaidar's reforms.

It is announced on the website message of TESIS X-ray observatory onboard the Russian satellite Coronas-Foton that the beginning of today's world financial crisis coincided in time with a period of abnormally low solar activity which should have changed by a rise about a year before. In the message it says "current solar minimum which is the deepest in this century coincides with a high precision with the development of financial crisis, the most large-scale in history, and world economy transition to a global state of recession."

Scientists note that possible solar influence on social processes can be explained by the fact that during solar maximums there were revolutions of 1905 and 1917 in Russia, the beginning of the Second World War (1939), and events of 1991 coinciding within a month with a maximum of 22-year solar cycle. Processes of social and economical activity slowdown are, on the contrary, connected with solar minimums, although physical mechanisms of solar influence on social processes are unknown yet.

As emphasized in Bogachev et al. (2009) we can consider the current solar cycle abnormal since about the middle of 2008—it is just the period (if to base on historic duration of cycles) when a solar activity decay phase should have been changed by a growth phase. It is obvious that it has not happened. Can we think about any parallels between this moment of solar transition to abnormal activity state and beginning of abnormal processes in the world economy? At present it is not known. However, the question is very interesting and can be the subject of more detailed statistical research.

A long period of solar calmness, the so-called *Maunder Minimum* from 1645 to 1715, was a period that had coincided with several disastrous events, including cooling of southern seas that were frozen and ice covering most European rivers during summer months. The death rate had a sharp increase because mass crop failures occurred.

Scientists think that such long activity decay can have serious consequences, as it is capable of destroying the climatic balance of our planet. Bogachev et al. (2009) reminded that a traditional method of definition of the solar activity level is based on Wolf numbers calculation. But based on this criterion it is impossible to understand if a new cycle has begun. Separate spots appear, among them there are even spots of new polarity (a cycle change is signalized by change of sunspot magnetic field polarity), but these events are so rare that it is impossible to understand a common tendency based on them.

Bogachev et al. (2009), Kuzin et al. (2009) has noted that such observations indicate that new conditions on the Sun during the present solar cycle (initiated in January 2008) has specific peculiarities—the magnetic field of the preceding solar cycle has been destroyed and one of the magnetic zones of the new activity cycle from an opposite polarity zone has been formed. The cycle itself has not begun, as usually, as a new belt. Either the field was too weak or the Sun needed formation of both activity zones for the cycle beginning.

It is clear that solar activity does not exert direct influence on economic and social processes (Abramov, 2001; Ivanov, 2002), which is different from climatological and biological processes. There is no a direct mechanism of influence, as claim some pretentious researchers. Nevertheless, if solar activity changes influence people and temperature, river inflows and precipitation amount, atmosphere circulation and solar radiation, (Babkin et al., 2004) all these impact on economy of individual countries and world economy as well (Bushuev et al., 2002).

3.2.10. Possible Mechanisms of Influence of Cosmophysical Factors on Climatological Processes

Although not all links of the chain of solar-terrestrial connections are equally researched, in general, a picture of solar-terrestrial connections is rather obvious. Quantitative solutions with badly known (or unknown) initial and boundary conditions are found difficult because of the absence of knowledge of concrete physical mechanisms that provide energy transmission between separate links.

Together with a search of physical mechanisms, an information aspect of solar-terrestrial connections are researched. Connections manifest themselves in two ways depending on if energy of solar disturbances inside the magnetosphere is redistributed smoothly or abruptly, in leaps and bounds. In the first case solar-terrestrial connections are manifested in a form of rhythmical oscillations of geophysical parameters (700-year, 22-year, 11-year, 27-day, etc.). In the second case, abrupt configurations are connected with a so-called trigger mechanism which can be applied to actions or systems in an unstable, close to critical state. In this case a small change of a critical parameter (pressure, current intensity, particle concentration, etc.) leads to qualitative change of a process or initiates a process.

For example, during formation of extra tropical cyclones during geomagnetic disturbances, energy of geomagnetic disturbance is transformed to energy of infrared radiation. The latter creates additional weak heating of the troposphere, in a result of which its vertical instability is developed. In this case, energy of developed instability can exceed energy of initial disturbance by two orders (Artekha et al., 2003).

According to the works of Ariel et al. (1986), Artekha and Erokin (2005), Gulinsky et al. (1992), Valdez-Galicia (2005), Dorman et al. (1987b), Libin and Jaani (1989), Dorman (1991), Pérez-Peraza and Gallegos-Cruz (1994), Chistiakov (2000), Avdiushin and Danilov (2000), Libin and Pérez-Peraza (2009), Kondratyev and Nikolsky (1982), Obridko and Shelting (2008),

Pudovkin (1992a,b), and others there are several probable mechanisms of the influence of heliophysical and cosmophysical factors on the lower atmosphere:

1. Mechanisms based on a solar constant change (astronomic and meteorological) (Nikolsky, 2009, 2010)
2. Change of atmosphere transparency under the influence of different extra terrestrial processes, in particular, under galactic cosmic ray action (Stozhkov et al., 2008; Svensmark, 2007, 2008)
3. Additional infrared radiation during magnetic storms
4. Influence of a solar wind on atmospheric electricity parameters
5. Condensation mechanism (see in the review Pudovkin and Raspopov, 1992a,b)
6. Ozone mechanism and intensity changes of ultraviolet radiation reaching the Earth surface because of an ozone layer configuration in the high latitudes as a result of accelerated particles influence on it
7. Hydrodynamic interaction of the upper and lower atmosphere
8. Solar dynamo model: Matter transmission from the subphotospheric layers of the Sun as a result of its rotation and convection. Interaction with the solar magnetic field triggers a dynamo process (an electric current generator transforming mechanic energy to energy of a magnetic field). When charged particles move together with flowing matter, a magnetic field connected with them moves too.
9. Infrasound (acoustical fluctuations of very low frequency): Infrasound appears in aurora regions in the high latitudes and spreads in all latitudes and longitude. So, it is a global phenomenon. Four to 6 hours after the beginning of world magnetic storm fluctuation, amplitude in the middle latitudes rises slowly. After its maximum it is miniaturized evenly during several hours. Infrasound is generated not only during auroras, but during storms, earthquakes, and volcanic explosions, so there are constant oscillations in the atmosphere which are in collusion with oscillations connected with a magnetic storm.
10. Micro pulsations or short-period oscillations of the Earth's magnetic field (with frequencies from several Hz to several kHz): Micro pulsations with frequency from 0.01 to 10 Hz work on a biological system, in particular on the human nervous system (2–3 Hz), increasing time of reaction to a disturbing signal, influencing psychics (1 Hz), arousing melancholy, horror, and panic without any reasons. Increase of diseases, frequency, and cardiovascular patients' complications is also connected with them.
11. Parametrical influence of solar activity on thermobaric and climatic characteristics of the troposphere: Zherebtsov et al. (2008) gives the analysis results of peculiarities and regularities of the troposphere temperature regime changes during the period of changing heliogeophysical activity and long-term changes of temperature and heat content of the troposphere. He has researched influence of atmospheric and oceanic circulation changes

on processes in the system Atmosphere-Ocean-Cryosphere—circulation in the ocean and energy exchange of the atmosphere with the ocean. Revealed regularities are fully explained in the framework of the model and mechanism of solar activity influence on the troposphere characteristics suggested by the authors earlier.

12. Directed spiral (whirling) radiation (Kondratyev and Nikolsky, 2005): In 1981 we noticed some cases of abnormal changes of the troposphere meteoparameters under the influence of solar activity (Kondratyev and Nikolsky, 1982), but only in 2003 analyzing the whole set of high-altitude observations of solar emission influence on atmospheric parameters did we identify manifestations of influence of not an electromagnetic or gravitational, but wavy, spiral (whirling) radiation flux from active regions (AR). Intensity of this specific radiation grows substantially during crossing an active region (AR) through the central part of the solar disk (a latitude zone $\pm 20°$ is regarded). Spiral (whirling) radiation (SWR) (as it comes out of its effects) influences mainly dynamic and structural parameters of the medium bringing (for example, in the Earth's atmosphere) additional energy to air vorticity, giving the air a comparatively small impulse, and a considerable moment of the impulse, because a whirling field from a local photospheric surface (for example, a spot with a geometrically correct form) can light 105 km^2 locally in different regions of the Earth simultaneously.

Effects of SWR influence on the atmosphere and, as it seems, other spheres should be regarded for the part of the Earth lighted by the Sun and the shadowy part separately, as SWR is characterized by a comparatively high penetrating capability and interacts with the solid lithosphere cover using it as a spherical lens. According to estimations (Kondratiev and Nikolsky, 1995a,b; Kondratyev, 2004a,b) and natural phenomena (e.g., craters), a focusing effect exceeds losses by several orders during movement of an SWR quantum field (spirons) along the lithosphere spherical waveguide. A high-speed (supersonic) concentrated whirl with a 30 to 50 m focus spot and high energy density on the whirl periphery up to 14 G/km^3 (after focusing) has been created in focus situated on the surface of the Earth spheroid or near the Earth before or after spheroid boundary. On the basis of data of high-altitude observations in cycles 21, 22, and 23 of solar activity, its influence on radiation, optical, microphysical, and meteorological atmosphere characteristic have been researched with an emphasis on the study of SA influence on dynamic (circulation) processes and changes of microphysical state of vapor molecular ensemble.

Contributions of spiral (whirling) radiation (SWR) and a proton flare flux to dynamic processes in the lower troposphere are illustrated by the example of the event of October 20, 1989. Observations of abnormal disturbances in typical daily movement of meteoelements, as well as movement of their synoptical periods, have led to clearing up the reason of appearance of power-intensive disturbances in synoptical processes in the troposphere. It seems that directed spiral (whirling) radiation and solar proton fluxes are basic and comparable force factors of solar activity influence the troposphere and the lower stratosphere; however, only the first factor influences effectively all the components—the atmosphere, hydrosphere, magnetosphere, lithosphere, biosphere and technosphere.

13. Cosmic rays (CR) and the atmosphere conduction (Stozhkov and Tikhomirov, 2008; Stozhkov et al., 2008): Ions formed by CR provide the atmosphere conduction. Current that flows in the atmosphere is one of the basic elements of a global electric chain which supports a constant negative charge of the Earth.

Lightning discharges of thunderclouds are the generator of electric charges of the atmosphere. Thunderstorm clouds are formed in the atmosphere fronts where formation and division of cloud charges take place. Positive and negative ions formed in the lower atmosphere by CR and natural radioactivity of the Earth is the source of thundercloud charges. These ions stick to aerosol particles, concentration of which is high in the atmosphere (more than 104 cm^{-3}). On charged aerosol particles water drops grow slowly as they climb up with ascending air.

According to Rusanov et al. (2009), negative charges are separated from positive ones due to the fact that water drops on negative centers grow ~10,000 times faster than on positive ones. As a result of this process the lower part of the cloud is charged negatively and the upper one positively. Lightning discharges appear when the so-called *extensive air shower* of up to 106 charged particles formed by a high-energy cosmic particle moves through the cloud.

Lightning charges happen on tracks of ionizing particles of extensive air showers. Thus, CR is a necessary constituent part of thunderstorm electricity and lightning discharges formation processes. Charged particles fluxes in the Earth's atmosphere intensify or reduce the process of cloud formation. During powerful flares with emission of high energy solar protons, the charged particle flux in the Earth's atmosphere increases, cloud density grows, and precipitation amount increases. During CR Forbush decreases, particle flux in the atmosphere decreases and precipitation level becomes lower.

In 1998 Danish scientists discovered a very interesting phenomenon using observations of clouds from satellites: area covered with clouds on our planet changes in correspondence with value changes of a CR flux that falls on our atmosphere. Every year a CR flux value becomes lower by 0.01 to 0.08%.

A negative trend is the result of explosion of a close supernova. This explosion happened several tens of thousands of years ago at the distance of several tens of parsec (1 parsec is 3.08×10^{13} km). Consequently, the area covered with clouds decreases slowly. This decrease should make temperature on our planet rise. We know that for the last 100 years temperature on the Earth surface has risen by ~0.5 °C. Thus, CR flux decrease can contribute to the effect of global warming.

In this case in practically all mechanisms a warmth flux from outer resources—solar flares, interaction of solar plasma and the Earth's magnetosphere, geomagnetic storms, geomagnetic convection, penetration of particles to polar regions, generation of additional quantity of nitrogen dioxide and ozone by solar and galactic cosmic rays in the lower stratosphere, solar activity influence on the atmosphere electric field, and so on—is a substantial factor of any discussed mechanisms.

In any case, a substantial source of global warming and climate change is situated outside the Earth.

3.3. MODELING THE INTERACTION OF COSMOPHYSICAL AND CLIMATOLOGICAL PROCESSES

Research devoted to problems of connection between the climate and changeability of solar activity and the solar wind can be divided into the following three groups.

1. Climatic changes during hundreds and thousands of years
2. Changes correlating with 22-year and 11-year solar activity cycles
3. Changing during several days or weeks

In any case, during the study of short-period and long-period atmospheric processes account of solar activity influence with applying data of indexes of geomagnetic activity, cosmic ray intensity, and so on, is very important.

In each case this influence depends greatly on the spectrum of electromagnetic and corpuscular solar radiation, on the current state of the solar and interplanetary magnetic fields, on the Earth heliolatitude, and a geographical region of atmospheric processes under study. That is why during modeling the solar activity impact on atmospheric processes it is necessary to account for any additional information, because using only Wolf numbers for establishing long-term interdependencies is not always justified.

The results of researches of cosmophysical and climatological (meteorological) processes mentioned in previous chapters prove the existence of cause-and-effect relations between solar activity and other processes. That is why we can admit that parameters describing atmospheric processes (in particular, for example, temperature $T(t)$) can be presented as the sum of preceding values of $T(t)$, solar activity $W(t)$, geomagnetic activity $\mathrm{K}p(t)$, solar radiation $R(t)$, and cosmic ray intensity $I(t)$, in an autoregressive model.

$$T(t) = \sum_{i=1}^{p} \alpha_i T(t-i) + \sum_{j=1}^{q} \beta_j W(t-j) + \sum_{k=1}^{s} \gamma_k I(t-k) + \sum_{l=1}^{m} \delta_l \mathrm{K}_p(t-l) + \xi_t \quad (3.10)$$

$T(t)$ is the predicted temperature value, where p, q, s, and m are the model order for each used series determining a back sight of each process for prediction of temperature estimation; α_i, β_j, γ_k, and δ_l are AR-model parameters.

In this case, while new data appear, one can observe renewal of autoregressive estimations and the possibility of predicting temperature one step further appears. (Predictions for the future consist of searching for a future value as weighed sums p of preceding $T(t)$ counting, q counting $W(t)$, $S(t)$ and $I(t)$ and m counting $K_p(t)$.)

There are two ways of building a model:

1. On the basis of a collected data set of temperature, solar activity, and cosmic ray intensity (a size of each is NO), the matrix for a system of linear equation (Eq. 3.10) is made up; vectors $\{\alpha\}$, $\{\beta\}$, $\{\gamma\}$, and $\{\delta\}$ are determined from its solution. But in this case it is necessary to take into account that autoregression coefficients can be determined practically for each accumulation interval. Using monthly average data for the period 1950 to 2004 it is possible to receive about 600 equations. Consequently, for yearly average data an equation number decreases up to $54 - k$, where k is the maximal autoregression order.
   If $k = 5$ an equation number makes approximately 50, and in this case we can try just to solve a kind of equation system (Eq. 3.10) assuming that noise is minimal.
   A system solution is actually a solution of system $N = \mathrm{NO} - \mathrm{k} - 1$ of Equation (3.11) given below, where Pi represents a surface temperature T, averaged over the region, if there are measures of several stations j. Thus, during estimating forecasts of temperature in Mexico, scientists used temperature values received from 30 to 40 meteostations in each of researched regions.
2. If the number equations (Eq. 3.10) is higher than an unknown $\{\alpha\}$, $\{\beta\}$, $\{\gamma\}$, and $\{\delta\}$ number and it is impossible to make a supposition about the minimal value, then a system solution adds up to solution of Equation (3.12), connected not with values of solar activity, temperature, geomagnetic activity, and cosmic ray intensity, but with their covariance functions Atj.

$$\begin{pmatrix} \mathrm{P_k \ldots P_{k-q}\; W_k \ldots W_{k-s}\; K_{pk} \ldots K_{pk-r}\; I_k \ldots I_{k-m}} \\ \vdots \\ \mathrm{P_{k+i} \ldots P_{k+1-q}\; W_{k+1} \ldots W_{k+1-s}\; K_{pk+1} \ldots K_{pk+1-r}\; I_{k+i} \ldots I_{k+1-m}} \\ \vdots \\ \vdots \\ \vdots \\ \vdots \\ \vdots \\ \vdots \\ \vdots \\ \vdots \\ \mathrm{P_N \ldots P_{N-q}\; W_N \ldots W_{N-s}\; K_{pN} \ldots K_{pN-r}\; I_N \ldots I_{N-m}} \end{pmatrix} \begin{pmatrix} \alpha_1 \\ \vdots \\ \alpha_q \\ \vdots \\ \beta_1 \\ \vdots \\ \beta_s \\ \gamma_1 \\ \vdots \\ \gamma_r \\ \delta_1 \\ \vdots \\ \delta_m \end{pmatrix} = \begin{pmatrix} \mathrm{P_{k+1}} \\ \vdots \\ \mathrm{P_{N+1}} \\ \vdots \\ \mathrm{W_{k+1}} \\ \vdots \\ \mathrm{W_{N+1}} \\ \mathrm{K_{pk+1}} \\ \vdots \\ \mathrm{K_{pN+1}} \\ \mathrm{I_{k+1}} \\ \vdots \\ \mathrm{I_{N+1}} \end{pmatrix} \tag{3.11}$$

$$\begin{pmatrix} A_{11} & \cdot & \cdot & A_{14} \\ \cdot & A_{22} & & \cdot \\ \cdot & & A_{33} & \\ A_{41} & & & A_{44} \end{pmatrix} \begin{pmatrix} \alpha \\ \beta \\ \gamma \\ \delta \end{pmatrix} = \begin{pmatrix} a_0 \\ b_0 \\ c_0 \\ d_0 \end{pmatrix} \tag{3.12}$$

It is clear in this case that, for example, for a region averaged temperature

$$P(t) = \sum_{J=1}^{N} T(t)_{i,j}/N$$

$$P(t-1)\cdot P^*(t-1) = \begin{pmatrix} P_{t-1}^2 & P_{t-1}\cdot P_{t-2}\ldots\ldots\ldots P_{t-1}\cdot P_{t-q} \\ P_{t-q}\cdot P_{t-1} & P_{t-q}\cdot P_{t-2}\ldots\ldots\ldots P_{t-q}^2 \end{pmatrix} \tag{3.13}$$

Analogous expressions are received for other parameters W, S, K_p, and I applied for building up a prognostic model. Covariance values A_{ij} and a_0, b_0, c_0, and d_0 can be considered as known—they are calculated from the real data. Thus, the system (Eq. 3.12) is a system of linear equations for defining unknown coefficients $\{\alpha\}$, $\{\beta\}$, $\{\gamma\}$, and $\{\delta\}$ of regression Equation (3.10), and can be solved only if the covariance matrix determinant differs from zero.

Noise exception is an advantage of the second way of an equation system solution, as mutual noise covariances with other processes are equal to 0.

So, setting oneself model orders for q, s, r, and m, temperature, solar activity, geomagnetic activity, and cosmic ray intensity respectively, it is possible not only to predict average values of temperature one step further (distance of this step will be determined only by a discretion data quantity Δt) but also to help to estimate contribution of this or that process to forecasted temperature. Note that, if values of one of the searched parameters $\{\alpha\}$, $\{\beta\}$, $\{\gamma\}$, or $\{\delta\}$ are small (much less than errors of their measurements), one can disregard the process itself in the model. Thus, during analysis of yearly average dependences between temperature and geomagnetic activity, the values were insignificantly small in comparison with errors and in absolute value.

A detailed analysis has shown that it is unacceptable to use only Kp-index to build up a prognostic model.

On the basis of values of averaged temperature in the northern part of Russia for 1950 to 2008 and Mexico for 1950 to 2004, solar activity for 1945 to 2008, geomagnetic activity for 1945 to 2004, and cosmic ray intensity for 1960 to 2007, scientists determined values of the model parameters and estimated $T(t)$ values for 2004, 2005, 2006, and 2007 which were later compared with the real $T(t)$ values for the same years. The analysis was based on monthly and yearly average values of $T(t)$, $W(t)$, and $I(t)$, and a two-year drift $W(t)$ relative to $T(t)$ was entered to yearly average data in advance (when using monthly average values a drift was not entered).

It is necessary to note that a number of predictors (the model order) should not exceed one tenth of a sample volume (as it is shown in Rozhkov (1979)), which can be explained by possible correlation connection between variables, on one hand, and sample limitation, on the other.

The received data show that α, β, and δ values for the whole period are practically not changed. Moreover, $T(t)$ values defined for 2004 to 2007 (in the framework of a prognostic model with coefficients α, β, and δ) differ from the real $T(t)$ values less than 12 to 25% for prognosis to 1 year, 27% for 2 years, and 30% for 3 years. A temperature forecast for 2008 given in 2005 was justified with a 40% error; the same forecast given in 2007 was justified with a 30% error.

Thus, the use of standard ARSS-models for working out forecast for temperature (and any other climatological parameters) based on preceding values and observation data of solar activity, geomagnetic activity, and cosmic ray intensity has good prospects.

The results of researches for the last years in this direction demonstrate rather good correlation of calculated and experimental values not only of $T(t)$, but of storminess $P(t)$, precipitation amount, and ice area in the Baltic and White Seas and in the Arctic. That is why further working out of such a model allows us to predict yearly average meteorological characteristics for one year further with 25 to 30% precision.

To get less precise estimations a two-parameter model is enough,

$$P(t) = \sum_{i=1}^{q} \alpha_i P(t-1) + \sum_{k=1}^{p} \beta_k \mathrm{W}(t-k) + \xi_t \tag{3.14}$$

Thus, the use of all existed spectral apparatus and comparison of results of various spectral calculations obtained by different researchers in different countries show that solar cyclic activity and its influence on the Earth's atmosphere are the source of the mechanism of impact on atmosphere circulation and water content of closed lakes. Thereby, behavior analysis of retardation between atmospheric processes and solar activity show the existence of constant drifts from 12 to 36 months between processes that correlate well with other technique calculation results.

It is necessary to emphasize that mutual autoregressive analysis for the whole data set and for even and uneven solar activity cycles separately result in an interesting picture. Brightly expressed 4- to 7-year, 10- to 12-year, and 80- to 90-year variations of ice area are observed in uneven cycles (on the backdrop of weaker expressed 400- and 720-year variations), 20- to 30-year, 80- to 90-year, and 400-year variations prevail in even cycles.

The analysis of surface temperature measuring in Mexico (Sonora, Tacubaya, Baja-California, and Sinaloa), Estonia (Tartu), Sweden (Stockholm), Lithuania (Kaunas), and Moscow for 1910 to 2007 demonstrates precise coincidence of separated frequencies (correlating to 2–4 year and 9–11 year periodicities) and good coincidence of 2- to 3-year retardations between processes. In this case, dynamics of oscillation behavior also coincides: 9- to

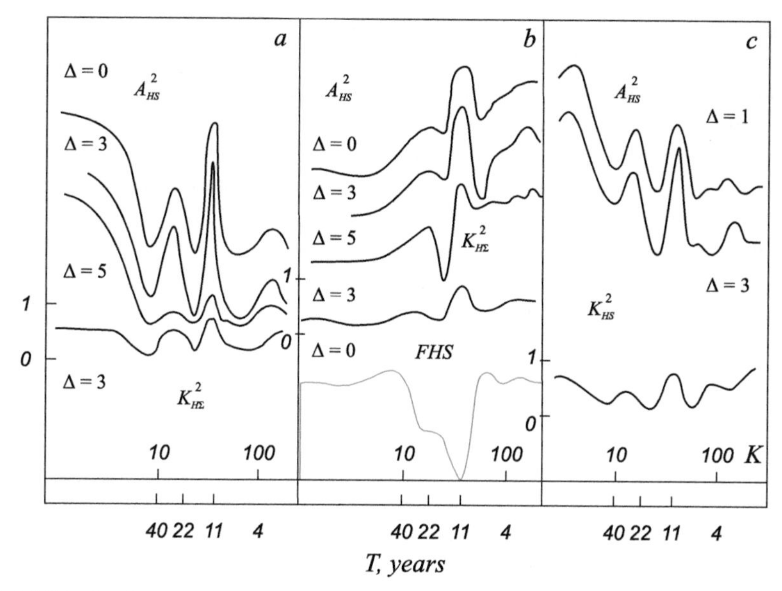

FIGURE 3.22 Power spectrums (A), coefficients of coherence (K2), and phase spectrums (FHS) for solar activity and atmospheric processes. *Source: Pérez-Peraza et al., 1997.*

11-year oscillations always exist, 2- to 4-year oscillations are less systematic, but also correlate well with the behavior of analogous solar activity oscillations (see Figure 3.22).

Analysis of observation data of precipitation in Estonia, Lithuania, and Russia for 1910 to 2007 and solar activity (sunspot area *S*) shows good coincidence of process behavior—the existence of 11-year and 22-year constituents and a quasi-biennial wave in the analyzed data.

On the basis of solar radiation, measurements in different points of the Earth surface in the period from 1950 to 2007 represent an attempt to estimate possible modulation of solar radiation by solar activity observed on the Earth. It is found out that in a broad frequency band in observation data of solar radiation on the Earth, oscillations with periods 2, 11, and 22 years tightly connected with solar activity are observed.

In connection with that, results of the analysis of solar activity and carbon dioxide contents in the Earth atmosphere are of great importance (using HL-index and the total sunspot areas). The autoregressive analysis made by the authors demonstrates precise coincidence of separated frequencies (correlating to 2–4 year, 9–11 year, 22–35 year, and 380-year periodicities) and good coincidence of 2- to 3-year retardations between processes. In this case, dynamics of oscillation behavior also coincides; if 9- to 11-year and 22-year oscillations of carbon dioxide always exist, 2- to 4-year oscillations are more casual which also correlate well with the behavior of analogous solar activity and atmosphere temperature oscillations.

Comparison of given results with similar researches of solar activity influence on the surface temperature, lake level, precipitation amount, and strong earthquake quantity show not only qualitative, but quantitative (to a precision of retardation) correlation. Besides, it is found that during joint analysis of temperature in different places of the Earth and solar activity, a choice of solar activity indexes does not play a decisive role. Thus, the authors consider sunspot area in the subequatorial zone of the solar disk the most acceptable index for calculations. That is why when solving problems of revealing mechanisms of large-scale processes in the atmosphere or trying to build up prognostic models of climatological or hydrological processes, it is necessary to take into account solar activity changes, processes in the interplanetary medium, and cosmic ray variations observed on the Earth.

In Rahmstorf et al. (2007) a group of scientists from different countries analyzed many forecasts of climatological characteristics made by different authors over the last 15 years and compared these forecasts with what had really happened during those 15 years. It was noticed that the changes of carbon dioxide contents in the atmosphere were correctly predicted, and temperature changes were predicted rather reasonably. All these indicators present a correspondence with the trends revealed earlier. But the average level of the World Ocean rose faster than it had been expected. From 1990 to 2005 it rose by 4 cm, although only 2 cm had been predicted. (It can be so that the reason of some deviation of observed values from predicted ones is the inner variability of the climatic system itself, a result of its component interaction dynamics unknown to us.)

Researchers from several international groups think that for prediction of future climatic changes, models based on autoregressive spectral analysis are very useful tools (for example, the authors of the book). These models are built on the basis of data that have been already observed during the previous years and on understanding interconnections of physical processes occurring on our planet surface. The forecast for the World Ocean level appeared to be the least satisfactory (the low panel of the diagram (Rahmstorf et al., 2007). For the last time this level has gone up much faster than it had been supposed by the IPCC model. The real increase (according to satellite measures) from 1993 to 2006 made up on average 3.3 ± 0.4 mm per year, while the most probable value of the model was less then 2 mm per year.

Rahmstorf et al. (2007) note that the rise of ocean level for the last 20 years (see Figure 3.23) was faster than for any 20 years during the preceding 115 years. Observed values correspond to utmost figures given in the model as unlikely and connected with the so-called "vagueness in the state of ice on land." And although simple heat expansion of water mass makes the basic contribution to the ocean level rise while global temperature increases, it seems that glacier melting plays an underestimated role. The authors come to the conclusion that we should deeply refine scientific prognoses of the climate change. Changes of carbon dioxide contents in the atmosphere and temperature

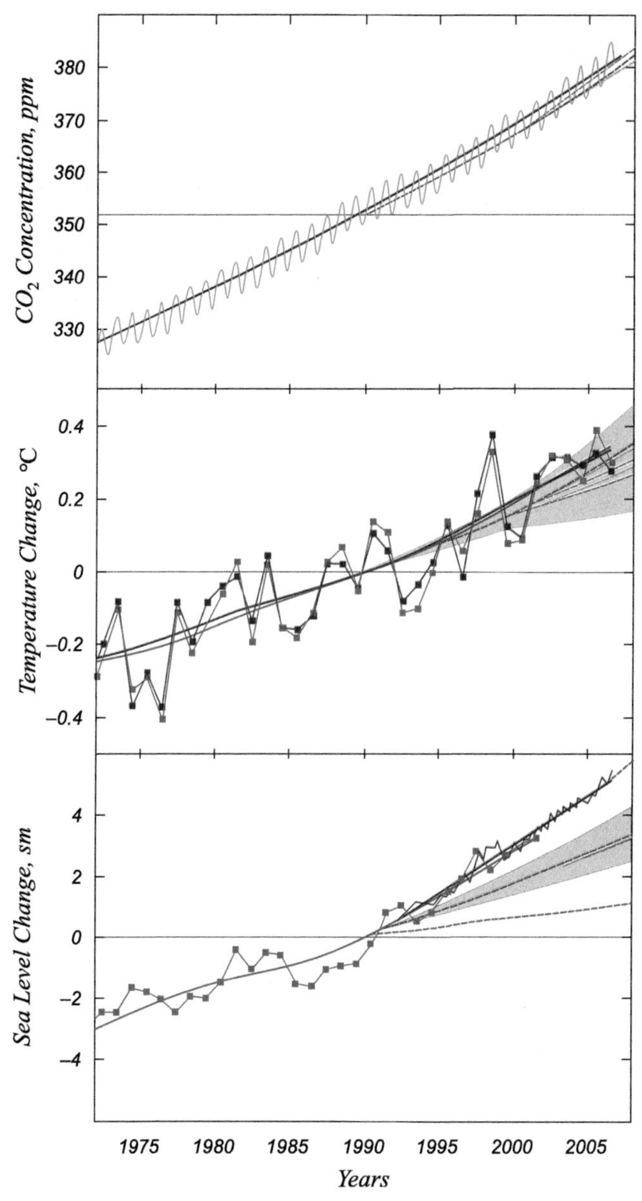

FIGURE 3.23 Changes of contents of CO_2 in the atmosphere (at the top of the graph), average temperature on the Earth surface (in the middle of the graph), and average level of the World Ocean (in the bottom of the graph) from 1973 to the present. *Source: Rahmstorf et al., 2007.*

changes were predicted rather well. As for the ocean level (the least satisfactory variant of the forecast), the reality appeared to be more severe than it had been predicted (Figure 3.23).

Nevertheless, the forecast results obtained in the whole series of works (and by the authors as well) are in agreement with the real data and fit themselves on the curves (see the figure) obtained in the work (Rahmstorf et al., 2007). These results allow us to use an autoregressive prognostic model for forecasting climatological processes (temperature, water content of closed lakes and seas, storminess, ice area, precipitation amount) rather reasonably taking cosmic ray intensity into account.

The thin solid lines represent the real data (see Figure 3.23), the thick solid ones represent the averaged real data which show the basic trend. The dotted lines are data of forecasts and given confidence intervals (areas colored with blue-grey). Temperature and the ocean level changes are given as deviation of a trend line in the place of its crossing the 1990 mark (taken as a zero).

Chapter 4

Estimation of Risks and Advantages Conditioned by Global Climatic Changes

Chapter Outline

4.1. RISKS TO HUMAN EXISTENCE

Existing risks are various and conditioned by several factors which characterize peculiarities of certain types of activity and specific traits of vagueness in conditions of which these activities are realized. Such factors are called *risk formation factors*. The term describes the processes and phenomena that promote appearance of a risk and determine its character.

Nowadays researchers' basic attempts are directed to specification of a list of risk formation factors for certain risk types and to working out techniques for estimation of influence of these factors on the dynamics of corresponding risks. For example, climatic risks and factors that influence them are hardly researched in Mexico or Russia. However, according to American specialists' research, in the U.S.A. climatic risks exert substantial influence on industrial production, which costs up to one trillion dollars (of the seven trillion dollar yearly gross product of the U.S.A.).

Makhov and Posashkov (2007) describe four types of strategic risks for the nearest 50 to 100 years, which humanity as a whole must regard.

Highlights in Helioclimatology. DOI: 10.1016/B978-0-12-415977-8.00004-6

1. **Cosmic risks**. This represents the danger of the Earth colliding with other cosmic bodies such as meteors, asteroids, or comets. Each is different in size, contents, physical peculiarities, and velocities. Meteors are stone and iron bodies, fragments of larger bodies such as comets or asteroids. They are several tens of meters in diameter. Impacting the Earth, free energy is produced from an explosion and a crater several times larger than the meteor diameter appears on the collision place. Asteroids are small planet-like bodies of the solar system. The most famous asteroid, which was 10 km in diameter, fell 65 million years ago in the region of the Yucatan Peninsula. According to one hypothesis, this collision was the reason of the extinction of dinosaurs. Comets are cosmic bodies that consist of a solid core, gaseous head, and a tail. The nucleus is tens of kilometers in diameter (that of Comet Halley was $16 \times 8 \times 8$ km), a head can be several hundreds of thousands of kilometers size across, a tail length can reach several hundreds of millions of kilometers. Collisions with cosmic bodies can lead to disasters of different scales. A meteorite of 1 km in diameter is enough to produce a global disaster; a stroke will annihilate everything within a radius of 1000 km of the impact point, conflagrations will cover wide territories, a great deal of ash and dust will be emitted to the atmosphere, and then it will precipitate during several years. Solar rays will not be able to penetrate to the Earth surface, and a sudden cooling will destroy many species of plants and animals, and photosynthesis will be halted. The Earth's magnetic field will also be disturbed, dynamics of tectonic processes will change, and volcanic activity will increase.
   Organization of measures to prevent a threat of collision with cosmic bodies includes two stages (Mikisha and Smirnov, 1999).
   a. Monitoring of objects (searching and identification of cosmic objects)
   b. Neutralization of dangerous objects with a help of the whole system (deviation of a threatening object from an orbit of meeting with the Earth, screening the Earth from collision with a threatening object, and, at last, annihilating of the threatening object)
2. **Resource risks**. These are risks connected with depletion and exhaustion of natural resources and with environmental pollution. When considering these risks it is convenient to divide resources in the following types (Mikisha and Smirnov, 1999).
   a. Fuel and energy resources (oil, coal, natural gas, nuclear energy, hydro energy, solar energy, wind energy)
   b. Raw material resources (metals, minerals, alloys)
   c. Natural resources connected with environment (land, water, forest, bioresources)
3. **Demographic risks**. These are mainly risks connected with today's global demographic transition, a process of transition from the state of high indicators of birth and death rates to the state of low coefficients of birth and death rates.

4. **Climatic risks**. These are risks connected with climatic changes and studied in this book. They are, first of all, extreme natural phenomena (climatic and weather abnormalities) and also risks connected with change of yearly average global temperature—a problem of global warming.

As mentioned in the previous chapters, there are two basic approaches to researching the Earth climate. The first approach is empirical-statistical. This means that on the basis of data about past climatic changes an attempt is made to restore the planet climate history in the past and predict these changes in future. The basic methods here are the following:

- Geological methods of stratigraphy (radiocarbon analysis of fossilization, rocks, glaciers, organic residues), paleotemperature methods
- Historical evidences (data records, chronicles, archeological finds)
- Direct instrumental observations and estimations (building up of statistic prognostic models)

The second approach is theoretical. It is connected with building of mathematical climate models.

There are three basic groups of natural climate formation factors (in correspondence with them it is possible to structure the model types):

1. *Heliophysical and astrophysical factors* are conditioned by influence of other solar system bodies on the Earth: solar luminosity, solar activity, gravitational influence, and tilt of the Earth axis to the orbit plane. As these factors are fundamental among mechanisms responsible for climate formation, they are taken into account in the majority of climate models to a lesser or greater degree.
2. *Geophysical factors* are connected with peculiarities of the Earth as a planet: form, size and mass of the Earth, velocity of its rotation round the axis, gravitational and magnetic fields, structure and inner processes (including tectonic activity), inner warmth resources (such as geothermal warmth flux and volcanism determined by them), and the atmosphere contents. These factors are studied separately, as a rule, with separate methods, and are accounted for in climate models as parameters.
3. *Circulation factors* are connected with processes inside the system consisting of the atmosphere, hydrosphere (ocean and glaciers), lithosphere (land), and biosphere: redistribution of energy and matter inside the system. It is thought that just circulation factors are responsible for appearance and frequency of extreme natural phenomena (Zhvirblis, 1987; Konstantinovskaya, 2001b; Kuzhevskaya and Dubrovskaya, 2007; Sherstiukov, 2007; Zherebtsov et al., 2005; Zveryaev, 2007; Zadde and Kijner, 2008).

For reliable estimations, common circulation models use, as a rule, two- and three-dimensional models for the atmosphere, ocean, and their joint system.

The kind of such models is the most frequent as besides the climate they are used for meteorological forecasts (weather forecast).

These models use equations of hydro thermodynamics describing transfer of energy and mass in the atmosphere and in the ocean, or movement of air and water masses participating in energy redistribution. Solutions of such equations are unstable locally; small disturbances of different nature (imprecision of initial conditions, inner and outer noise impact, etc.) grow in time or in space very quickly and, at least, make a long-term forecast inconsistent.

In models of the ocean circulation besides hydro thermodynamic equations, the equations describing state and change of sea ice (crystallization, melting) are used. The role of the ocean in climatic changes is high because of its inertness, which is larger than that of the atmosphere; that is why during research of slow processes from several years to several decades the ocean process modeling is of a great importance.

Because of a great number of factors (and huge number of parameters) it is impossible to build up a general climate model today. That is why all existing models account only for a part of factors, the rest are just parameterizations of high usefulness as support in the obtention of results of modeling and forecasts made with a help of such models. It must be pointed out here that the roles of some factors, processes, and interactions are not yet fully studied. A lot of processes cannot subject to adequate generally accepted explanation, and some phenomena and interdependences are hidden from observers.

Every year a number of new observations and factors grow, but vagueness is kept and even grows, as climatic models are very sensible to small changes. It makes a problem of prediction of natural processes more complicated, because the climate is a complex nonlinear self-organizing system of nonstationary nature. Similar systems usually demonstrate an unstable character, which predetermines the existence of unpredictability horizon that restricts the development of more adequate models. All these affect the limits of theoretical physical climate models and the results obtainable from them.

4.2. RISKS AND ADVANTAGES CONNECTED WITH GLOBAL WARMING

Figure 4.1 shows a detailed picture of temperature fluctuations over the last 2500 years. What were the main climatic events during this period? First, it was unusually cold; similar to the time of early antiquity. Other outstanding climatic events were warming during the Roman period and a significant cooling of the Great Migrations.

Finally, the peak at the turn of the first and second millennia—the Warm Medieval Climatic Optimum. This gained fame, in particular, through the development of the Normans in Greenland. After that is a clear trend of a decrease in temperature. The coldest periods were noted in the fifteenth and at the end of the seventeenth century. In July 1601, Moscow was built on a sleigh.

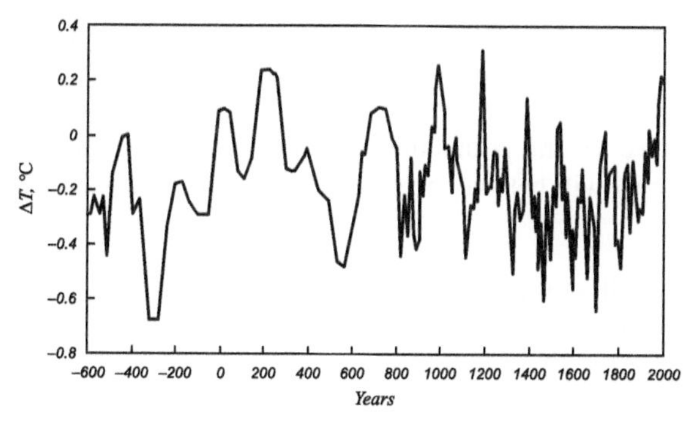

FIGURE 4.1 The history of climate in recent 2500 years. *Source: Klimenko, 2009a,b.* www.polit.ru/article/2005/11/02/climate/

An even colder period came in the late seventeenth century. The seventeenth century was the coldest period in the last few thousand years. The population of Finland, Estonia, Livonia, northwestern Russia, Scotland, Denmark, and Northern Germany during this time was reduced by 30 to 40%. After that, the temperature gradually increased (Libin and Pérez-Peraza, 2009).

History and projections of the global temperature anomaly (relative to 1951−1980) and forecasts from various scientists until 2200 are shown in Figure 4.2.

Since the end of nineteenth century until the present time the temperature has increased by 0.5 to 0.6 °C. Certainly, it depends on what period we compare. If we take the temperature of the last 5 years, it's an increase of 0.8 °C (the last 5 years were very warm). But comparing the value of 20 years, it was actually 0.6 to 0.7 °C. It is important that the entire planet awaits warming.

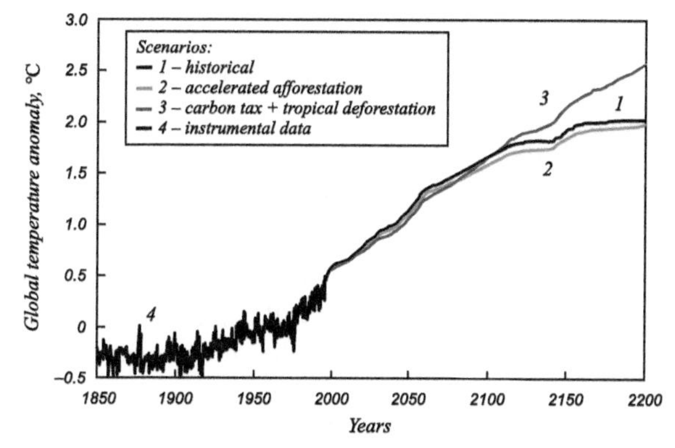

FIGURE 4.2 History and projections of global temperature anomaly (relative to 1951−1980) and forecasts from various scientists until 2200. *Source: Klimenko, 2009a,b;* www.polit.ru/article/2005/11/02/climate/

In total, over the last 150 years, a positive trend of rising of the Earth's yearly average temperature has been observed. Up to the present point, it has risen by $0.8 \pm 0.2\,^{\circ}C$.

Scientists have several points of view concerning yearly average temperatures and how they will behave in future. The two predominant theories are as follows.

- Temperature will continue to rise, at least, up to 2100.
- Temperature will have stopped rising by the middle of the twenty-first century, after this it will drop.

This opinion discrepancy is due to an absence of a stable position about the reasons of the warming; and this is where the arguments and debates occur. Some researchers think that warming is connected with the impact of human activity, others believe it is entirely due to natural reasons, and a third group insist on the necessity to consider both natural and anthropogenic influences.

Existing methods do not reveal a precise reason for the warming. It seems that the main constituent of yearly average temperature rise is natural; it is responsible for 70 to 90% of the temperature rise value, and anthropogenic activity can be responsible for 10 to 20% with 10 to 20% error and vagueness (Libin and Pérez-Peraza, 2009).

Debates about consequences of warming for the Earth as a whole, and for humanity in particular, are even more fervent. Opinions are antipodal, from utmost pessimistic to neutral and even optimistic ones. It is understandable; economical and political interests of countries and private groups are often behind these positions. And it is not easy to determine which research of warming problems are objective and which ones are subjective. That is why we will try to separate those possible consequences which are not well argued or objective (Shnol et al., 2000; Malkov et al.; 2005, Makhov and Posashkov, 2007; Vasiliev and Dergachev, 2009; Libin and Pérez-Peraza, 2009).

- Melting of sea ice and retreatment of continental glaciers
- Rise of the World Ocean levels on account of glacier melting and heat penetration of the upper ocean layer
- Possible latitude drifts of climatic zones
- More intensified warming of the land and, as a result, more intensified heat penetration of soil

These are direct results of warming, and there is some debate since it is not yet clear in what way these factors will influence humankind and our economic activity, and what will be the ecological and economical consequences.

As noted in the work of Makhov and Posashkov (2007), besides negative consequences, reports of the Intergovernmental Panel on Climate Change (IPCC) consider positive ones, including consequences for separate regions.

(Apparently, climatic models of common circulations, radiation models, and biological models were used in research supporting that point of view).

In IPCC reports the forecast and analysis of possible risks between now and 2100 are given. It is suggested that the yearly average temperature of the Earth surface will rise by 1 to 4 °C, and that the ocean level will become 15 to 95 cm higher by the end of the twenty-first century.

Risks and possible consequences for human beings include the following.

1. Influence on ecosystems
   a. Change of geographical locations of ecosystems and biocenoses (latitude drift to the North)
   b. Change of species composition of plants and animals (as a result of conflagration, diseases, plant pests increase)
   c. Possible change of bioproductivity—productivity of some plants and animals will fall, that of other ones will rise
   d. Changes of soil and land characteristics (bog reclamation, salinization, over wetting erosion, desertification, congelation melting)
2. Influence on hydrology and water resources
   a. Growth of drought number in arid zones can induce water deficit
   b. Frequent floods in the middle and moist zones will lead to sinking of territories and infrastructure destructions
   c. Rise of the ocean level will lead to flooding of some territories (shelves, small islands, coastal zones)
3. Influence on agriculture
   a. Quantity growth of insects—pests and agent of diseases will lead to drop of productivity of some crop types
   b. Rise of carbon dioxide concentration can lead to a rise of productivity of some crop types and to a drop of productivity of other crop types
4. Influence on human health
   a. Growth of number of inflectional diseases (fever, malaria, cholera, plague, etc.)
   b. Thermal load
   c. Intensity of air pollution
5. Social and economical consequences
   a. Removal of negative consequences will demand expenses
   b. Migrations from regions with unfavorable climatic changes to more favorable regions are possible
   c. Warming in the middle and cold territories will lead to economies in energy on domestic heating
   d. Retreat of Arctic ices will open up the possibility to research the Arctic and open new trade routes (the North seaway) (Alexeev, 2006; Gorbatenko et al., 2007; Gritsevitch et al., 2008; Stern, 2006)
   e. Damage from losses of some tourist zones is unavoidable

6. Political consequences
   a. Some states will be in a more advantageous position than others; changes in alignment of forces can take place
   b. Struggle for territories and resources can be intensified

All these consequences have a conditional and probabilistic character—they reflect only one of several possible scenarios of development of events; the real situation could be better or worse. In a certain sense this scenario will enable the estimation of the scale of possible changes in connection with climatic oscillations, and give us an idea about the position of some international scientific organization on this question.

G.D. Oleynik, chairman of the Committee of the Federation Council for the North and Small Ethnic Communities, said in his speech during the First World Summit of Regional Governments devoted to problems of the climate changes (Sant-Malou, France, October, 29–30, 2008):

The climatic changes are worldwide independent on reasons arousing them. They result in necessity of international cooperation of these forces and mechanism which makes the country to take corresponding measures in the framework of such cooperation, especially in northern regions... Dynamics, scales and character of the climatic changes and their consequences for economy are different with substantial vagueness conditioned by stochasticity of natural phenomena themselves, their interaction with economic and social systems as well as by discrepancy of economic effects of global warming. The climatic changes can be favorable for some territories and, on the contrary, unfavorable for others; and this situation can change to the opposite one in course of time or according to other scenarios. So, solving a problem of smoothing the climatic changes demands working out some special national strategy of stable economical development of northern regions, oriented to life improving (in a wide sense of the word) which should be a source of means and mechanisms of economic complex and population adaptation to the climate changes and to reducing risks of such changes.

Source: http://www.severcom.ru/cooperation/inter_contacts/item64-1.html

4.3. POSSIBLE REASONS FOR GLOBAL WARMING CONNECTED WITH SOLAR ACTIVITY VARIATIONS

It is important to remember that we began our discussion of global warming after the first estimations of surface temperature behavior in the high latitudes and solar activity (Libin and Pérez-Peraza, 2009). In a year, 15 trillion kilojoules of solar energy reach the Earth. Solar radiation is partially absorbed and reflected by the atmosphere. As result, about 5000 kilocalories (20,000 kilojoules) of energy covers each square kilometer of the Earth surface; 30% of this energy is reflected to the cosmos, the rest of the energy determines meteorological processes.

To be exact, solar energy (solar rays) is absorbed not by the atmosphere, but mainly by the ocean and land. Warming of the Earth surface induces convection—ascending warm air is replaced by descending cool air. During vapor condensation, cloudiness appears in the ascending air.

Because of rotation of the planet, the Coriolis forces act on moving air. They twist ascending air counterclockwise forming cyclones and descending air (anticyclones) is twisted clockwise in the Northern Hemisphere. In the Southern Hemisphere rotation directions are opposite. Convection covers the whole Earth's troposphere, and on its upper border a part of heat energy goes to the cosmos as infrared radiation with a maximum on millimeter wavelengths.

The height at which the atmosphere becomes transparent for heat radiation (thickness of the troposphere) changes from 4 km near the poles to 12 km near the equator. Exact values depend on moisture (vapor concentration) and also on contents of other gases absorbing infrared radiation, primarily carbon dioxide and methane. The vertical energy flux existing everywhere is aided by a slower transmission of warmth from tropics to poles, as the planet luminosity by solar rays is maximal at the equator. It changes each year regularly in each latitude of the planet. It is followed by the weather change.

Weather is determined by many parameters (pressure and air moisture, velocity and direction of the wind, cloudiness), and surface temperature is the most important among them. Surface temperature has been measured reliably all over the planet for more than a century.

An average temperature (based on many-year data) for each season is a rather stable climate characteristic. Yearly change of the average temperature is one month delay in phase from the luminosity sinusoid as a result of thermal inertia of the ocean and land. Diverting our attention away from these regular seasonal temperature changes we have analyzed the character of their abnormalities or weather fluctuations of temperature relative to its climatic average. Statistical processing of temperature variations resulted in a lot of surprises: the basic results of our works ones are briefly described in this book.

Analysis of 380-year variations of solar activity, temperature, and CO_2 made by the authors leads to the tendency that these three processes are situated on a branch of 400-year variation growth. Maybe it is global warming! Scientists of Pulkovo astronomic observatory accentuate the importance of this problem; they also considered solar activity influence on a series of events, including global weather and climate variations, and concluded that it is a proven and real fact.

The term cosmic weather which has been used during the last decades characterizes the whole complex of external (relative to the Earth) geo-effective factors, the basic of which are solar magnetic field variations and resulting phenomena induced by these changes. Fluxes of high-energy particles appearing during solar flares and coronal mass emissions can destroy radio communication, hamper radio navigation, lead to power supply failures, and damage equipment of cosmic apparatuses in a short-term aspect. Besides, these fluxes are dangerous for cosmonauts and high latitude air travelers.

Solar activity modulates a flux of galactic cosmic rays which influences cloud formation on the Earth and its reflectance to the energy flux coming from the Sun. This can induce long-term trends of the Earth climate and lead sometimes to great weather abnormalities. Timely observation and prediction of solar activity changes as well as terrestrial phenomena induced by it can reduce economical risks and allow us to work out an optimal strategy for prevention of natural disasters.

That is why the project of Constant Cosmic Solar Patrol (PKSP) that started at the Roscosmos is very important and will be performed on the basis of orbit satellites by 2015 within the framework of the Federal Cosmic Program. The constant cosmic solar patrol has to provide monitoring of solar ionizing radiation changes in the soft X-ray and utmost ultraviolet spectra.

There is one more thing to note. Today there is consensus among reputable geologists that the level of the World Ocean periodically became lower or higher, resulting in water covering of large parts of continents, except mountains, or, ebbing again. Such global floods are called the *thalassocratic phase* of the Earth development (from Greek, “thalassa” means sea and “kratos” means power). The last such flood is considered to have happened about 100 million years ago around the epoch of the dinosaurs. Characteristic discovered in the inland regions proves that North America (from the Gulf of Mexico to the Arctic) was flooded by the sea. Africa was divided into two parts by a shallow strait crossing the Sahara. So, each continent was shortened to a large archipelago size.

A curious book *Our Future* under the editorship of Ms. Gro Harlem Brundtland, the Prime Minister of Norway at that time, was published in the middle of the 1980s. The authors of some of the articles, famous scientists, predicted that at the beginning of the twenty-first century, because of global warming, about half the glaciers of the Antarctic and Greenland would melt and the risen ocean would flood several tens of world ports and coastal fertile or industrial low places, and several great rivers would overflow. Is this the next disaster? What real threats should we expect from global warming, even knowing that natural phenomena repeat periodically?, Unfortunately, this knowledge does not lead to most of people to feel better.

4.4. CLIMATE CHANGES AND NATIONAL SECURITY (AN EXAMPLE FOR THE SPECIFIC CASE OF THE RUSSIAN FEDERATION)

Leading Russian scientists V.M. Kattsov (Kattsov et al., 2003, 2007a,b), B.M. Meleshko (Meleshko et al., 2004), and S.S. Chicherin (Kattsov et al., 2007b) write in their report “Climate Change and the National Security of the Russian Federation” (Kattsov et al., 2007b):

Global warming creates a situation for the Russian Federation (RF) (with account of its geographical position, peculiarities of economic potential, population problems, and geopolitical influence) where it is necessary to be aware of national interests relative to the climate variations and threats to national security connected with that and also to work out corresponding home and foreign policy. Neglecting the problem of global climate variations, inactivity justified by references to insufficient study is unreasonable and can lead to serious risks for stable development and security of the country.

One of the bright examples of transformation of scientific problems to political ones is the Arctic which is of great interest for RF and other countries situated outside the Arctic. Over the last decades rather quick changes of the Arctic climate have been observed (more variations are expected in this twenty-first century, see more detailed information below). These variations can radically aggravate existing or give birth to new intergovernmental problems connected with search and production of energy resources, use of transport seaways and bioresources, delimitation of the continental shelf, the environment, applying of maritime law, and so on, and become a factor of maritime activity destabilization in this region (Kattsov et al., 2007a; Kovalishina, 2009).

A univocal estimation of consequences of expected warming for RF (profitable or not, as a whole) is impossible with account of interaction of different factors on its huge territory. Meanwhile, in Stern's report (Stern, 2006) the author says that 2 to 3 °C warming will be a favorable factor for Canada, RF, and Scandinavia, situated in the high latitudes, due to intensity of crop capacity, reduction of death rate from low temperatures, shortening of heating period, and, possibly, growth of tourism.

It is necessary to make all-round analysis of possibilities and resources of adaptation of any country to the climate variations, to consider existing results of home or foreign researches, and to stop discussing if global warming exists or not, at least, at the state level.

Over the last decades substantial climate changes connected with global warming have been observed on the Earth. Global climate change creates an increase in quantity and magnitude of natural disasters, exerts substantial impact on life conditions of population, and does great harm to national security. It is enough to say that natural disasters occurring on the Earth during the last decade have done more than $800 billion harm and more than 200 million people have been their victims.

Taking into account of a worldwide character of the problem of global climate change, only one year has it been twice discussed at the highest political level—on the General Assembly of United Nation Organization (UNO) in February 2008 and on the summit of the European Community in Brussels in March 2008. It was accentuated that global climate change can lead to geostrategic variations: appearing of new economical interests connected with struggle for access and control of energy resources in connection with possible improvement of access to hydrogen resources in the Arctic and growth of possibility of serious potential conflicts on this matter.

The observed global warming on the planet makes up such a situation in which it is necessary to estimate results of possible nonanthropogenic (and, possibly, anthropogenic as well) disasters and to work out programs of complex security of the territory and population considering these phenomena. Again, for the particular case of the Russian territory, it has been established, by instrumental observations, that in the twentieth century and in the beginning of the twenty-first century temperature grew faster than at any time in the last 200,000 years: a yearly average temperature which was ~15 °C over many thousands of years has a risen by more than 1 °C since 1900 including 0.5 to 0.6 °C in the last 10 years. (It should be noted that observed growth rate of an average temperature in the northern latitudes is twice as high as in the middle latitudes.)

The prognosis of the climate change made by the authors shows that an observed warming trend will be kept at least until 2060. Observed growth rate of an average temperature in the northern latitudes being twice as high as in the middle latitudes, by our estimations a yearly average air temperature in the northern latitudes may rise by 2 to 3 °C by 2060.

Tsalikov (2008) gives some generic data on condition changes for the Russian territory occurring during the twentieth century connected with global warming on the Earth. He discusses consequences connected with permafrost recession (permafrost covers up to 65% of the country area). A lot of cities and villages of Western and Eastern Siberia, pipelines and gas pipelines, highways and railways, and transmission and communication lines are found on the permafrost. If the average temperature increases by 4 °C irreversible changes will take place in the permafrost regions (see Figure 4.3).

In Russia the common area of permafrost extension is ~10.7 million km^2, which makes up about 63% of the country's territory (see the figure). In the region of permafrost extension on the territory of the RF there is more than 80% of discovered reserves of oil, about 70% of natural gas, huge deposits of coal and peat are concentrated, and a ramified infrastructure of fuel and energy companies' operations are built.

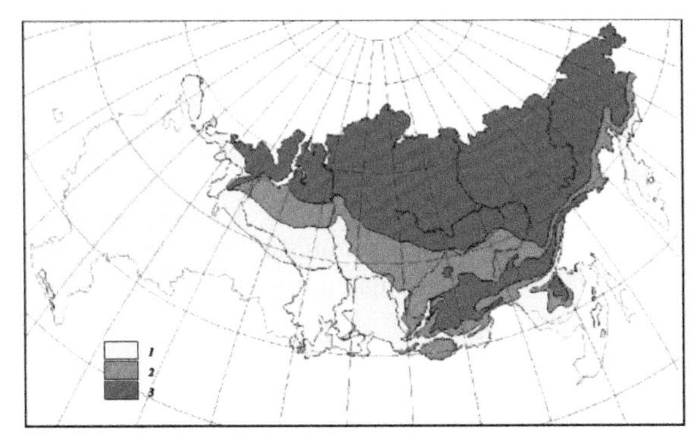

FIGURE 4.3 Permafrost zones (dark ones). *Source: Anisimov and Lavrov, 2008.*

As described in the works of Anisimov and Nelson (1997), Anisimov and Belolutskaya (2002), and Anisimov and Lavrov (2008), during the twentieth century a rise of temperature of upper many-year frozen soil and increase of depth of seasonal fire-setting have been observed. Moreover, during the last three decades these processes have been accelerated. Since the beginning of 1970s temperature of frozen soil has become 1 to 1.5 °C higher in central Yakutia and 1.0 °C higher in Western Siberia, while the air temperature has become 1.0 to 2.5 °C higher. There is little doubt that these changes are conditioned by global processes, as in North Alaska warming also occurred, even more intensified (Osterkamp and Romanovsky, 1999). Since the beginning of the twentieth century up to the 1980s the temperature of upper frozen rocks became 2 to 4 °C higher and, on average, by 3 °C more during the next 20 years up to 2002. In northwestern Canada (Majorowicz and Skinner, 1997) the upper layer of permafrost has become 2 °C warmer over the last two decades.

Especially great interest in information “anomalous” areas where the background of global climate warming prevailed tendencies for a long time cooling (up to mid 90s). Are among them the north-east of Canada. It is noteworthy that since the middle-1990s (after a sharp increase in global warming), in this region the temperature of the upper layer of permafrost is not decreased, but has increased by about 2° C. This confirms the view that even in the abnormal regions the changes are caused by global warming.

Nowadays in West Siberia intensive melting of frozen soil is observed (up to 4 cm per year). Melting of frozen soil will lead to growth of nonanthropogenic and emergency anthropogenic cases related to building and construction instability and communication destruction. When a yearly average air temperature becomes 2 °C higher, pile capacity becomes 50% lower, and about one forth of standard domestic houses built in the period 1950 to 1970 in such cities as Yakutsk, Vorkuta, and Tiksi could be under threat of destruction. Risk for infrastructure objects is especially great where frozen soil contains a great quantity of ice. Such regions are a substantial part of the Lena valley, West-Siberian Plain, the Chukchi Peninsula, and a large part of European north island territories where large oil and gas complexes, transmission lines, and the Bilibinskaya NPS are situated.

According to estimations (Tsalikov, 2008), it is thought that by 2060 the permafrost zone will have moved by 150 to 200 km, proven by the results of forest advance to the north on the territory of Russia and Canada (Porfiriev, 2005a,b,c; Porfiriev, 2006 a,b,c; Libin and Pérez-Peraza, 2007, 2009). Nowadays in West Siberia, because of permafrost degradation, about 35,000 thousand pipeline and gas pipeline breakdowns happen every year, connected with loss of stability of bases, mounting deflection, and pipelines breaking. Permafrost melting will result in landslides on melting slopes and a slow thawed soil flow, and also in substantial gaps on account of soil consolidation and carry-over of it with water.

Tsalikov writes in his work (2008):

It is evident that operating helipads and airfields which are necessary for food, post, petroleum and lubricant delivery to northern regions, medical aid, and people rescue will be useless. Preparation of transport infrastructure of northern territories in new climatic conditions is becoming a very real problem. Destruction of underground storage walls will result in substantial problems. Mining operations (oil, gas, metals) have been organized on northern territories for decades. A great amount of base oil was lost during accidents and leakage of pipelines, but not distributed in the soil and so it stayed in the ground frostbound. During permafrost melting new biocenoses can be poisoned with oil. These situations are called timely chemical bombs. *This term means retardation of harmful effect. These bombs can be of a metal character; a great amount of harmful heavy metals is contained in wastes and trades of mining production on permafrost. In the north we have already met the problem when agricultural fertilizers and pesticides washed away during the thaw were found in the surface waters. Permafrost melting is especially dangerous in Novaya Zemlya in the zones of radioactive waste storage locations and on the Yamal peninsular in the region of perspective oil production.*

During research of precipitation amount on the Earth over the last 80 years the authors found one more problem (Pérez-Peraza et al., 2005; Libin and Pérez-Peraza 2009). According to the authors' results, global warming will condition 6 to 7% growth of yearly average precipitation amount relative to nowadays on the Russian territory in a cold period, and 3% growth on the Mexican territory (Pérez-Peraza et al., 1999, 2005).

As a result of forecasted temperature and precipitation variations, a substantial rise of yearly river inflow in northern regions and lake levels are expected by 2050 (Pérez-Peraza et al., 1999, 2005; Libin and Perez-Peraza (2009)). That is why problems connected with floods and inflow which are predominant today among all natural disasters, according to total yearly average damage, will become an important consequence of the climate variations for northern territories of Russia. Precipitation and inflow increase will create serious problems with regard to the protection of people and territories from flood (Kononova, 2008).

Such floods have already been observed lately. Among them there are unusual floods such as that which almost destroyed the city of Lensk in May 2001; partial sinking of several large European cities in the summer of 2002; and the greatest and longest floods in Western Europe in the summer of 2003.

Tsalikov (2008) demonstrated in connection with global warming a serious problem that appears conditioned by Arctic ice melting.

It is shown on navigators' ice maps of the nineteenth century that in 1890 solid Arctic ices covered the whole Barents Sea and spread to the Iceland coast blocking the Fram Strait. On the satellite photo of the same region made in 2003 one can see that waters reach the Spitsbergen archipelago and stretch in the Barents Sea up to the coast of Novaya Zemlya.

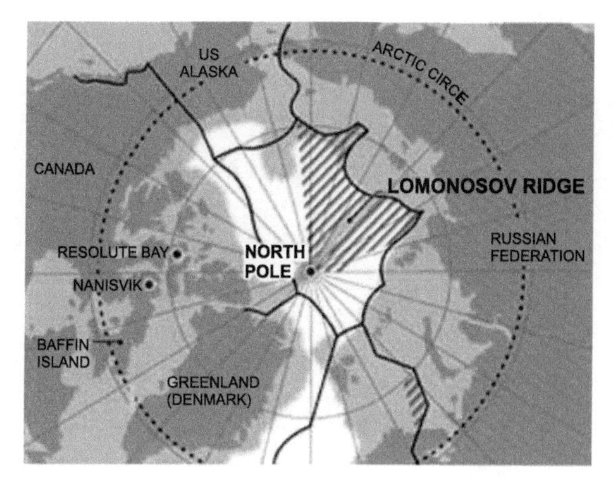

FIGURE 4.4 In the Arctic, almost disappeared old ice. *Source:* http://nsidc.org/arcticseaicenews; www.wunderground.com/climate/SeaIce.asp.

It means that for the last 100 years the ice cover in this part of the Arctic decreased by almost one third. The same satellite photos show cogently that Arctic ice cover grew by 25% between 1979 and 2003. Estimation shows that by 2050 the northern seaway will have been opened 100 days per year instead of the 20 of today, and by 2070 the Earth may be fully deprived of the northern icy cap.

Of great importance and significance for the RF are the problems of increase of navigation scale in the northern latitudes, quantity of the Russian Northern Fleet ships, and the problem of safety on water in these regions.

Unprecedented speed of Arctic ice melting can place survival of northern local people in jeopardy, lead to sinking of large territories, disappearing of certain biological species, and destruction of national communities' infrastructure. However, it will open a new way between Asia and Europe, easing the access to fuel resources on the sea shelf of the Arctic Ocean (Figures 4.4 and 4.5).

4.5. CLIMATE OSCILLATIONS ARE THE REALITY

Though in the greater scheme of things, global warming is only just beginning, the international community should prepare to more serious testing. In fact, climatologists believe there is nothing abnormal about today's heat. If we review archives, we learn that the heat and drought in June through August of 2010 were predicted the previous year. NASA's James Hansen writes in his article published in December 2009,

The following summer in Europe will be the hottest in the whole modern history of meteorological observations. For the last 12 months, a daily average temperature of the

FIGURE 4.5 The Arctic Ocean today. *Source:* http://sp.rian.ru/neighbor_relations/20101005/147696353.html.

Earth atmosphere has been the highest for the last 130 years. Established hot weather will become a consequence of warm flow movement changes in the central zone of the Atlantic Ocean, where similar to the Pacific phenomenon La Niña an area of cool water appeared last year. Next year the Atlantic Ocean will release much more warmth than usual, which will result in heating of the lower atmosphere. Mathematical models show that temperature will rise during the following months, and we, living in the Northern Hemisphere, will have to survive the hottest summer for the whole history of modern meteorological observations.

Russian scientists agree with American colleagues' opinion. In winter of 2009 their estimations also showed a sudden rise of temperature for the summer of 2010. However, long-term forecasts of Hydrometeoservice of Russia for spring and summer of 2010 did not promise either such drought or sudden temperature rise. This was not a disadvantage of modern science, but demonstrated principal difficulties of the Earth's climatic system which enters its own, sometimes very substantial, amendments to our theoretical estimations. In this case, according to Hydrometeoservice data, there is a very small possibility of repeating such an abnormal summer 2012. Even at the expense of global warming, which affects in Russia to a lesser extent (due to the geographical position of Russia), such dry summer (consecutive several years) may be repeated only once or twice over the next 100 years.

In some senses, what happened over the European part of Russia in the summer of 2010 is a direct continuation of what happened the previous winter. Specialists of Hydrometeoservice of Russia attribute it to a reduction of western transfer over the Atlantic. Large air masses formed over the ocean usually penetrate freely to the territory of Europe and Russia and bring warm rainy weather in winter and cold rainy weather in summer. But in 2010 we observed a rare phenomenon: intensifying of cyclonic activity over the Atlantic Ocean and long-term maintenance of anticyclones over the European part of Russia. To put it another way, cyclones from the ocean became blocked by powerful air fluxes which resulted in stable anticyclone weather over the European territory. In summer air was formed not over the ocean as usual, but over the land. Such stable dynamic balance can last for a long time. In winter such processes bring stable frosty weather with minimal precipitation amount; in summer the result is bright dry weather with short breaks when a small cyclone manages to run a blockade, producing phenomena like when Eastern Europe became flooded by heavy showers. Such weather combinations set in with a certain cyclicity. We observed similar phenomena in 1972 and in the 1930s. More often this weather type leads to heavy droughts and conflagrations.

Abnormal situations such as those described above appear relatively often. There is nothing unexpected in the process of such phenomena appearance for specialists of Atmospheric Physics. In the absence of clouds, anticyclones are very stable. Sun warmth freely penetrates the Earth surface and warms the land; in its turn air is warmed from the land and feeds the anticyclone, which has grown to a gigantic size that summer from below. The situation becomes more complex if a cyclone of the same power is near an anticyclone, as happened in Russia.

A problem is that science could not predict 10, 20, or 30 years ago, and cannot predict today how long such situations can last. A principal difficulty is that small changes are accumulated in atmospheric processes which are difficult to fix, but one day when a critical mass is accumulated these changes result in qualitative drift. So, the presence of a stable anticyclone is not surprising. However, its scale (area and duration) is surprising. There has not been such a powerful and long anticyclone over the central region of Russia for the whole history of instrumental observations.

The situation formed in the summer of 2010 in Russia is not a direct consequence of global warming. Warming appears on account of natural and, to a lesser degree, anthropogenic reasons. When it gets warmer, the process of heating in the tropics goes more slowly than in the polar latitudes. Correspondingly, temperature difference becomes less, and less energy is left in the atmosphere to resolve a stagnant situation in Europe. So, global warming really could have contributed to abnormal heat duration.

Specialists in paleoclimatology (Bradley, 1985; Shaviv, 2002; Carslaw et al., 2002; Drummond and Wilkinson, 2006; Rosemarie et al., 2007; Sahney and Benton, 2008; Borrero et al., 2008; Gillman and Erenler, 2008) say that there have been such abnormalities earlier. According to Russian chronicles, between 1350 and 1380 Russia survived a whole series of terrible summer seasons with the same conflagrations of peat bogs. A chronicler wrote in 1371, "Darkness was lasting for two months. Darkness was so great that one could not see one meter behind," (Shaw, 1976; Cheshihin-Vetrinsky, 1879). The abnormal weather of the summer of 1841 created a monthly average temperature rise of 4.6 °C, or practically by the same as in summer of 2010. After that temperature got back to normal gradually. Today many scientists also hope, based on past examples, that the climate will get back to normal in the near future.

Today the Earth has entered another cycle of atmospheric circulation with durative periods of relative stability. Over the last few years in Russia either long-lasting frosts or long-lasting rains have been repeated in winter (earlier, weather lasted for three to five days, then it changed). Based on past meteorological observation data and some authors' conclusions, we can predict confidently that there will not be any climatic drifts over the next few decades. Autumn with its characteristic increase of precipitation amount, changeable weather, and rather early frosts will be consistently on time (or may arrive a little late because of the intensity of heating of the surface). We do not expect any serious trends or deviations of the climate (besides a slowly rising yearly average temperature). However, in Europe and in the European part of Russia we have entered a cycle of longer-term periods of atmospheric circulation, which is why periods with long-lasting single-type climatic conditions are expected. Over the ocean, as a result of intensified evaporation, huge damp air masses were accumulated in summer 2010; these masses will move to Europe as soon as a border between cyclone and anticyclone is broken. So, the autumn of 2010 was rather rainy and long-lasting, especially during October and November. Predicting estimations made in 2010 had shown that January and February of 2011 would be cold and severe, what was completely confirmed. Predicting estimations made in 2011 had shown that January and February of 2012 also need to be cold and severe. Past January and February confirmed our forecast.

The academic co-society is rarely unanimous. Global warming exists and it is impossible to stop it with either limitations to greenhouse gases emission or any other measures. NASA analytics support that,

"The climate in the first half of the twenty-first century will become warmer at a record pace. In the last century warming was 0.06 °C per decade. Up to the middle of this century this velocity will have been 2 to 2.5 times higher. After that the climate will begin to settle down. Warming maximum is expected in the first half of the next, twenty-second century on the level of 1 to 1.5 degrees higher then today. In the nearest decades many people and countries will find themselves in climatic conditions in which they have never lived before and have no experience of such living."

Source: a) http://www.earth-policy.org/indicators/C51/global_temperature_2010; b) http://www.pewclimate.org/docUploads/101_Science_Impacts.pdf; c) http://www.treehugger.com/climate-change/2011-year-weather-extremes-more-come.html.

Global warming will touch especially the central regions of Russia, which are the most sensitive to climatic changes. Because of peculiarities of the geographical position of Russia, when a yearly average world temperature rises by 1 °C, a yearly average temperature of Central Russia rises by 2.5 °C.

Also 1 °C of yearly average temperature rise is equivalent to 600 km forwarding of climatic conditions to the southwest. In the climatic relation, Moscow is in the Ukrainian climatic zone, and if climatic tendencies are invariable in the beginning of the 2030s the climate of Russia will correspond to that of Spain or North Africa. However, unlike the Mediterranean regions, there is no sea in Russia to cool night temperature. As we saw in the summer of 2010, in general, minimal difference between day and night temperatures influenced growth of deaths from heat. The Spanish have a rest from heat with the help of the sea, but in Moscow temperature did not deviate from a 27 °C mark even at night, which is considered to be dangerous, especially for old people and the sick.

There is also the issue of whether modern economy is able to endure these new natural conditions with long-lasting periods of heat and cold. Issues of economic development connected with climate variations along with the problems of food supply and financial crisis are now challenges on a global scale. Solutions lie in the framework of integrating the problem of reducing climatic risks to a strategy of stable economic development. Such risks can be illustrated, for instance, by constant flooding all around the world (see Figure 4.6).

On the one hand, it is supposed that economical development oriented to life improvement (in a wide sense) is a source of means and mechanism of economical complex and people adaptation to the climate variations and of reducing risks of such changes (Porfiriev, 2006a,b,c,d,e, 2008, 2009a,b, 2010a,b). On the other hand it is supposed that risks of climatic variations are considered and estimated together with other risks of stable economic and society development, and the problems of global warming in the series

FIGURE 4.6 Royal Leamington Spa, Warwickshire, U.K. on April 10, 1998. The town center is under water and many cars are abandoned in roads. These floods affected a large swathe of central England and caused losses of £350 million. They illustrate a possible impact of global warming. Rainfall and torrential rain will rise, thereby increasing susceptibility to flooding. *Source: Saunders, 2009.*

FIGURE 4.7 Julius Schnorr von Carolsfeld "World flood" (160 illustrations of the Old Testament) *Source: www.paskha.net/christianity/bible-illustrations/016_noah-builds-the-ark.htm.*

of challenges to this development should be determined only on this comparative basis.

Man has gone overboard with his anthropogenic activity, but he has not yet done a serious harm to nature; as long as man's activity doesn't escalate to serious levels, nature will manage with these consequences.

The Sun has cycles that influence the climate. We are lucky, as we are witnesses of its very intensive phase. Today we do not need ice breakers in the north as much as was needed for northern navigation and economy 30 years ago. It is not a human fault. Nature is a self-controlling phenomenon, and certainly not self-destructing. The Sun is a decisive player in all of these processes and will give us a lot of new puzzles.

It is generally accepted that for some time we will observe temperature, lake level, and precipitation amounts rise. Is the next Deluge in store for us? No existing theory gives the final answer.

Journalist Andrey Zavolokin (Zavolokin, 2002) writes,

In Noah's Biblical life story it is said that God left a covenant to the patriarch that all living beings on the Earth would never be annihilated by the Deluge. But there was no a covenant that we would never be slightly wet.

Water flooded the Earth for forty days, and as it rose it lifted the boat off the ground (Figure 4.7). The water continued to rise, and the boat floated on it above the Earth. The water rose so much that even the highest mountains under the sky were covered by it. It continued to rise until it was more than twenty feet above the mountains… And the waters continued to cover the Earth for one hundred fifty days".

The Old Testament. Flood. Genesis 7:17-24

Part II

Illustrative Examples of Helioclimatology Applications

Chapter 5

The Case of Hurricanes vs African Dust

Chapter Outline

Highlights in Helioclimatology. DOI: 10.1016/B978-0-12-415977-8.00005-8

5.1. THE LINKS BETWEEN GEOEXTERNAL FORCING AND TERRESTRIAL PHENOMENA: THE CASE OF HURRICANE GENESIS

The links between space weather and meteorological weather have often been discussed, not only in the last century (Mason and Tyson, 1992; Mazzarella and Palumbo, 1992), but also for several centuries (Rodrigo et al., 2000) and even some thousands of years ago (Neff et al., 2001). A great deal of effort has gone into clarifying the mechanism of all complicated interconnections between the cosmophysical and climatic phenomena of the Earth, some of them recently summarized by Benestad (2006), Kanipe (2006), Haigh et al. (2005), and Fastrup et al. (2001).

Over the last few years, more and more investigations have shown that solar activity has a noticeable impact on meteorological parameters (Ney, 1959; Gray et al., 2005; Kristjansson et al., 2002; Laut, 2003; Tinsley, 1996, 2000; Tinsley and Beard, 1997) and cosmic rays (Marsh and Svensmark, 2000; Kudela et al., 2000; Gierenes and Ponater, 1999; Dorman, 2006; Mavromichalaki et al., 2006; Raisbeck and Yiou, 1980; Yiou et al., 1997). Some indications show that several purely meteorological processes in the terrestrial atmosphere are connected with the changes in the CR intensity, and that they are influenced by solar activity and magnetosphere variations (Kudela and Storini, 2005; Marsh and Svensmark, 2000; Kristjansson et al., 2002).

One of the main goals of space climate research is to know how and when the periodicities of space phenomena modulate terrestrial climatic changes. Some insights have been obtained; for example, the solar Hale cycle (20–25 years). Changes in solar activity for the last 500 years have been studied (Raspopov et al., 2005), with the aim of revealing a possible contribution of solar activity to climatic variability. On the other hand, quasi-periodic climatic oscillations with periods of 20–25 years have been revealed in the analysis of parameters such as ground surface temperatures, drought rhythm, variations in sea surface temperature, precipitation periodicity, and so on (Ol', 1969; Cook et al., 1997; Pudovkin and Lyubchich, 1989; Pudovkin and Raspopov, 1992a,b; White et al., 2000; Roig et al., 2001; Raspopov et al., 2001, Khorozov et al., 2006).

To understand the involved physical mechanisms, confident observational or experimental facts are required. Many researchers have attempted to clarify the mechanism of all the complicated interconnections between the cosmophysical phenomena and climatic phenomena at Earth; for instance, Fastrup et al. (2001), Haigh et al. (2005), Benestad (2006), Kanipe (2006).

One of the principal difficulties in quantifying the role of the space phenomena on climate changes has been the absence of long-term measurements of both the climatic and space phenomena. Consequently, people often recur to the use of proxies. In the last years more and more investigations have shows that the solar activity (Tinsley, 2000; Kristjansson et al. 2002) and cosmic rays, have noticeable impact on the meteorological parameters. However, the influence of cosmophysical phenomena on climatic phenomena is currently debated (e.g., Haigh, 2001; Shindell et al., 2001).

Previous work by means of a correlational analysis (Elsner and Kavlakov, 2001; Kavlakov et al., 2008a,b; Pérez-Peraza et al., 2008a), mentioned here below (Section 5.3), seems to indicate that certain extraterrestrial phenomena could have some kind of relation with the occurrence of Hurricanes. It is even speculated that such correlations could seat the basis of deeper studies to use the results as indicators of hurricane precursors. To give those results a higher meaning, it is convenient to carry out spectral studies of the different involved time series to delimitate with more preciseness the existence of those potential relationships.

That is, to find incident cosmophysical periodicities that may modulate terrestrial phenomena. Though the AMO (Atlantic Multidecadal Oscillation) has been linked with the frequency of Atlantic hurricanes, however, in the present context, little attention has been given to such a large scale climatic phenomena. For instance, the question about the role of the Sun in modulating these phenomena has not been clarified, it requires further assessment. In Section 5.7 we describe the behavior of the main periodicities presented by the AMO in relation to solar activity phenomena and galactic cosmic rays (GCR). Special mention is made of the links between geoexternal forcing and hurricanes.

5.2. NORTH ATLANTIC HURRICANES

Hurricanes are considered one of our plant's most astonishing meteorological phenomena. Strong winds, clouds of great size, and intense storms unite to advance from the ocean and to reach mainland, razing everything in their passage. Fallen trees, damage to buildings, changes in the natural landscape, and eliminating life are some of the consequences that these unpredictable events can generate. Due to the great intensity that they reach, with winds that can surpass 350 km/h, hurricanes are classified as true natural disasters: whole towns disappear under the force of the impetuous winds. There is no way to counteract the force of a hurricane; man is merely a spectator. Hurricanes have always been associated with damage that translates to human and material losses, but it is necessary to point out that they also bring benefits such as an increase in precipitation in regions where the agricultural development depends on precipitation, as well as the recharge of dams and bodies of water, vital for the development of the populations.

The word hurricane has its origin in indigenous religions of old civilizations. The Mayan named their god of storm *Hunraken*. Taino people, a culture of the Caribbean, gave the name Hurricane to a God they considered malicious. Nowadays, hurricanes are not considered wicked forces but due to their great force and potential for loss of lives and material damages, they are considered one of the most powerful phenomena in nature.

Tropical hurricanes are the only natural disasters that are given individual names. These names are known well before they occur, as are their possible effects, contrary to other natural phenomena such as earthquakes, tornados, and floods. Gilberto, Katrina, Mitch, and Isidoro, to name some of the most recent, are examples that recall vivid images, for the severe damages that caused. These phenomena present common characteristics, although each one has its own particular features.

The destruction caused by the hurricanes in the Caribbean and Central America has modified the history and the future of these regions. The danger comes from a combination of factors that characterize these tropical cyclonal storms: elevation of sea level, violent winds, and strong precipitation. The Yucatán Peninsula, for example, is affected in a direct or indirect way by most of the hurricanes that are formed in the Western Caribbean. The hurricanes can have a diameter as long as the peninsula itself, so that practically any hurricane that is formed affects the coast of the peninsula (Mulokwa and Mak, 1980).

To give an idea of hurricane's power, the energy concentrated in the vortex system is estimated to be $>10^{16}$ J. If we consider that the air over a surface with a diameter of ~800 km has a mass of ~2×10^{12} tons, turning with average velocity of ~15 to 20 m/s, we could easily calculate an energy of ~10^{11} KWh. This corresponds to the energy released during the explosion of more than 2000 atomic bombs of the Hiroshima type. That explains the devastating effect of a hurricane when it hits a populated area. North Atlantic hurricanes frequently strike the Caribbean islands, Mexico, and the United States. A single hurricane hitting over these coasts could take hundreds of human lives and can cause damages of billions dollars. Practically every year, one or two such hurricanes devastate these regions. They rank at the top of all natural hazards (Elsner and Kara, 1999).

One famous example is Hurricane Katrina (Figure 5.1), which destroyed not only the city of New Orleans but vast areas of the states of Louisiana and Alabama. Tropical hurricanes are sometimes driven by weak and erratic winds that makes it difficult to predict them. Published warnings have substantially reduced the number of deaths.

FIGURE 5.1 (a) Pictures of hurricane Katrina, taken from a satellite (NOAA). At the right is the hurricane eye. Katrina was the deadliest Category 5 hurricane to strike the United States since 1900. (b) Walls of the hurricane eye and spiral bands of Hurricane Katrina.

Today much effort is devoted to better understanding hurricane formation and intensification to unveil connections with other physical processes; for instance, the fact that there has been a low Atlantic hurricane activity in the 1970s and 1980s compared to the past 270 years, but increasing destructiveness over the past 30 years. Also, even if, since the beginning of the 1990s, there is a general upward trend in the frequency of tropical cyclones, there have been some years that do not follow such a tendency, as it was the case in 2006. It is hoped that the conglomeration of different research with very different focus contribute, overall, to the task of improving prediction methods of the complicated trajectories for a better forewarning of hurricanes, prediction of their probable devastation, and to warn with enough time the threatened population.

Hurricanes are perturbations that take place in tropical regions where the waters of the ocean are relatively warm (temperatures around the 26–27 °C). They are characterized by a large center of low pressure, around which the air rotates at great speed embracing an extension of several hundreds of kilometers. Hurricanes have a certain anatomy and their classification depends on the intensity of the winds, on the atmospheric pressure, and on the potential damages they may cause. Powered by intensive solar heating and producing fast evaporation in the second quarter of each year, large upward hot high velocity circular wind streams are born over the warm equatorial waters, with a velocity higher than 60 km/h and reaching rotational velocities beyond

350 km/h. *Tropical cyclone* is the scientific term for a closed meteorological circulation of enormous mass of atmospheric air rotating intensely, that is developed on tropical waters. These systems of great scale, non frontals, and of low pressure happen in areas of the world known as tropical basins of hurricanes.

Therefore, the tropical cyclone is a low-pressure system that is located over hot waters of tropical oceans (between the tropics of Cancer and Capricorn and at least 4–5 ° away from equator). The intensive heating, low pressure, and resulting powerful evaporation quickly increase the rotational wind speed. This huge system moves from east to west and slightly to the north, but deviations to the east are also possible—dangerous especially for the west coast of Mexico and U.S.A. Generally these cyclones are known as *hurricanes* if they are formed over the Atlantic and northeastern Pacific Oceans. We use, hereafter, indifferently the terms *cyclone* and *hurricane*. If they are formed over the western Pacific Ocean, they are called *typhoons*. Because of the Earth rotation, typhoons rotate counterclockwise in the Northern Hemisphere and clockwise in Southern Hemisphere.

In the first moments of the formation of a tropical cyclones, when the circulation of the closed isobar reaches a speed of 18 m/s, (i.e., <34 kt or 61 km/h), the system is denominated as *tropical depression* (TD). This is considered a tropical hurricane in formative phase. If the sustained speed of the wind ranges from 18 to 32 m/s (34 till 63 kt, i.e., 62–115 km/h) it is called a *tropical storm* (TS) and then given a name. Likewise, when the speed of the wind exceeds 119 km/h (or $\geq$33 m/s; $\geq$ 64 kt) the system is called a hurricane (or a typhoon). That is the speed that defines the start of a hurricane over the Atlantic and a typhoon over the Pacific. They have a defined nucleus of pressure in very low surface that can be inferior to 930 hpa. Every year an average of 10 tropical storms develop in the Atlantic Ocean, the Caribbean, or the Gulf of Mexico, and about 6 of those end up becoming hurricanes. In a 3-year period, the North Atlantic coasts receive an average of 5 hurricanes, 2 of those considered bigger hurricanes. In general, tropical depressions and tropical storms are less dangerous than hurricanes; however, they can still be fatal. The winds of depressions and tropical storms are not normally considered dangerous.

The intense rains, floods, and the severe natural phenomena of tornados, are the biggest threat. Hurricanes can then be described as turbulence phenomena caused by a current of hot air that is formed in the summer in the tropic and that goes to the North Pole compensating the difference in temperature between the Equator and the Pole. One countercurrent of the north to the south compensates the difference in pressure. This circulation of winds north to south and south to north in the Northern Hemisphere, together with the daily circulation of the Earth that causes the trade winds, are the main factors from the point of view of the winds to create situations that can form hurricanes. Another condition for the formation of a hurricane is the temperature of the surface of the ocean, an energy source that gives form to the phenomenon, which should be ≥26 °C. Under these conditions, it is the column of hot and humid air originated in the ocean that becomes the nucleus around which rotate the winds and later form the so-called "eye" of the hurricane.

The adjacent air is gradually involved in the rotation and the diameter of the whole vortex spreads to between 500 and 1000 km. With the further increase of the circular velocity, reaching sometimes 150 to 160 kt (80 m/s), the whole vortex spreads out to a gigantic ring with a diameter of several hundred kilometers. In its center is the relatively calm "eye" region of the hurricane. Around it, the rotational velocity is the greatest and decreases out of the center. With the increase of the circular velocity, the whole vortex spreads out to a gigantic ring with a diameter of several hundred kilometers (Table 5.1). In the east-west motion the whole system sweeps a path of about 1000 km wide. It gradually intensifies its rotational wind velocity, simply cooling the hot oceanic surface (e.g. Kerry, 2006), lingering over the ocean sometimes 20 to 30 days and forming complicated trajectories. The loss of energy of the phenomenon usually happens when the hurricane moves inside coastal areas and onto the continent.

The energy that requires a hurricane to maintain its activity comes from the liberation of heat that takes place in the process of condensation of the vapor of water that it evaporates from the surface of the ocean, forming nebulosity and

TABLE 5.1 Basic Hurricane Parameters

Parameter	Range	Average Unit
Diameter (D)	200–1300 km	500 km
Eye	6–80 km	50 km
Rotational velocity (V)		0 m/s
Duration	1–30 days	8 days
Kinetic Energy	4–8 Twh	6 Twh
Surface winds	> 33 m/s	
Energy Source Latent Heat Release		
Equivalent Energy	2000 atomic bombs (Hiroshima type)	
Lives (North Atlantic)	200,000 from year 1700	
Damages (North Atlantic)	$1180B from year 1900	

intense precipitation. When a hurricane enters the continent it loses intensity quickly when stopping the process of strong evaporation from the surface. The hurricane works like a vapor machine, with hot and humid air providing its fuel. When the Sun's rays heat the waters of the ocean, the humid air warms, expands, and begins to rise as it forms globes of hot air. More humid air replaces that air and that same process begins again. The rotation of the Earth eventually gives a circular movement to this system, the one that begins to rotate and to move as a gigantic spinning top. As in all hurricanes, this turn is felt clockwise in the Southern Hemisphere and counterclockwise in the Northern Hemisphere.

All the tropical depressions that develop into hurricanes originate practically under the same conditions, and they retain the same meteorological characteristics throughout his life. The physical differences that can be pinpointed from one event to another reside in the speeds that each event can reach and the length of time that they remain.

Recent studies on the formation of hurricanes highlight cause, the violent circulation of air, and the transformation of the liberated caloric energy when it condenses the vapor of water contained in the air that ascends from the surface in a very extensive area. Such a condition implies having an appropriate provision of latent heat, and of some mechanism that triggers and maintains the upward vertical movement required to produce the condensation of the vapor, and with it the liberation of that latent heat. These requirements are satisfied when the temperature of the seawater in a specific area is equal to or higher than 26 °C, when its distance to any coast or island is more than 400 km, and when inside that same region, convergence associated with any perturbation exists, be it tropical wave, polar water-course, inter-tropical convergence line, or area. The conditional instability is an atmospheric state that favors the formation of a hurricane in a potential region; it is a clear relationship between the presence of the instability and the favorable months for the formation of the tropical hurricanes.

The temperature in the ocean and the high relative humidity in the stocking and low troposphere are also requirements for the development of the hurricane. Figure 5.2 shows a map of the superficial temperature of the sea for the summer in the Northern Hemisphere. The yellow, orange, and red colors demonstrate the temperatures of the quite hot water to sustain hurricanes.

Another necessary condition of the organization of the circulation inside the region in which ascents of air and the liberation of latent heat of vaporization takes place, is that they happen in a superior latitude at 5 °, since in an inferior latitude the organizing effect of coriolis (rotation of the Earth) has very low values. It is for this reason that hurricanes are formed and are intensified when they are located on tropical or subtropical oceans in both hemispheres where the force of rotation of the Earth is sufficiently strong, so that the rotation movement begins around the center of low pressure and whose temperatures of water at surface level are around 26.5 °C or warmer. Figure 5.3 shows the main development region for tropical cyclones, the basin bounded by 25 and

FIGURE 5.2 Hurricane Katrina and the *sea superficial temperature* (SST).

60 °W longitude and by 8 and 23 °N latitude where effects of ocean heat content on hurricane genesis are almost constant.

Depending on the rotational velocity, which at the extreme can exceed 165 knots (>300 km/h), hurricanes themselves are classified in several ways, generally based on the vortex wind velocity and their destructive power. In 1969, the Organization of United Nations requested the evaluation of the damages generated by the passage of hurricanes in a certain type of housing. The North American engineer Herbert Saffir and the then director of the National Center of Hurricanes of United States, Robert Simpson, developed a mensuration scale to qualify the potential damages that can be caused by a hurricane, considering the minimum pressure, the winds, and the tide after its passage. This is now know as the Saffir–Simpson scale and consists of seven categories: tropical depression, tropical storm, and five categories of hurricanes going from hurricanes categories of 1 up to 5 (Table 5.2).

Independent of hurricane category, the damages are more intense when the translation speed is small or almost zero, provided they stay for a longer time over one location.

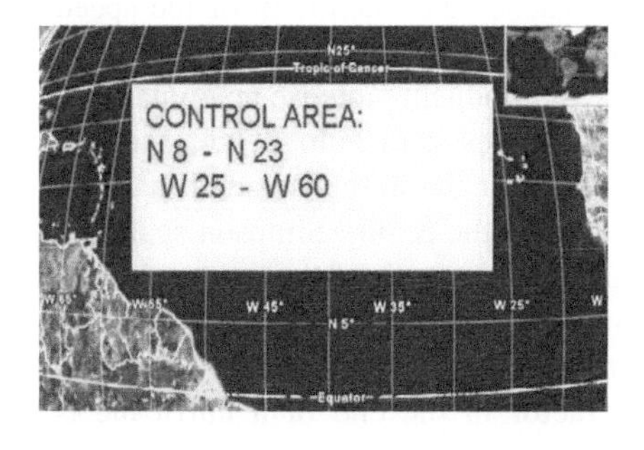

FIGURE 5.3 Control geographic basin, where the water surface temperature is practically constant.

TABLE 5.2 Saffir–Simpson Scale Maximum Rotational Wind Velocity

Storm Type	Knots	km/h	m/s
TD	30–34	56–62	15–17
TS	35–64	63–118	18–32
H1	65–82	119–153	33–42
H2	83–95	154–177	43–49
H3	96–113	178–209	50–58
H4	114–135	210–249	59–69
H5	>135	>249	>69

According to the Saffir–Simpson scale, hurricane evolution is as follows:

- *Birth* (tropical depression): First it is formed as a peculiar atmospheric depression because the wind begins to increase in surface with a maximum speed of 62 km/h or less; the clouds begin to be organized and the pressure descends until near 1000 hectopascales (hpa).
- *Development* (tropical storm): The tropical depression grows and it acquires the characteristic of a tropical storm, which means that the wind continues increasing to a maximum speed of 63 to 117 km/h; the clouds are distributed in hairspring form and a small eye begins to form, almost always in circulate form, and the pressure decreases to less than 1000 hpa. It is in this phase that it receives a name corresponding to a list formulated by the World Meteorological Organization (Committee of Hurricanes). Formerly, each hurricane was denominated with the name of the saint of the day in that it had been formed or it had been observed. If a hurricane causes a devastating social and economic impact on a country, the name of that hurricane does not appear in the list again.
- *Maturity* (hurricane): The tropical storm is intensified and it acquires the characteristic of hurricane—the wind reaches the maximum of the speed (up to 370 km/h) and the cloudy area expands obtaining its maximum extension between 500 and 900 km of diameter, producing intense precipitations. The eye of the hurricane, whose diameter varies from 24 to 40 km, is a calm area free of clouds. The intensity of the hurricane in this stage of maturity graduates by means of the scale of 1 to 5 of the Saffir–Simpson scale.
- *Dissipation* (final phase): The pressure in the center of the system begins to increase and the winds fall gradually accompanied by a weakening of the system. In this stage the hurricanes that penetrate land become extra-tropical hurricanes. A central factor in the end of a hurricane is

the lack of energy sustenance provided by the warm waters. Another is that when arriving to Earth, the friction with the irregular surface of the land provokes cloudy expansion of the meteor and it causes its detention and dissipation in strong rains. An additional factor is that the hurricane meets with a cold current.

In Figure 5.4 we present some images that were taken over a period of 6 days, during the stages of development of a hurricane.

Series of images of the Hurricane Floyd, 1999. (a) A good tropical disturbance that favors a tropical depression; (b) 12 hours later, we see a tropical depression that continues the escalation; (c) Hurricane Floyd has been intensified to a tropical storm; (d) one can already observe Floyd as a hurricane of Category 1; (e) Hurricane Floyd observed as Category 4.

Tropical cyclone intensification depends on many factors (DeMaria and Kaplan, 1999) including oceanic heat content and proximity to land. Hurricanes are formed and are intensified when they are located on tropical or subtropical oceans in both hemispheres where the force of rotation of the Earth (Coriolis) is sufficiently strong so that the rotation movement begins around the center of low pressure and whose temperatures of water at surface level are quite warm. The main regions are not stable as for their location, since this obeys the position of the centers of maximum marine heating, that are in turn influenced by the cold currents of California and the equatorial warm countercurrent in the Pacific Ocean, as well as by the drift of the warm current of the Gulf Stream. Also, they do not stay on land, independent of the superficial temperature.

An analysis of the trajectories of tropical hurricanes shows that there is no coastal area of Mexico that is free from the threat of the tropical depressions that arrive in many cases at hurricane intensity. In the Gulf of Mexico and in the Pacific, the Mexico's coast is vulnerable to the effects of tropical storms, although their behavior on both coasts is different. The depressions that are generated in the southeast of Mexico, specifically in the Bank of Campeche, generally move northward, while those of the Caribbean travel toward the west until touching the coasts of Central America, or those of the Yucatán Peninsula. When they cross it, they vanish, but not enough to be annulled, due to the

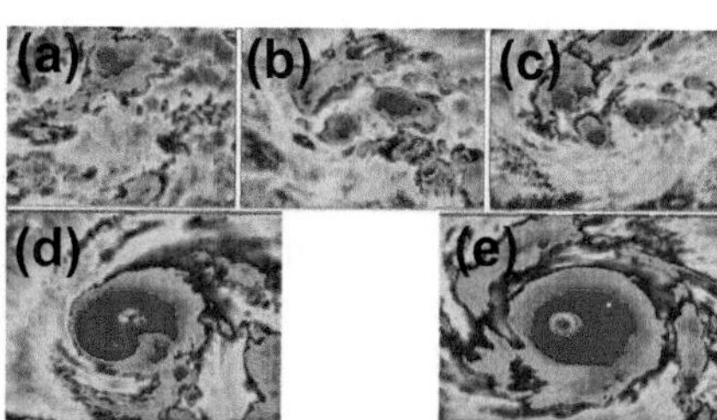

FIGURE 5.4 Series of images of the Humicate Floyd, 1990.

narrowness of the peninsula, so when arriving at the Gulf of Mexico they find the warm water again that reefed them, recovering their fury and continuing their devastating path.

In a study on the activity of the depressions in the North Atlantic during the first half of last century, some investigators found that more than 78% of those that occurred in the Gulf of Mexico took place starting from 1932, and only 36% have reached hurricane force. The duration of these depressions has been 4.4 days and of the hurricanes, 2.2 days. The closed form of the Gulf conditions their short duration and low frequency, since the storms reach the land rapidly and then vanish. The Yucatán Peninsula of is the most affected by the depressions and, of the total previously mentioned, 46% affected the peninsula. In the last two decades the frequency and intensity of the hurricanes in this region has increased; the stand-outs being Gilberto in September 1988 and Mitch in October 1998.

The season of hurricanes emerges when the climatic equator moves in a direction of the poles carrying with it high temperatures that heat the air and the seawater, giving rise to the emergence of an area of low pressure. This generally happens between the months of May and November.

For a hurricane to be formed there need to be present the following elements:

- *Pressure*: A preexistence of a convergence zone in the low levels and low superficial pressure of synoptic scale.
- *Temperature* > 80 °F: At this temperature, the water of the ocean is evaporating at the required quick level so that the system is formed. It is that evaporation process and the eventual condensation of the vapor of water in the form of clouds that liberates the energy that gives the force to the system to generate strong winds and rain. And, since in the tropical areas the temperature is usually high, they constantly originate the necessary elements.
- *Humidity*: As the hurricane needs the evaporation energy like fuel, there must be quite a high humidity, which happens more readily on the sea, so the advance and increment in energy happens there more easily, weakening when arriving to mainland.
- *Wind*: The presence of warm wind near the surface of the sea allows a lot of evaporation and it begins to ascend without big setbacks, originating a negative pressure that crawls to the air in hairspring form toward the inside and up, allowing the evaporation process to continue. In the high levels of the atmosphere the winds should be weak so that the structure stays intact and the cycle is not interrupted.
- *Gyre (or Spin)*: The rotation of the Earth gives a circular movement to this system, which begins to rotate and to move as a gigantic spinning top. This in turn is carried out counterclockwise in the Northern Hemisphere and clockwise in the Southern Hemisphere.

Finally it is worth mentioning that Meteorologists have records of North Atlantic hurricanes that date back into the nineteenth century. Over the last half-century, these records are based on a wide range of measurements including ship and land reports, upper-air balloon soundings, and aircraft reconnaissance. More recently, radar imaging and satellite photographs have been included. The geographical position of the eye and the rotational velocity is measured and published every 12, and more recently, every 6 hours.

5.3. CORRELATIONAL STUDY OF NORTH ATLANTIC CYCLONES[1]

Correlational works between solar and climatic parameters give interesting results (e.g., Chernavskaya et al., 2006). Within this context, in a series of works (Elsner and Kavlakov, 2001; Kavlakov et al., 2008a,b; Pérez-Peraza et al., 2008a,b) several efforts have been addressed to find possible interconnections between the appearance and development of Atlantic hurricanes and changes in geomagnetic activity and sharp cosmic ray (CR) decrease, namely Forbush events (FE), which in turn are known to be related to solar activity changes (SS index).

The authors tried to examine the eventual connection of CR, SS, AP, and K_p changes with the processes developing in the atmosphere, far before the formation of North Atlantic hurricanes. Their main hypothesis is that any specific changes of the collateral parameters during the days preceding the cyclone appearance could be used as an indicative precursor for an approaching hurricane event. Their efforts were especially concentrated not only in North Atlantic hurricanes, but in particular those which struck the east coast of Mexico. All such hurricanes, recorded in the period 1950 to 2007 were analyzed.

To reveal any possibility for immediate (not delayed) relationship between hurricanes and cosmo-geophysical parameters, simultaneous data was used to find statistical dependencies between the specific sharp changes in the geomagnetic field and cosmic ray intensity and the corresponding values of the hurricane intensification. For instance, investigating the daily intensification of Katrina (Figure 5.1), it has been noticed that a strong geomagnetic change (through changes in K_p and AP indexes) was recorded 5 days before its maximum value (Figure 5.5). Some repeatedly observed coincidences between the hurricane appearances and preceding sharp cosmic rays (CR) decreases Forbush events (FE). For example Hurricane Abby in 1960 and Celia in 1970 (Figures 5.6 and 5.7) suggested a possibility for closer relationship between the FE and hurricane intensification.

---

[1] In collaboration with Professor Stilian Kavlakov of the Bulgarian Academy of Sciences, Copernic, Sofia, Bulgaria.

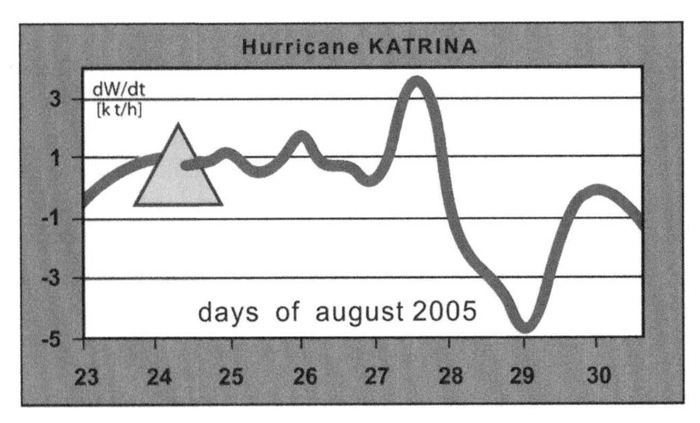

FIGURE 5.5 A strong geomagnetic disturbance (shown with a triangle) was recorded 5 days before the maximum intensification.

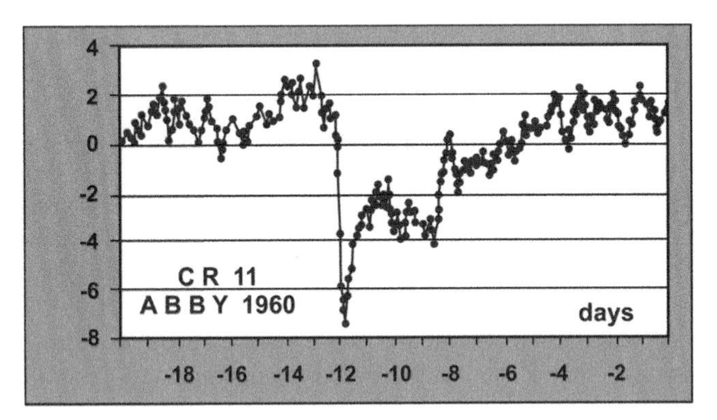

FIGURE 5.6 A Forbush event (7.6%) 12 days before the start of Hurricane Abby, 1960.

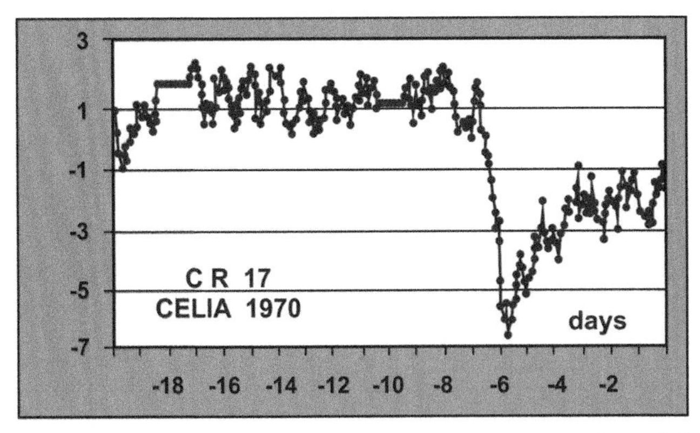

FIGURE 5.7 A Forbush event (6.57%) 6 days before the start of Hurricane Celia, 1970.

These facts provoked the interest of the previously mentioned authors for a detailed study of such collateral phenomena and their statistical comparison with hurricane phenomena. Within that context they were looking for signatures of coadjutant hurricane forecasting to conventional meteorological models, well before the period at which those models usually produce their predictions: the behavior of the (CR) intensity, sunspots (SS), geomagnetic indexes (AP), and (K_p) had been analyzed in long intervals preceding the development of the North Atlantic cyclones, that is, before the first observations of the just born cyclonical system. Two approaches have been done to attack the problem, as described in the following two subsections (Section 5.3.1 and Section 5.3.2).

5.3.1. Parameterization of Hurricane Activity in Terms of Their Intensity (Maximum Rotational Velocity)

In Elsner and Kavlakov (2001) variations in geomagnetic activity in the magnetosphere have been statistically linked to hurricane intensity over the North Atlantic. Based on daily hurricane intensity, a positive correlation between the averaged K_p index of global geomagnetic activity and hurricane intensity as measured by maximum sustained wind speed was identified. Later Kavlakov (2005) analyzed all "major hurricanes" (Categories 3, 4, and 5), that is, those with a maximum rotational velocity of over 95 knots (170 km/h), during the period 1954 to 1999. To avoid any overlapping influences over the investigated parameters spread in front of the hurricanes, only those with a time separation between them exceeding 35 days were selected. Only 39 hurricanes were found filling those conditions, in that 45-year period. Cosmic ray intensity data was taken from 14 neutron monitor CR stations encircling the Atlantic Ocean and situated in Europe and North and South America, characterized by their continuity and stability. The general trend of CR intensity changes are well intercorrelated for all nearby stations. During the periods when some of the stations were not working, a corresponding statistical weighting of the data was applied. The full set of daily sunspot numbers was obtained from the website of the National Geophysical Data Center in Boulder, Colorado, U.S.A. The AP and K_p indices, which describe the status of the geomagnetic field, were taken from the website of GeoForschungsZentrum, Potsdam, Germany, as a sum of the 8 absolute 3-hourly values each day over the same 45-year period.

From the whole data set for CR intensity, SS number, AP, and K_p indices, only those included in the 34-day intervals preceding the day when every one of the examined hurricanes reaches its maximum rotational velocity, namely the M-day in that work, were extracted. Data was overlapped and averaged separately for all these variables. Results based on daily hurricane intensity are summarized in Figures 5.8 through 5.10. Changes of the CR in the intervals of 34 days before the M-day are depicted in Figure 5.8: the statistical error is in the limits of the size of the points. Figures 5.9 and 5.10 show the

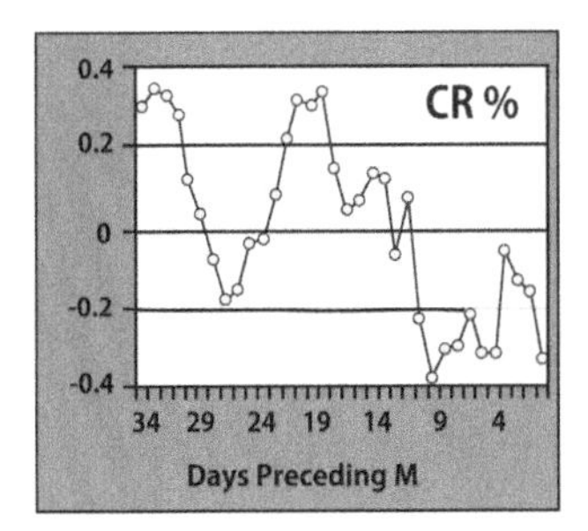

FIGURE 5.8 Changes of CR intensity preceding the M-day.

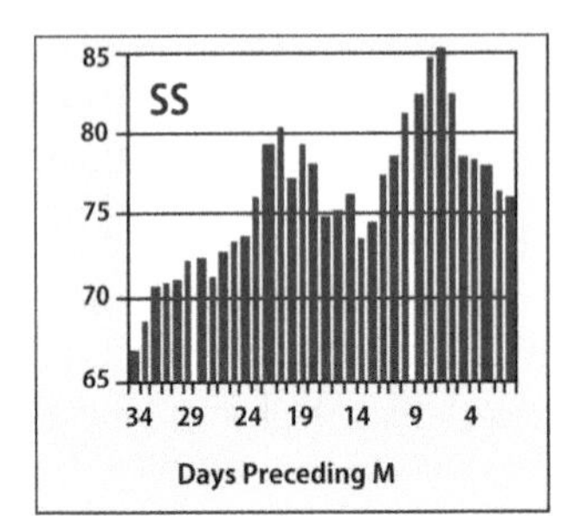

FIGURE 5.9 Values of the SS index.

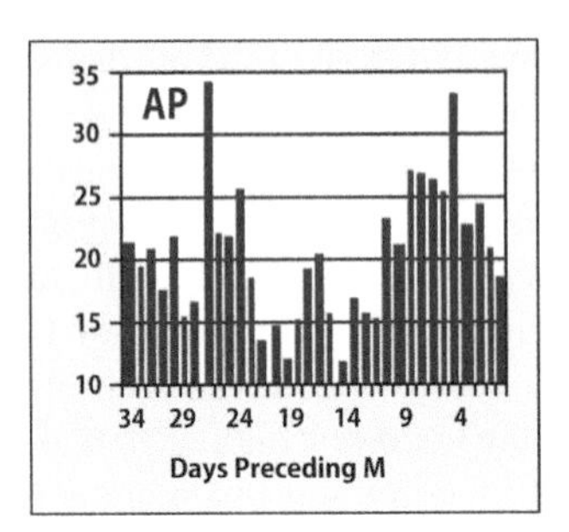

FIGURE 5.10 Values of the AP preceding the M-day.

changes of the SS and AP indices (K_p index follows sensibly the same trend of the AP one).

The obtained results show that specific changes in the solar activity, CR intensity, and geomagnetic field precede the appearance of these major North Atlantic hurricanes. It can be seen from Figures 5.6 and 5.7 that there is a change of intensity of the FE-type just 8 to 12 days before the time of hurricane maximum intensity. The corresponding amplitude of these changes is often well noticeable even in the case of separate individual events. SS and AP indexes show an abnormal trend around 8 to 12 days before the M-day.

In particular it is found that 2 to 3 days before and after the maximum K_p values, the hurricane intensification is much higher than the average.

It is important to emphasize that these changes are persistently accompanying the days preceding the major hurricane maximum intensity. The existence of stable precursors for the chosen types of hurricanes, exceeding Category 2, has been evidenced in that work. The obtained specific changes of these parameters should be taken in consideration, when complicated processes in the upper atmosphere are used to determine the hurricane formation. Potentially, that should contribute to hurricane development forecasts.

The combination of the special changes of SS and AP with a Forbush decrease of the global CR intensity, as shown in Figure 5.8, must draw attention to the places where generally hurricanes are born. That is especially true for the summer period, when such simultaneous changes could be taken as an early indication for elevated probability of a major hurricane formation. Of course, the basic observations of all meteorological parameters should be taken into consideration first.

5.3.2. Parameterization of Hurricane Activity in Terms of Their Intensification (First Derivative of the Rotational Velocity, Assuming that Changes in the Rotational Speed Reflects any Energy Impute in Hurricane Dynamics (Kavlakov et al., 2008a,b,c))

The earlier work described in the previous section is expanded on by focusing on hourly intensification rates rather than daily hurricane intensity. Results appear to be more general in that there is no need to separate the tropical cyclones by category. Intensification (or intensification rate) is the change of intensity with time. The terms *intensifying* and *deepening* refer to positive intensification while the terms *decaying* and *filling* are used for negative intensification.

Here, the interest is the study of hurricane intensification rates around the time of a major geomagnetic disturbance and whether there is, on average, a statistically higher rate during these disturbances; the K_p index is widely used in ionospheric and magnetospheric studies and is recognized as measuring the magnitude of worldwide geomagnetic activity. A similar study is described in Section 5.3.4 with sharp cosmic ray decreases (FE).

The authors considered the maximum wind speed (intensity) for all tropical cyclones (hurricanes and tropical storms) over the North Atlantic, which includes the Atlantic Ocean, Gulf of Mexico, and the Caribbean Sea, during the period 1951 to 2005. The data are derived from the hurricane database (HURDAT or best-track) (Neumann et al., 1999) maintained by the National Hurricane Center (NHC). HURDAT consists of 6-hourly positions and intensities. The 6-hourly values were converted to 1-hourly values using cubic spline

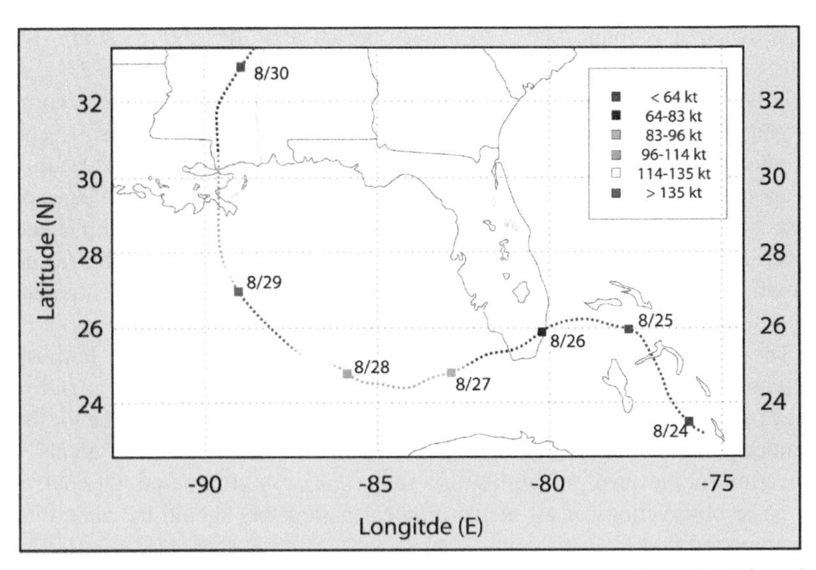

FIGURE 5.11 Track of Hurricane Katrina. The positions are every hour (interpolated from the mostly 3-hour reports) beginning at 2100 UTC on August 23, 2005 over the southern Bahamas. Color denotes storm intensity as indicated by the maximum sustained wind speed (kt). The date (month/day) is shown at the 0000 UTC time.

interpolation. As an example, Figure 5.11 shows the 1-hour position and intensity for Hurricane Katrina in 2005.

Tropical cyclone intensification is a time-derivative quantity. While it was tempting to use a simple finite difference procedure to approximate the derivative, the order of the error on this approximation procedure is commensurate with the derivative value. Then, the hourly intensification rate was estimated from an asymmetric 6-point (3 left, 2 right) 3-degree Savitzky–Golay first derivative filter (Savitsky and Golay, 1964) that reduces the error. Hourly intensification rates are obtained for all tropical cyclones (hurricanes and tropical storms) for a total of 105,638 hours over the period 1951 to 2005. Figure 5.12 shows a time series plot of the hourly intensification for Hurricane Katrina.

In the work described above (Kavlakov, 2005) the behavior of K_p was investigated and some other parameters before the start day of the hurricane. In subsequent works (Kavlakov et al., 2008a,b,c) the basic position was rearranged: it was taken as "0" days, not the day when a hurricane starts, but the opposite—the day with a high K_p peak; thus, they registered all the hurricane activities around it.

Data of K_p were obtained from the U.S. National Oceanic and Atmospheric Administration (NOAA) website. Interest is on days surrounding a K_p maximum event (called here a K_p day). A K_p day is defined as one in which the daily K_p index exceeds 420, or more than 70% above the long-term average. In this way we

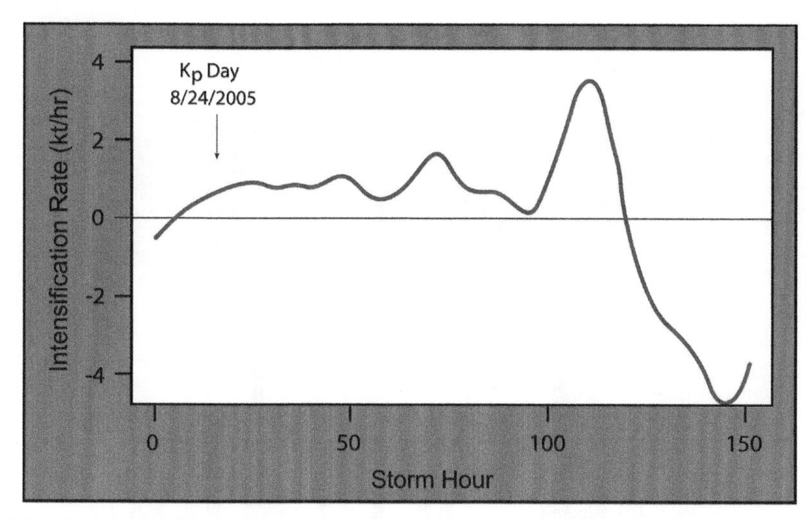

FIGURE 5.12 Intensification rate for Hurricane Katrina. The intensification rate in kt/h is given every hour for the lifetime of the tropical cyclone until it passes north of 33 °N latitude. The arrow locates a K_p day. The location is 0900 UTC on August 24, 2005 when the K_p index reached a value of 8.7 steps.

identify 224 K_p days during the hurricane season months of May through November over the period 1951 to 2005, overlapping active storm hours.

The monthly distribution of K_p days (Figure 5.13) showing how many of the 224 K_p days fall in each of the months is fairly uniform with a maximum during September and a minimum in June (for example, 35 occurred during the month of May and 9 occurred in 1951). The annual distribution shows the well-known 22-year solar cycle. There is no significant long period trend in these counts. In contrast, the hurricane season is strongly peaked around the month of September and there is an increasing trend over time in the number of tropical cyclone hours (Figure 5.14).

5.3.3. Cyclone Intensification around K_p "0" Days (Days of K_p Max)

We statistically analyze the relationship between geomagnetic disturbances and hurricane intensification by averaging intensification rates over 5 days centered on the K_p day and comparing this mean intensification with the overall average intensification. From all 105,638 hours of tropical cyclone activity, the mean intensification rate is +0.0342 kt/h which equals 4.1 kt over any 5-day period. This compares with a mean intensification rate of +0.0713 kt/h or 8.56 kt over the 5-day period based on 10,995 hours of intensification (108 separate tropical cyclones) ±2 days of the K_p day. We note larger intensification rates surrounding the K_p day, on average, for tropical cyclones weaker and stronger than hurricane intensity (64 kt). Results are shown on Table 5.3.

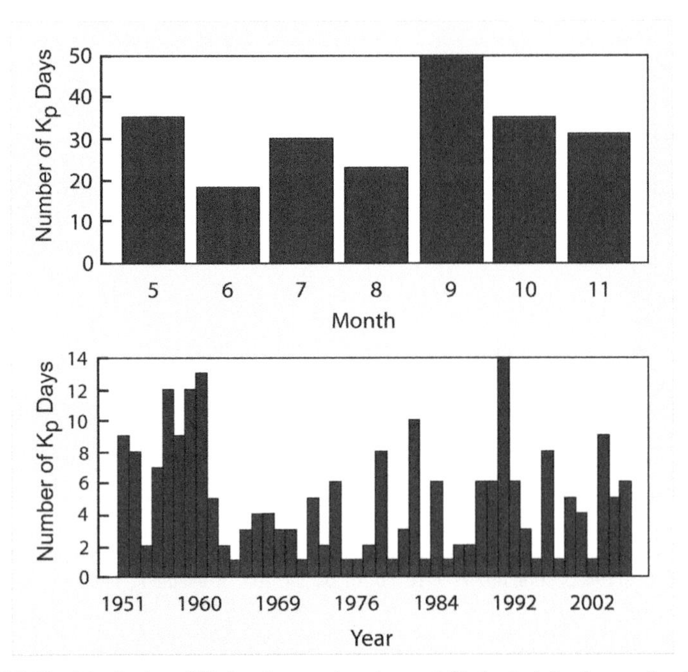

FIGURE 5.13 Distribution of K_p days by month and year. A K_p day is defined as one in which the daily K_p index exceeds 420. By this definition there are a total of 224 K_p days in the period 1951 to 2005.

To test the significance of these differences we randomly assign days as K_p days and compare the mean intensification rate (bootstrapped rate) over the 5 days centered on these random dates. We repeat this procedure many times (200–1000) and count the number of bootstrapped rates that exceed $+0.0713$ kt/hr. The number of times the rate is exceeded divided by the total number of bootstrapped rates is the p-value. We find a p-value of 0.12 for all cyclones, 0.13 for weak cyclones, and 0.10 for strong cyclones. The obtained number distribution is shown on Figure 5.15.

We note larger intensification rates surrounding the K_p day, on average, for tropical cyclones weaker and stronger than hurricane intensity of 64 kt. While suggestive, the results are inconclusive regarding the relationship between geomagnetic disturbances and hurricane intensification.

Tropical cyclone intensification depends on many factors (DeMaria and Kaplan, 1999) including oceanic heat content and proximity to land. These factors will confound our ability to identify a significant geomagnetic signal in the data. To provide some control, authors repeated the previous analysis using only cyclones confined to the open waters of the tropical Atlantic. In this way we consider only storm hours far from land over a fairly uniformly warm part of the basin. The chosen control region is part of the main development region for tropical cyclones and is bounded by 25 and 60 °W longitude and by

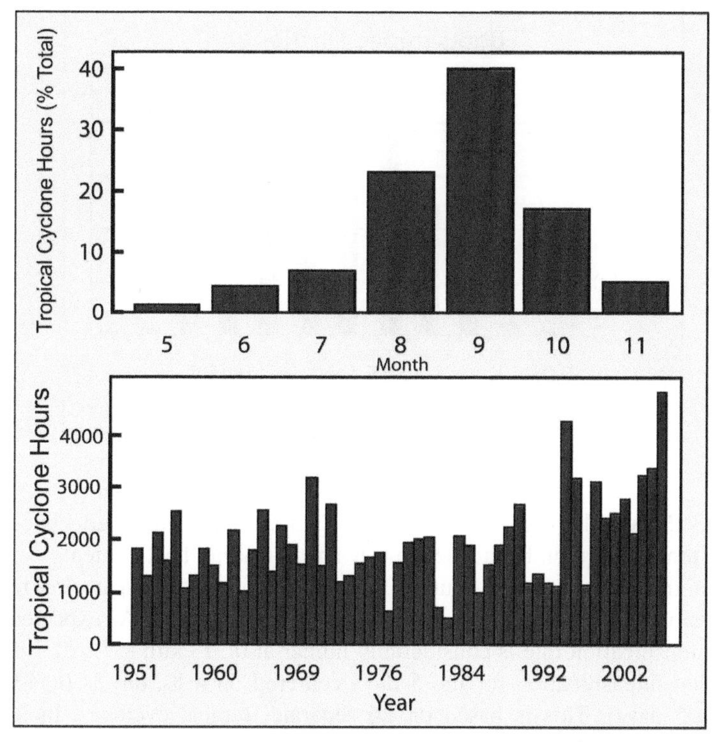

FIGURE 5.14 Tropical cyclone hours. The monthly distribution and annual counts of tropical cyclone hours for the North Atlantic over the period 1951 to 2005.

TABLE 5.3 Average Intensification Rates Over 5 Days Centered on the K_p Day and Comparison with the Overall Average Intensification

	"0" days	Cyclones n	Cyclones hours	Average (dW/dt) [kts/hour]	Average (dW/dt) [kts/5days]
	Over whole Atlantic region				
All		603	105638	0.0342+/-0.006	4.10+/-0.07
KP_{max}	224	108	10995	0.0713-/-0.008	8.56+/-1.02
FE	166	96	7691	0.0546-/-0.0095	6.5+/-1.1
	Only over hot waters				
All		131	17579	0.313+/-0.04	37.6-/-1.9
KP_{max}	224	26	2230	0.543+/-0.12	65.1+/-14.4
FE	166	25	2104	0.363+/-0.216	43.6+/-25.9

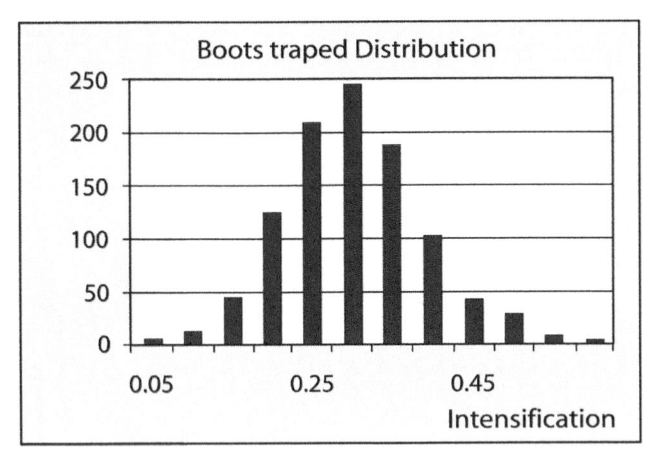

FIGURE 5.15 Distribution of "bootstrapped" rates.

8 and 23 °N latitude (Figure 5.3). In this way it is controlled for effects of sea-surface temperature on hurricane genesis and for ocean heat content.

In contrast to the 105,638 hours of the whole basin, within the control region there are only 17,579 cyclone hours over the 55-year period. As expected, the mean intensification rate is considerably higher at 0.313 kt/h (37.6 kt/5 days). The mean intensification for the 5 days centered on a K_p day is 0.543 kt/h (65.1 kt/5 days). This is based on 26 separate tropical cyclones, including Tropical Cyclone Dog in 1952 and Tropical Cyclone Iris in 2001. We repeat the bootstrap procedure as described above for determining the statistical significance. Figure 5.16 shows a histogram of mean intensification rates for 1000 bootstrapped rates. The actual rate is noted with an arrow. The p-value is 0.007 indicating a significant increase in intensification around K_p days relative to the average. Similar results are noted for tropical cyclones greater than and tropical

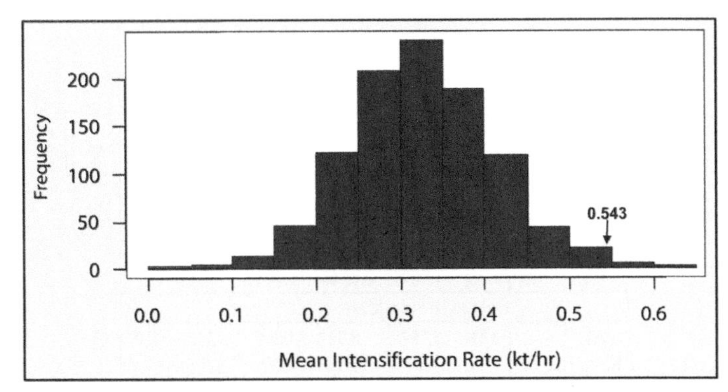

FIGURE 5.16 Histogram of bootstrapped intensification rates over the control region. The mean intensification rate ±2 days of the K_p day is 0.543 kt/day, which is greater than all but 7 of the 1000 bootstrapped rates.

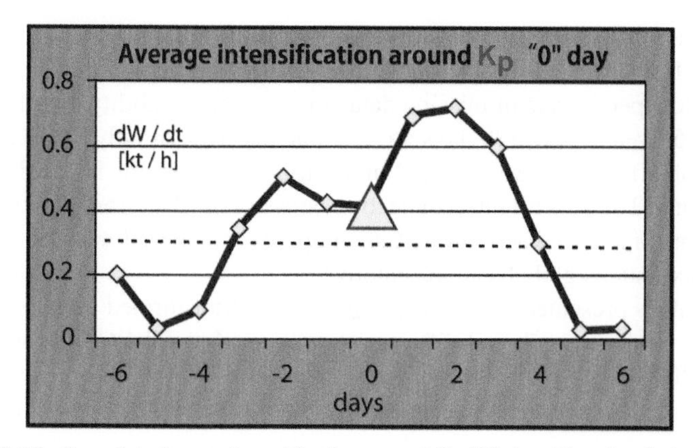

FIGURE 5.17 Lag plot of mean intensification around K_p "0" day. The triangle indicates the position of the K_p "0" day. The influence of the geomagnetic disturbance appears most pronounced over a 7-day period centered on the K_p day.

cyclones less than 64 kt, although the significance is more pronounced for the weaker cyclones.

The 2-day window surrounding the K_p day is arbitrary, so we also consider the mean hurricane intensification for storms before, during, and after the K_p day. Figure 5.17 shows the mean intensification rate as a function of lag time from the K_p "0" day. We note that the effect appears most pronounced for lags from −3 to +3 days, with a peak at +2 days.

Summarizing, it was found in the analyzed work that intensification is related to the geomagnetic disturbances mainly in the North Atlantic higher latitudes. It is understandable why the K_p effect is less pronounced over the regions with overheated surface water, where more of the cyclones are born (Table 5.3). There, the dominant factor is the energy extracted from the water surface. That reduces all other accompanying factors participating in the cyclone formation. Over the whole basin, where generally the primary creating and supporting effect of water surface temperature is reduced, these other factors became more active.

What must be kept in mind from this work is that it was found that 2 to 3 days before and after the maximum K_p values, the hurricane intensification is much higher than the average.

5.3.4. Cyclone Intensification around FE "0" Days

Works described in Section 5.3.1 were based on data from several neutron monitors (NM) situated around the Atlantic Ocean. But the gain for the statistics achieved in this way was suppressed by the difficulties of combining together the different data, available in different intervals. So, we accepted that the use of only one long-running continuous CR station could be much more suitable. That is why we took the whole neutron monitor data set received on

Climax CR station (39.37 °N, 106.18 °W, alt. 3400 m, and 2.97 GeV cut-off rigidity). They covered the period 1951 to 2005 with negligible instrumental changes, low percentage of missing data, and very high stability. For the whole period of 55 years, only 399 days are without any data, or only 2.02%. That is 97.98% of effective measured CR intensity. We carefully interpolated the missing data. The general interconnection between the data from practically all NM stations, measured on different geographical places, allow us to consider the Climax data as globally representative.

The values presented in counts per hour were transformed in daily percent deviations from the general 55-year average value (394,600 counts/hour, or 9,470,400 counts/day). The statistical error then is 0.032% for a single day. In most cases we averaged over many days, and the error is generally below that value.

Analogically to the K_p data, all Forbush effects were extracted from the whole set of Climax CR data, and defined as an FE "0" day when the sudden decrease in CR intensity (or Forbush event) is below −3% (Figure 5.18). It could be seen that these events are equally spread throughout the year (Figure 5.18). That means there are no significant monthly changes in their number. As we mentioned before, in contrast, the hurricane season is strongly

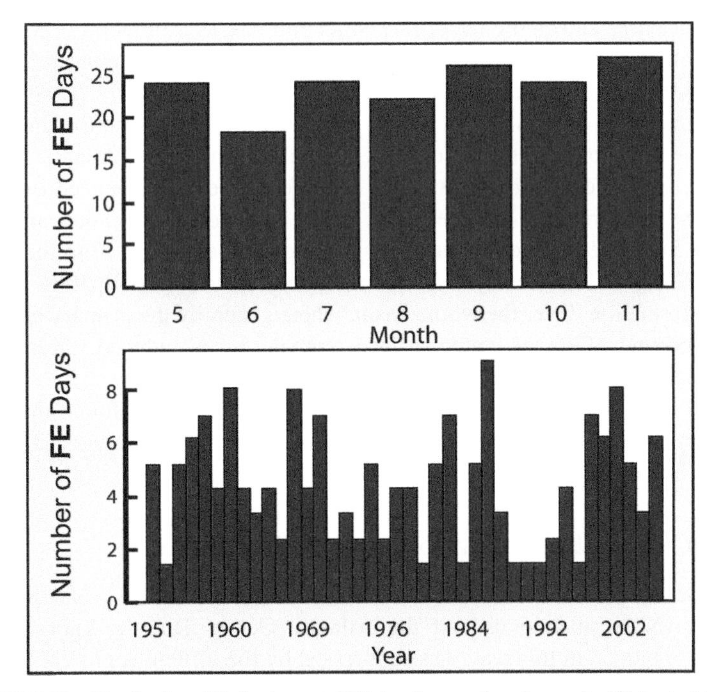

FIGURE 5.18 Distribution of Forbush event (FE) days by month and year. An FE day is defined as one in which cosmic ray intensity drops by at least 3%. The annual distribution shows the well-known 22-year solar cycle. There is no significant long period trend in the frequency of Forbush days.

peaked around the month of September and spread practically only over the months from May till November. In these 7 months in long hurricane seasons over the period 1951 to 2005, we identify 166 FE "0" days. The annual distribution shows the well-known 22-year solar cycle. There is no significant long period trend in the frequency of Forbush days.

The relationship between dW/dt and FE was statistically analyzed by averaging intensification rates over 5 days, centered on the FE "0" day, and comparing this mean intensification with the overall average intensification. We used the same geographical area as previously to classify the hurricane intensification over the whole Atlantic area and over hot waters with practically constant temperature. Results are shown in Table 5.3 on the corresponding rows to FE. It could be noticed that about 7 to 10% of the hurricane hours are in the interval of 5 days around the FE "0" days. From all 105,638 hours of tropical cyclone activity, the mean intensification rate is +0.0342 kt/h, which equals 4.1 kt over any 5-day period. This compares with a mean intensification rate of +0.0546 kt/h or 6.5 kt over the 5-day period based on 7691 hours of intensification (96 separate tropical cyclones). We note a weak rise of the intensification rates surrounding the FE "0" day, on average, for tropical cyclones. There are 17,579 cyclone hours within the hot water control region over the 55-year interval (1951–2005).

As expected the mean intensification rate is considerably higher at 0.313 kt/h (37.6 kt/5 days). The mean intensification for the 5 days centered on an FE "0" day is 0.363 kt/h (43.6 kt/5 days). This is based on 2104 hours of 25 separate tropical cyclones.

To test the significance of these differences it was randomly assigned days as FE "0" days and applied the same bootstrapped rate method. The mean intensification rate over the 5 days centered on these random dates was estimated with 17% accuracy. The 2-day window surrounding the FE "0" day is arbitrary so we also consider the mean hurricane intensification for storms before, during, and after the FE day. Correspondingly to the average intensification around K_p "0" day shown in Figure 5.17, it is shown in Figure 5.19 the mean intensification rates as function of lag time from the FE "0" day.

While suggestive, the results are inconclusive regarding the relationship between cosmic ray decreases and hurricane intensification. The difference with geomagnetic activity can be noticed easily. While for cosmic rays after the "0" days there is a rise of the intensification in both cases, generally the distribution around FE "0" day does not show any significant difference from the average. In contrast, the distribution of the average hurricane intensification around days with high geomagnetic disturbances (K_p "0" days) is significantly greater than the average.

Summarizing, it was found a slight rise of the hurricane intensification around the FE "0" days over both investigated regions. But that rise as well as the form of the curve on Figure 5.19 shows a statistically insignificant

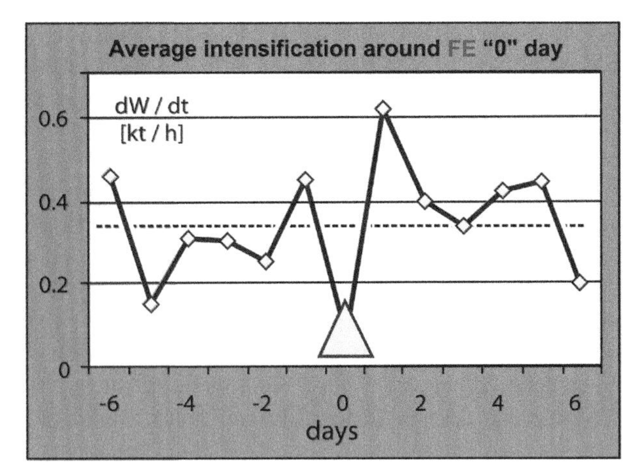

FIGURE 5.19 Lag plot of mean intensification around the average FE "0" day. The triangle indicates the position of the FE "0" day. The influence of the cosmic rays on tropical cyclone intensification appears less pronounced than that of geomagnetic variations.

relationship between cosmic ray sudden decreases (the Forbush effects) and hurricane intensification in the interval of ±5 days.

It appears that the sharp cosmic ray intensity decreases have predominantly long time range influences: taking into consideration the persistent appearance of FE from 5 to 20 days before the hurricane start (Figure 5.6 and Figure 5.7) we could deduce that any suggested mechanism that relates the FE with the hurricane formation needs considerable time for activation. As could be seen here the direct impact is negligible.

At present, an accurate forecast of cyclone activity cannot be claimed only on the basis of preceding K_p and CR; Forbush effects data changes, however, looking at these results could be strongly alerted if a sharp rise in K_p index is recorded during the hurricane season, or a Forbush event appears at the end of the summer. Then investigating all the parallel atmospheric data we could be closer to a true prediction.

5.3.5. Hurricanes Hitting the Coasts and Landing in Mexico

Following the same hypothesis, that any specific changes of the collateral cosmophysical parameters during the days preceding the cyclone appearance may be used as an indicative precursor for an approaching dangerous event, a correlational analysis was done (Pérez-Peraza et al., 2008a) using not only the maximum rotational velocity as in Section 5.3.1, but also the estimated total energy. Data were used for all recorded cyclones in the Atlantic Ocean and in the Pacific Ocean from January 1, 1950 till December 31, 2004 that fill the following requirements: during their development their maximum rotational velocity V_{max} reaches at least 35 knots (which means we include in our investigation not only the hurricanes, but also the tropical storms, as defined in the Saffir–Simpson scale), and during their displacement they have touched either the Mexican coast or the Mexican borders. After careful examination of the track of every cyclone in that 55-year interval, 119 cyclones born over the Pacific and 59 cyclones born over the Atlantic Ocean were found obeying these conditions. The main characteristic of a subset of 24 hurricanes (among the most popular of them) is displayed in Table 5.4.

On the Internet there are a lot of cyclone data (e.g., *http://www.aoml.noaa.gov/hrd/hurdat/Documentation.html*; *http://stormcarib.com/climatology/#links*; *http://weather.unisys.com/hurricane/atlantic/index.html*). We carefully examined data chosen from them. Solar activity is characterized by different indexes, among one of them is the daily number of sunspots number (SS). A full set of daily sunspot numbers for the period 1950 to 2005 was obtained from the website of the National Geophysical Data Center in Boulder, Colorado, U.S.A., *http://www.ngdc.noaa.gov/stp/solar/ssndata.html.*

For cosmic ray (CR) data, instead of using data from several neutron monitors (NM) situated around the Atlantic Ocean, since the gain of the statistics, achieved in this way, was suppressed by the difficulties of sticking together the different data available in different intervals, it was decided that only one, but long-running NM with more than 50 years continuous measurements would be much more suitable. That is why the whole set of Climax NM was considered, as described in the previous section. It must be added that the map of cosmic ray intensity isolines at height about 200 km (e.g. Khorozov et al., 2006) shows that the concerned region in the North Atlantic is crossed practically by the same isoline, which includes Deep River, Climax, Mexico, and other NM stations.

The general interconnection between the data of practically all NM data, measured on the same isoline, permits us to consider the Climax data as globally representative. The Climax data were taken from http://ulysses.sr.unh.edu/NeutronMonitor/00ClimaxCorr.html. The values presented in counts per hour were transformed in daily percent deviation from the general 55-year average value (394,600 counts/h or 9,470,400 counts/day). The statistical error then is 0.032% for a single day. In most cases we averaged over many days, and the error generally is below the size of the point, presented on graphs.

The daily values of geomagnetic activity indexes K_p and AP characterizing the geomagnetic activity used in this work were taken from the web page of GeoForschungsZentrum, Potsdam, Germany and compared with those of National Geophysical Data Center in Boulder, Colorado, U.S.A. Full data set for our 55-year period were available on both websites (potsdam.de/kp_index/index.html; http://www.geomag.bgs.ac.uk/data_service/data/magnetic_indices/apindex.html).

TABLE 5.4 Parameters of a Typical Subset of Hurricanes

Table4

0	1	2	3	4	5	6	7	8	9	10	11	12	13	14	15	16	17	18	19	20	21	22	23	24	25	26	27	28
Number of CR	Type	HURRICANE	Tropical depression					L	D	U_{av}	V_{max}	V_{av}	U/V	These Max and Min are for 20 days interval, prior to the corresponding shown date (column 2,3,4)														
stations used here		NAME	Beginning of Trop.Depr.			Position		Whole Track	Whole Duration	Aver. linear	Rot.Veloc. max	aver.	Correl. Coef.	CR Max	CR Min	Diff Max-Min	SS . Av.	SS Max	SS Min	Diff Max-Min	AP . Av.	AP Max	AP Min	Diff Max-Min	KP . Av.	KP Max	KP Min	Diff Max-Min
			yy	m	d	LAT gedree	LONG	km	hours	km/h	knots		%	%	%													
1	0	FLORENCE	1954	9	11	20,9	-94,7	326	42	8	65	47,9	-0,80	0,402	-0,618	1,0199	4	18	0	18	15	27	8	19	22	30	14	16
1	0	GLADYS	1955	9	4	20,6	-94,1	903	60	15	80	50,0	-0,14	0,8201	-0,606	1,4256	41	89	11	78	8	20	2	18	14	26	3	23
1	0	JANET	1955	9	21	13,2	-54,3	4921	210	23	150	101,0	0,21	0,7498	-1,019	1,7685	53	89	7	82	11	22	4	18	18	28	9	19
1	0	ANNA	1956	7	25	20,6	-92,7	662	42	16	70	50,7	0,53	1,1	-1,5	2,6	128	216	65	151	11	25	4	21	17	30	7	23
3	0	CARLA	1961	9	3	12,5	-77,0	7417	306	24	150	67,8	-0,51	0,6	-0,6	1,2	60	108	33	75	11	37	2	35	16	35	4	31
9	0	INEZ	1966	9	21	9,9	-35,1	7969	486	16	130	76,2	-0,02	2,8	-2,9	5,8	47	89	18	71	23	112	3	109	23	46	6	40
9	0	BEULAH	1967	9	5	14,0	-57,0	5327	420	13	140	76,8	0,46	0,9	-0,5	1,4	117	130	77	53	10	25	4	21	17	30	7	23
12	0	EDITH	1971	9	5	11,4	-58,0	2516	306	8	140	52,0	0,10	1,7	-1,1	2,9	57	108	19	89	9	32	2	30	15	33	2	31
9	0	CARMEN	1974	8	29	16,8	-55,8	5454	294	19	130	66,2	-0,05	1,6	-6,0	7,6	39	69	8	61	19	46	4	42	23	39	9	30
9	0	FIFI	1974	9	14	15,3	-65,0	4241	198	21	90	52,0	-0,68	0,8	-3,3	4,2	36	80	8	72	15	32	4	28	21	33	9	24
9	0	CAROLINE	1975	8	24	22,4	-69,8	3178	198	16	100	38,9	-0,57	0,9	-1,1	2,0	48	104	7	97	11	29	4	25	17	32	7	25
9	0	ALLEN	1980	7	31	11,0	-30,0	8148	276	30	165	97,1	-0,16	2,3	-5,3	7,6	156	241	81	160	12	46	4	42	17	31	9	22
8	0	GILBERT	1988	9	8	12,0	-54,0	8025	276	29	160	74,5	-0,48	1,7	-3,5	5,2	102	164	21	143	11	24	3	21	18	28	7	21
9	0	GERT	1993	9	14	10,6	-80,7	3036	174	17	85	37,8	-0,15	0,5	-0,5	1,0	25	53	9	44	16	91	3	88	18	48	4	44
7	0	ROXANNE	1995	10	7	14,0	-82,1	3230	324	10	100	53,0	0,24	0,8	-0,7	1,5	12	30	0	30	14	57	3	54	17	41	5	36
7	0	DOLLY	1996	8	19	17,3	-80,2	501	144	3	70	41,0	0,47	1,2	-0,8	2,0	16	23	9	14	7	15	4	11	13	23	8	15
7	0	MITCH	1998	10	22	11,6	-76,1	11957	456	26	155	63,6	-0,19	1,1	-1,2	2,3	64	105	19	86	14	62	2	60	17	43	4	39
1	1	ALICE	1954	6	24	22	-94	1182	60	20	70	47,0	0,12	0,5	-0,4	0,9	0	0	0	0	6	13	3	10	11	20	5	15
12	1	ELLA	1970	9	8	15,3	-83,5	2181	120	18	110	52,6	0,13	1,3	-1,2	2,5	113	136	91	45	10	23	4	19	17	26	7	19
10	1	ANITA	1977	8	29	26,9	-88,4	1601	120	13	150	72,0	-0,03	1,2	-1,1	2,4	33	42	15	27	13	29	5	24	20	32	9	23
9	1	BARRY	1983	8	23	26,0	-76,0	2706	144	19	70	38,8	-0,33	1,3	-1,2	2,5	74	105	40	65	15	62	2	60	18	34	3	31
9	1	DIANA	1990	8	4	13,2	-79,5	3178	138	23	85	42,4	-0,60	1,7	-2,5	4,2	136	204	57	147	17	102	4	98	19	48	8	40
10	3	FERN	1967	10	1	20,3	-93,0	823	78	11	75	51,0	0,22	0,9	-1,7	2,5	59	79	36	43	19	85	3	82	21	48	4	44
11	3	LAURIE	1969	10	17	17,3	-85,1	2953	252	12	90	44,0	0,02	1,6	-2,5	4,1	89	123	42	81	19	90	3	87	19	46	5	41
7,208	0	Average	1974	8,63	15,38	16,463	-73,17	3851	214	17	109,583	58,1	-0,09	1,1935	-1,746	2,9397	63	100	28	72,2	13,2	46,1	3,5	42,6	17,8	34,6	6,46	28,13
12	3	Max	1998	10	31	26,9	-30	11957	486	30	165	101,0	0,53	2,8351	-0,381	7,627	156	241	91	160	23,2	112	8	109	23,1	48	14	44
1	0	Min	1954	6	1	9,9	-94,7	326	42	3	65	37,8	-0,80	0,402	-6,044	0,9077	0	0	0	0	5,76	13	2	10	10,7	20	2	15

5.3.5.1. Data Processing

Every cyclone was characterized with:

1. Origin: Atlantic (A), or Pacific (P)
2. Start (D0) [year, month, day], when the circular wind velocity (V), measured on the ocean surface has reached 35 nodes
3. Geographical position [B(/1,k1)] of the cyclone eye center, when the start occurs
4. Maximum rotational speed V_{max} reached during the cyclone development
5. Hurricane rank after the Saffir–Simpson Scale, corresponding to the V_{max} value
6. Duration in days (L)
7. Total cyclonal energy (CE)

The daily averaged rotational velocity was taken as a characteristic of the cyclonal energy. The sum of these daily values was accepted as proportional to the total cyclonal energy (CE).

To obtain a basic view for the yearly changes of our parameters, we introduced a yearly cyclonic energy [CE(y) = CE1 + CE2 + _ _ _ + CE*n*], characterizing the total energy of all cyclones. CEi is the estimated energy of "i" cyclone appearing in the year (y). The CE(y) and the CE(y) (smoothed), plotted together with the average yearly values of the parameters CR, SS, K_p, and AP are depicted in Figure 5.20 The correlation coefficients between these parameters are shown in Table 5.5.

If the start of the next cyclone occurs less than 20 days after the start of the preceding one, the less powerful of them is defined as *overlapped.* The overlapped cyclones were not included in the calculations. So, 78 from all 119 Pacific and 44 from all 59 Atlantic cyclones (122 from all the 178 hurricanes) were classified as not overlapped in the 55-year period. Depending on their Saffir–Simpson rank (V_{max}) they were subdivided in: 6 separate groups (rank 5, 4, 3, 2, 1, TS); 3 compound groups (rank 5 + 4, 3 + 2, 1 + TS), and 2 compound groups (rank 5 + 4 + 3, 2 + 1 + TS) (Table 5.6). In Figure 5.21 the rank distributions of the Atlantic and the Pacific cyclones are shown.

Obviously the Pacific cyclones are mostly low ranked. Their places of birth in the Pacific are dispersed along the Mexican west coast while the Atlantic cyclones are spread rather far from the Mexican east coast. From that it is easily understandable why the Atlantic cyclones are predominant in the highest ranks, while the Pacific cyclones are abundant in the lowest ranks. To have a general view on the behavior of all parameters (SS, CR, AP, K_p) 35 days before and 20 days after the cyclone start, every one of the 122 cyclones were investigated separately, creating in this way 488 graphs.

Since it is not possible to present all them, therefore, on the basis of such a vast amount of data, it was decided to evaluate the averaged time between the appearance of these specific CR, SS, and AP changes and the cyclone formation

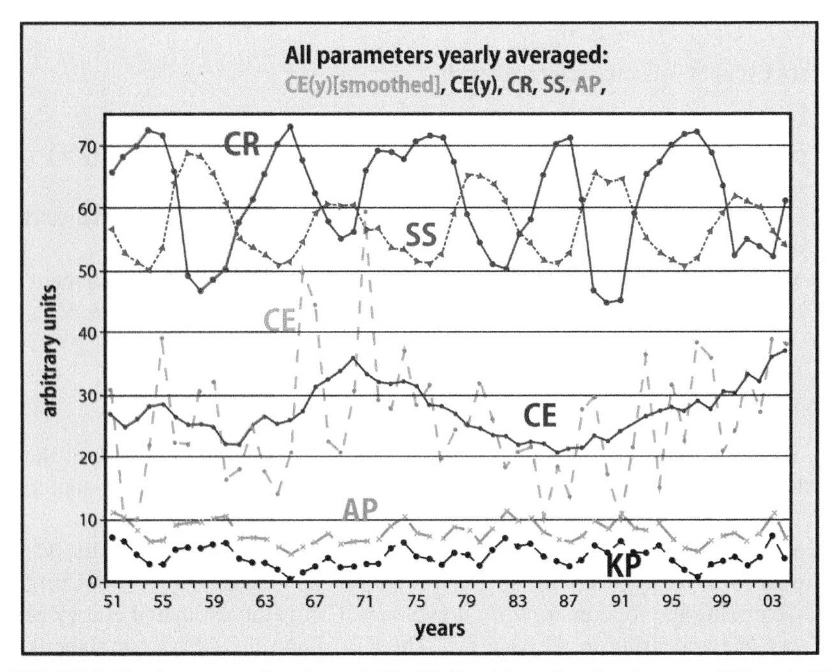

FIGURE 5.20 Average yearly values of CR, SS, K_p, AP, total cyclonal energy (CE), and CE smoothed.

and then to determine the size and the time distribution of these changes and their interconnection. The behavior of the CR, solar activity, and geomagnetic status are examined in the intervals of 35 days before and 20 days after the start of every cyclone. So, from our large amount, we chose three graphs, presenting the changes of CR, SS, and AP for the most powerful cyclones (Allen, Gilbert, and Mitch) in the 55-year period (Figure 5.22). There is a remarkable disturbance in CR intensities, appearing 5 to 20 days preceding the cyclone starts. It is accompanied by one or two clearly pronounced Forbush decreases.

It can be seen in Figure 5.22, in the middle row that SS changes reach 150 to 200 SS units, sometimes even with more of 20 days in advance of the day of the hurricane occurrence. It can also be appreciated that the Ap values show consecutive picks, reaching 100 to 120 Ap units while, spreading over the whole period of 30 preceding days to the cyclone day (the K_p graphs repeat closely those

TABLE 5.5 Correlation Coefficients

CE	SS	CR	AP	K_p
	−0.13	0.30	−0.32	−0.31

TABLE 5. 6 Cyclones Sorted Depending on V_{max}

DISTRIBUTION	Saffir-Simpson Scale		Rank	H5	H4	H3	H2	H1	TS	All
		V_{max}	(Knots.)	>135	114–135	96–113	83–95	65–82	35–64	>35
		V_{max}	(km/h)	>249	249–210	209–178	177–154	153–119	118–63	>63
ALL CYCLONES	6 groups	Cyclone	quantity	7	17	15	17	34	32	122
ATLANTIC	6 groups	Cyclone	quantity	6	7	7	4	11	9	44
PACIFIC	6 groups	Cyclone	quantity	1	10	8	13	23	23	78

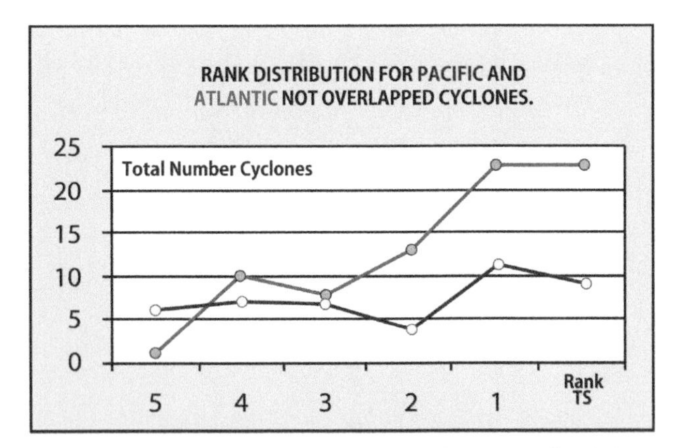

FIGURE 5.21 Distribution of Atlantic and Pacific cyclones according their categories.

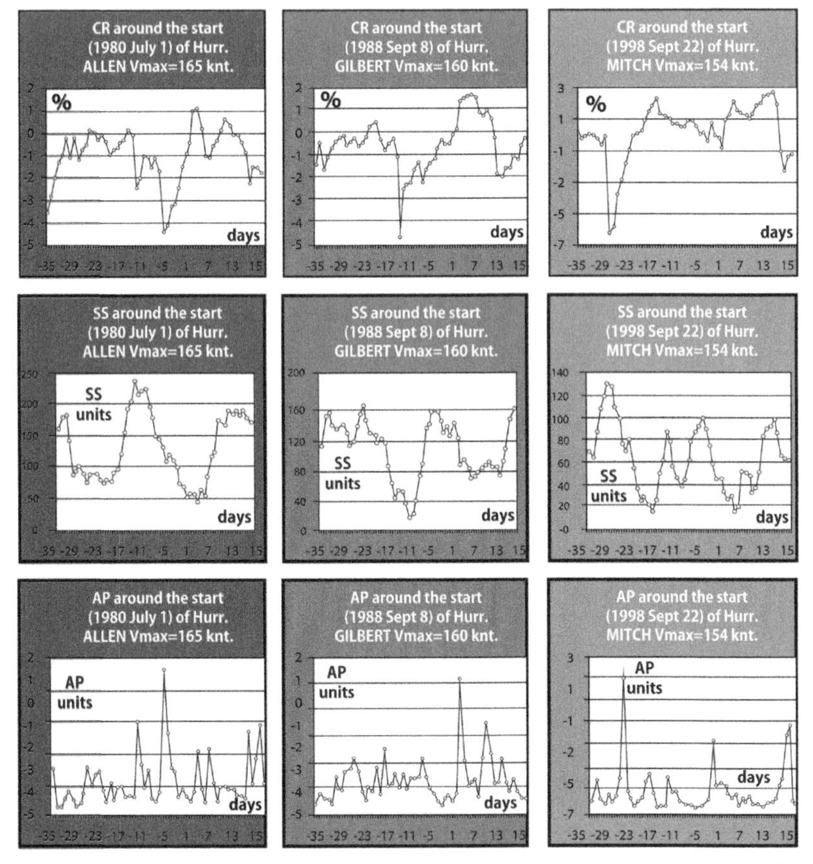

FIGURE 5.22 Time behavior of CR, SS, and AP during three intense hurricanes: Allen, Gilbert, and Mitch.

of AP, which is why they are not presented). Looking at the shape of these graphs, it can be seen that these parameters fluctuate considerably for every single cyclone not only in size, but also in appearance. To understand better these changes, an averaged curve over all years has to be obtained. This is a complex task.

A rough estimation of our parameter changes preceding the appearance of the cyclones was done dividing the whole 30 days of the preceding period in three continuous intervals (each of 10 days). The deviation (the difference between the maximum and minimum values of SS, CR, AP, K_p) in every interval was averaged for all 122 cyclones together. It was decided to separate the hurricanes into two groups (Table 5.6) for all 39 major cyclones M ($M = H5 + H4 + H3$) and for all 83 minor (small) cyclones S ($S = H2 + H1 + TS$) (Table 5.6). The results are shown on the graphs of Figure 5.23. Generally the highest SS disturbances appeared 30 to 21 days before the cyclone start, while for CR, these disturbances systematically reach their maximum in the 20- to 11-day interval. To reveal the fine structure of these changes all parallel averaged daily values of CR, SS, AP, and K_p were correspondingly arranged. The cyclones were subdivided in groups as shown in Table 5.6.

A *running average method*, smoothing the obtained curves over nine adjacent values was used. An averaged parameter over all hurricanes did not enhance the peaks, but mostly reduced them, though more pronounced disturbances could be seen before the appearance of more powerful cyclones.

The graphs obtained for hurricanes sorted depending on their category and smoothed, averaging over nine adjacent values, are shown in Figures 5.24 through 5.26. Some of the results, obtained for CR changes around the cyclone start, are shown in Figure 5.24. The same behavior of the changes for SS is observable in Figure 5.25. Investigating separately all the four parameters averaged for every one group, we obtained 84 new averaged graphs.

It is interesting that the precyclonal wave amplitude in AP and K_p values remains relatively stable before all cyclones and does not depend much on their power. But this wave slightly changes its phase with the cyclonal power. The corresponding changes of AP for three groups of cyclones and for all of them

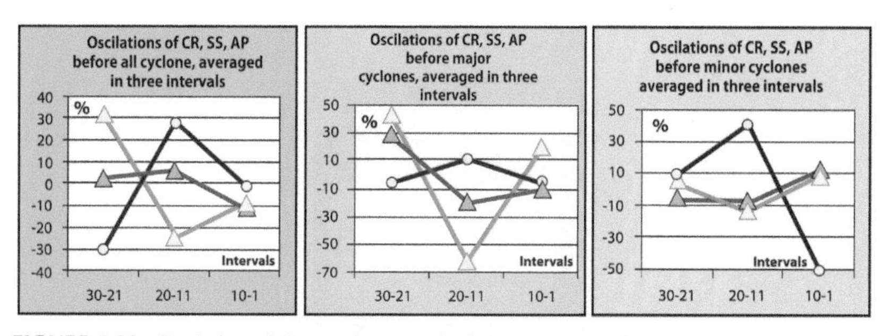

FIGURE 5.23 Deviation of the maximum and minimum values of SS (green lines), CR (blue lines), and AP (red lines) in intervals of 10 days for the average of all cyclones (122), the major cyclones (39), and the minor cyclones (83).

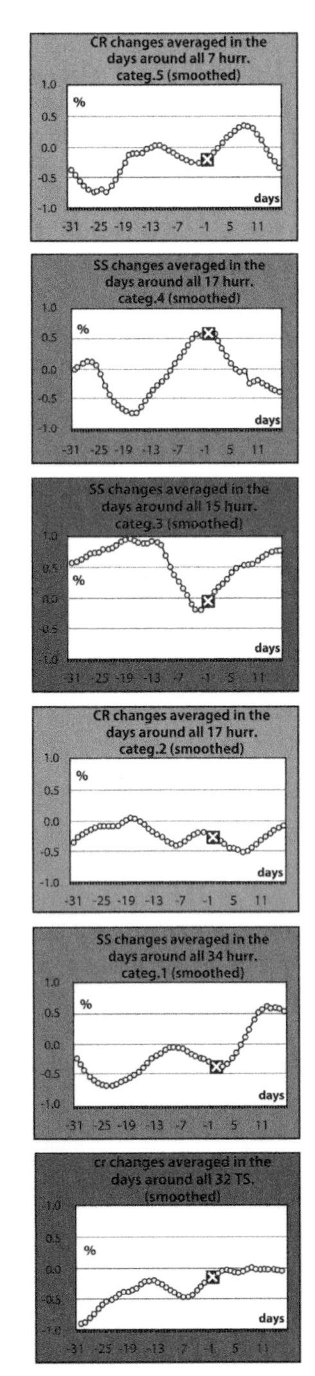

FIGURE 5.24 A running average smoothing of the obtained CR curves over nine adjacent values for hurricanes of Categories 5, 4, 3, 2, 1, and TS.

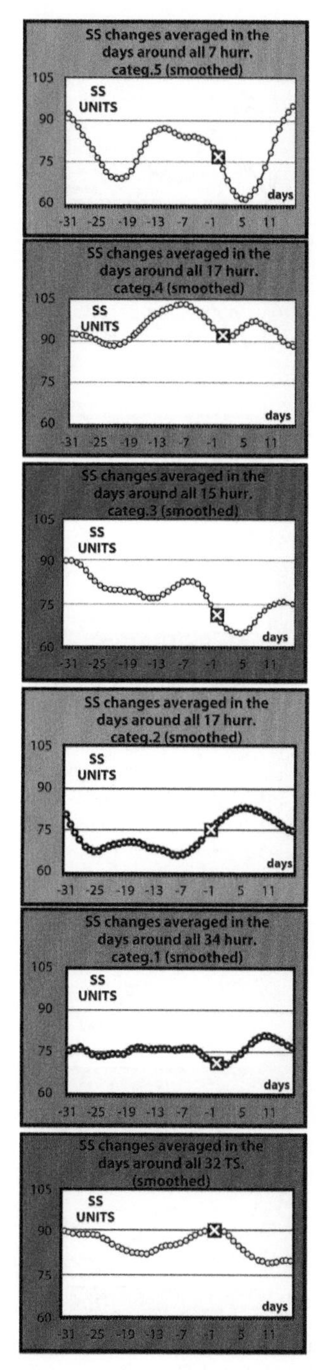

FIGURE 5.25 A running average smoothing of the obtained SS curves over nine adjacent values for hurricanes of Categories 5, 4, 3, 2, 1, and TS.

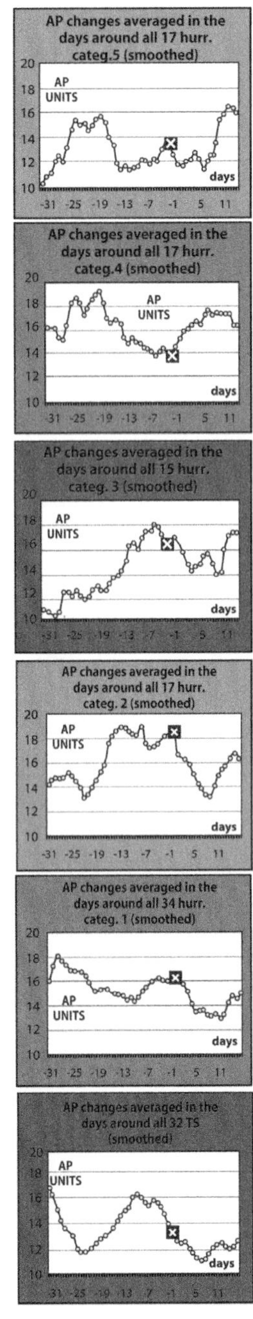

FIGURE 5.26 A running average smoothing of the obtained AP curves over nine adjacent values for hurricanes of Categories 5, 4, 3, 2, 1, and TS.

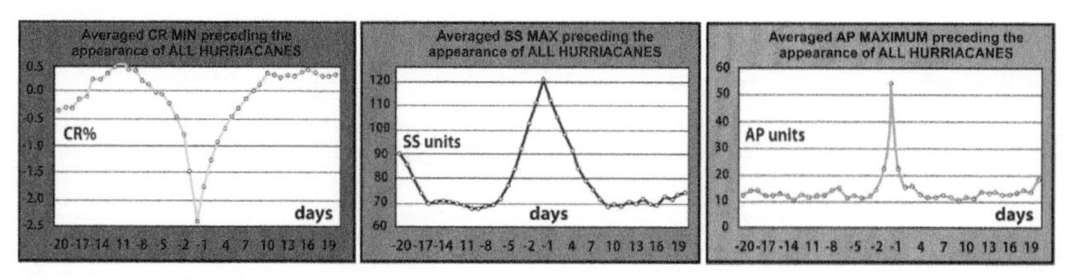

FIGURE 5.27 Overlapping of the CR, SS, and AP peaks of hurricanes of Categories 2, 3, 4, and 5 for all the 122 hurricanes in the 20-day interval before the cyclone appearance to visualize the size of the peaks measured from their basis.

together are depicted in Figure 5.26 (the changes of K_p are similar). To find the dependence of the peak parameters on the hurricane category we overlapped the highest peaks of CR, SS, and AP in Categories 5, 4, 3, 2, and in the 20-day intervals before the cyclone appearance. The graphs are shown in Figure 5.27 combining all 122 hurricane cases together, where the concept of relative maximum or minimum has been introduced to measure the size of the peaks measured from their backgrounds. Thus, on the graphs we have: for CR $-2.5 - -0.3 = -2.8\%$, for SS $120 - 70 = 50$ SS units, and for AP $55 - 12 = 43$ AP units. In Figure 5.28 these peaks were obtained separately for hurricane categories 5, 4, 3, and 2; from the graphs it was estimated the dependence of the relative peak size on the hurricane category.

The results are shown on the graphs of Figure 5.29. There, the relative peak size in the Categories 5, 4, 3, and 2 shows a well-expressed change with the

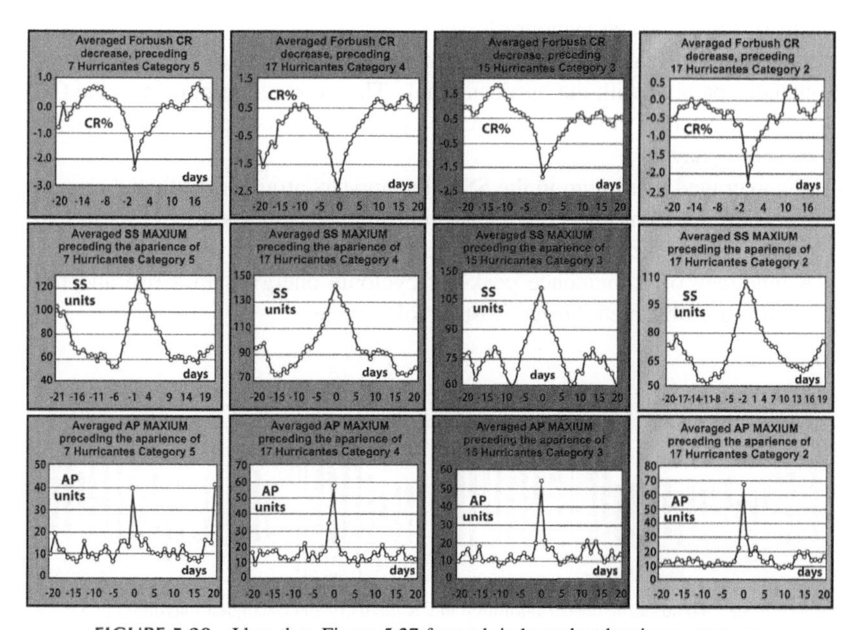

FIGURE 5.28 Idem than Figure 5.27 for each independent hurricane category.

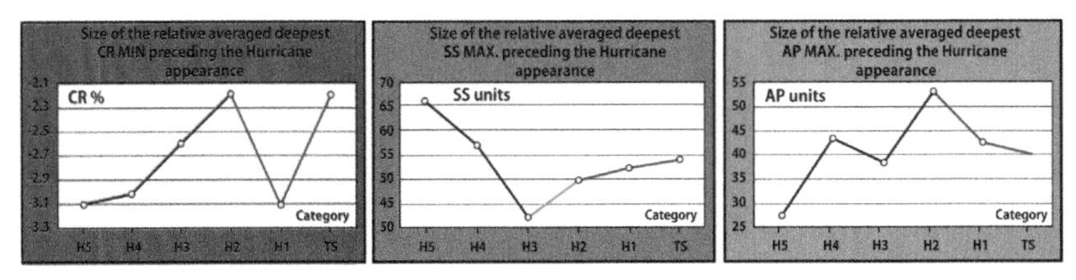

FIGURE 5.29 Dependence of the relative peak size on the hurricane category.

hurricane category. That fades away for the lowest Category 1 and for the thunderstorms (the lighter gray lines in the graphs). The trend deeper minimum in CR intensity values as well as higher maximum in SS for higher hurricane categories is within the expectation. For the analysis of the time distribution of the preceding peaks, the place of the highest maximum (for SS and AP) was located and the deepest minimum for CR in the interval of 20 days before the start of the cyclone. The distributions for these extremes are shown in Figure 5.30 in the combined graphs for all elaborated cyclones. It is interesting to notice the predominant minimum for CR at the beginning of this interval around the 18th to 19th day. The maximum for SS and AP are concentrated in two peaks around 5th and the 16th preceding days.

What is seen on the presented graphs can be summarized as follows:

- The presence of specific peaks in the CR, SS, and AP values, measured in the 20-day interval, preceding the hurricane start was confirmed. This is especially valid for the major hurricanes, but it is noticeable in considerable part of lower ranked cyclones.
- It was shown that for high ranking hurricanes the size of these parameter changes parallel with the hurricane rank. The averaged minimum in the CR values preceding the entire hurricane categories is located around the 19th preceding day.
- The averaged maximum in the SS value appears around the 5th and around the 16th preceding days. The maximum for AP appears also in these two places around the 5th and around the 16th preceding days.
- A tendency of coincidence peaks of cyclonic energy with a rise and fall periods are observed during most of solar cycles.

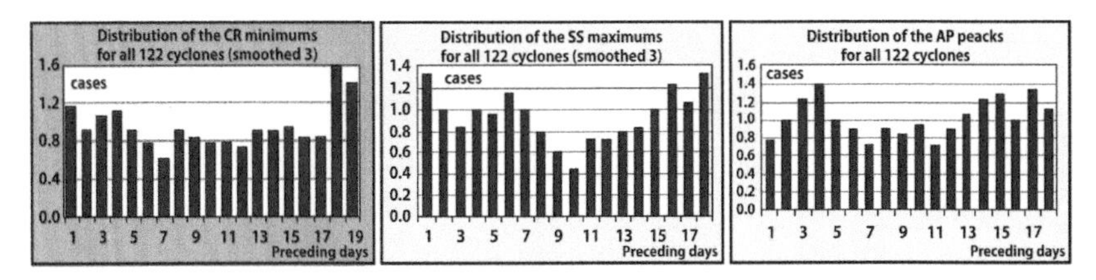

FIGURE 5.30 Trend of the deeper minimum in CR intensity and the higher maximum in SS, for higher hurricane categories.

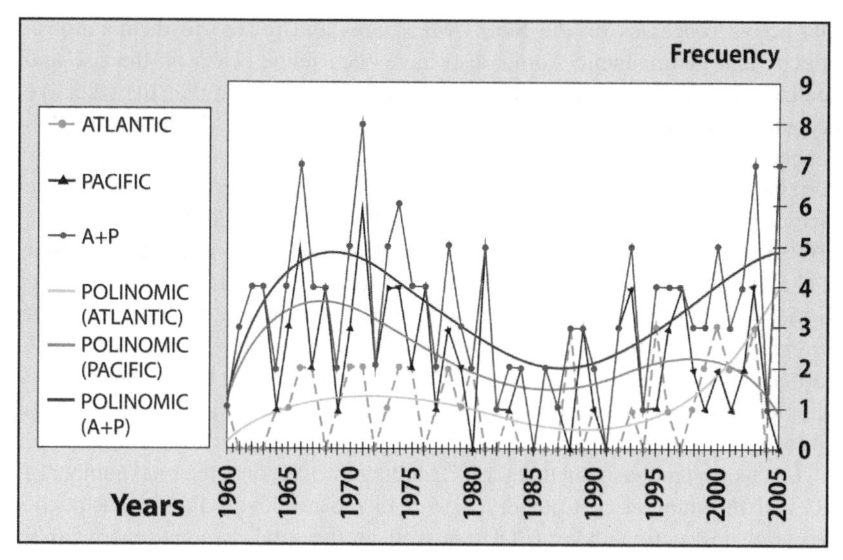

FIGURE. 5.31 Cyclical behavior of the TS landing into Mexico, 1960 to 2005. Source: *Pérez-Peraza et al., 2008a.*

In Figure 5.20, the curve presenting the yearly change of total energy (CE and CE smoothed) released from all the cyclones has an interesting form. It could be well approximated with two overlapping sinusoids: one of them with a period approximately 3.3 years, the other one about 30 years. Furthermore, based on worldwide analysis of accumulated cyclone energy (ACE), it has recently been shown (Klotzbach, 2006) the existence of a large increasing trend in tropical cyclone (TC) intensity for the North Atlantic basin and a considerable decreasing trend for the northeast Pacific. In the particular case of TC landing on Mexican coasts, Figure 5.31 shows the same tendency: a frequency increase in the Atlantic and a decrease in the northeast Pacific over the last decade.

The sum of the two basins is dominated by the Atlantic trend. It can be seen from Figure 5.20 that a similar increasing trend is followed by the total cyclone energy CE during the last years. Also, it should be noted that in both curves (Figure 5.20 and the CE smoothed curve in Figure 5.31), there is a cyclical behavior dominated by a quasi 30-year wave. It is worth mentioning that an analysis of the total number of TC and the number of Category 4 to 5 (Saffir–Simpson scale), in the northwest Pacific basin, also shows such a cyclic behavior (Webster et al., 2006), as it does the other index of TC intensity, namely the potential destruction index (PDI) (Chan, 2006).

Furthermore, the regularity exhibited between the outstanding peaks of a cyclonic energy and the rise and fall periods of the 20th solar cycle can be seen in Figure 5.20, as well as the tendency of coincidence peaks of cyclonic energy with rise and fall periods during other solar cycles. It should be noted that active processes on the Sun: CME, flares and related to them Forbush effects, and geomagnetic storms also have occurrence peaks on the rise and decline phases of a solar cycle. One would expect, then, that the selective account of cosmophysical factors would allow obtaining better correlation dependences. However, the yearly averaged SS, CR, AP, and K_p have a low correlation factor with CE, as was shown in Table 5.5. Nevertheless, among the factors that may in principle have a physical connection with tropical cyclone (TC) development, a very weak anti-correlation of -0.3 is found (Webster et al., 2006) between storm intensity and sea surface temperature (SST), similar to those of our parameters of Table 5.5. In contrast, the number of TC of Category 4 to 5 shows positive correlations, ranging from 0.55 to 0.65 with vorticity (rotational flow), vertical wind shear, and the lower tropospheric moist static energy (amount of available thermodynamic energy in the atmosphere) (Webster et al., 2006).

It is worth emphasizing that the TC landing into Mexico, the total number of TC and the number of Category 4 to 5 in the northwest Pacific basin, the potential destruction index (PDI), as well as the total cyclonal energy (CE smoothed) show a similar cyclic behavior of 30 years.

5.4. AFRICAN DUST

Great quantities of dense dust supply from the great North African and Asian deserts are often carried over huge areas of the Caribbean, the tropical North Atlantic, and the temperate North Pacific and Indian Oceans during much of the year, with different effects in those regions (see Figure 5.32). The environmental conditions of Earth, including the climate, are determined by physical, chemical, biological, and human interactions that transform and transport materials and energy. This “Earth system” is a highly complex entity characterized by multiple nonlinear responses. One important part of this system is *soil dust*, which is transported from land through the atmosphere to the oceans, affecting ocean biogeochemistry and hence having feedback effects on climate and dust production (Jickells et al., 2005).

Dust production arises from saltation and salt blasting, when winds above a threshold velocity transport soil grains horizontally, producing smaller particles, a small proportion of which get carried up into the atmosphere for long-range transport processes. These processes depend on rainfall, wind, surface roughness, temperature, topography, and vegetation cover, which are interdependent factors linked to aridity and climate in a highly nonlinear way. Such a production depends on the supply of wind-erodible material, which ironically usually requires fluvial erosion, often from adjacent highlands, followed by subsequent drying out and the loss or absence of vegetative protection. Desert dust aerosol is dominated by particles of diameter 0.1 to 10 μm, with the mean size being around 2 μm. Such aerosols have a lifetime of hours to weeks, allowing long-range transport over scales of thousands of kilometers

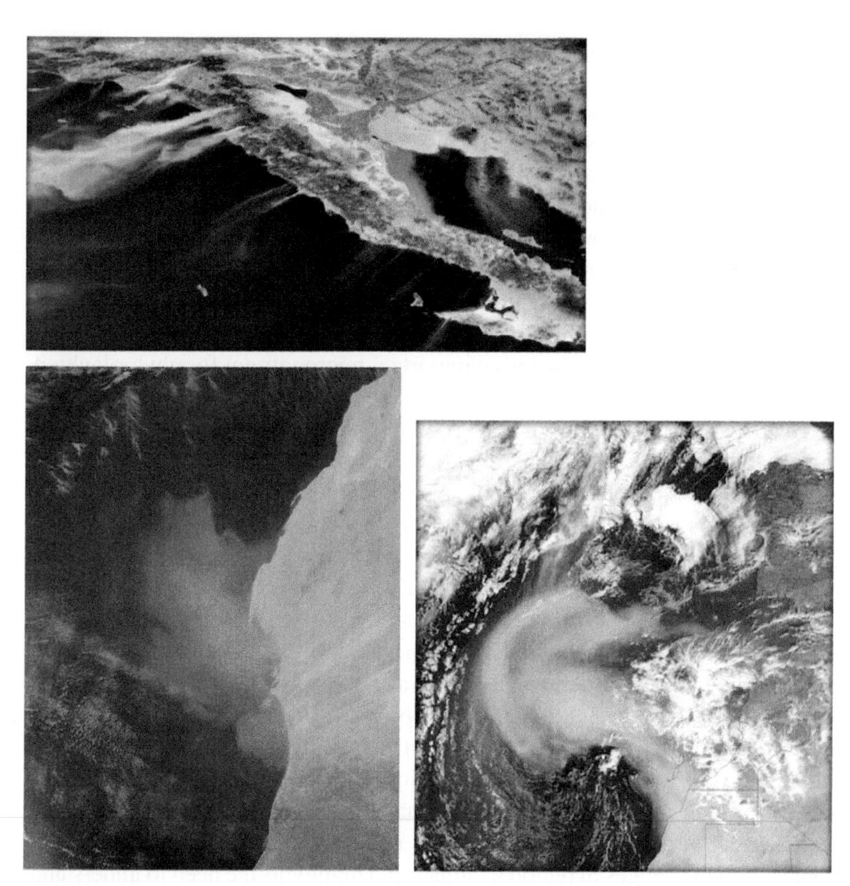

FIGURE 5.32 Typical African dust Storms of Western Africa. Source*: Earth Observatory, NASA.*

(Duce, 1995; Ginoux et al., 2001) but producing strong gradients of dust deposition and concentrations that vary substantially on time scales of ~ 1 day.

Dust winds show large interannual changes that are highly anti-correlated with rainfall in the Sudan-Sahel east African drought regions. The annual emission budgets over Sahara (north of 21.25 °N) and Sahel (south of 21.25 °N) in North Africa indicate that those from Sahara are twice the corresponding emissions from Sahel, and represent 64% of North African and 42% of global emissions. For both regions, the year 1996 has the lowest annual emission. At global scale it is know that dust concentrations were sharply lower during much of the twentieth century before 1970, when rainfall was more normal. The interannual variability of African dust transport over the north tropical Atlantic is monitored using *in situ* surface concentration measurements performed at Barbados since 1966, along with the Total Ozone Mapping Spectrometer (TOMS) and Meteosat dust optical thickness (DOT) records covering the last two decades.

Much of the transport of dust occurs at altitudes of several kilometers, with subsequent removal by wet deposition. Dust deposition estimates are in the order of 1.7×10^{15} gr year^{-1} varying substantially from year to year, with almost two-thirds from North Africa and 26% of the dust reaching the oceans. Dust production, transport, and deposition to the oceans depend on climatic factors, particularly atmospheric structure, which regulates uplift, and wind speed and precipitation, which influence removal. Over large areas of the Earth, the atmospheric aerosol composition is dominated by mineral dust. Dust storms and dust plumes are the most prominent, persistent, and widespread aerosol features. The great variability of African dust transport has broader implications. Iron associated with dust is an important micronutrient for phytoplankton (Falkowski et al., 1998).

Thus, variations in dust transport to the oceans could modulate ocean primary productivity and, consequently, the ocean carbon cycle and atmospheric CO_2. Dust could play a positive role in reducing global warming by greenhouse gas CO_2. Carbon fixation by phytoplankton in the oceans acts as a sink for CO_2. Aeolian dust deposition is the primary source of bio-available iron in the iron-limited open oceans and effectively controls phytoplankton blooming (Martin and Gordon, 1988). Another important effect of dust particles is their role in the photochemical production of tropospheric ozone by reducing the photolysis rates by as much as 50% (e.g. Dickerson et al., 1997; Liao et al., 1999; Martin et al., 2002) and by providing reaction sites for ozone and nitrogen molecules (e.g. Prospero et al., 1995; Dentener et al., 1996). Additionally, dust particles affect air quality (Prospero, 1999) and are potential vectors for long-range transport of bacteria.

The great variability in dust transport demonstrates the sensitivity of dust mobilization to changes in regional climate and highlights the need to understand how dust, in turn, might affect climate processes on larger scales; mineral dust, emitted by wind erosion of arid and semiarid areas of the Earth, is thought to play an important role in climate forcing. However, it has been difficult to quantify because of the relatively complex and highly uncertain effect of dust on radiative forcing (Intergovernmental Panel on Climate Change (IPCC), 2001; Sokolik et al., 2001). Because of the great sensitivity of dust emissions to climate, future changes in climate could result in large changes in dust emissions from African and other arid regions that, in turn, could lead to impacts on climate over large areas.

To understand the forcing involved in past climate trends and to improve estimates of future dust-related forcing, it is necessary to characterize the variability of dust emissions in response to climate-change scenarios and to distinguish between natural processes and human impacts. Aerosols, including mineral dust, can affect climate directly by scattering and absorbing solar radiation and indirectly by modifying cloud physical and radiative properties and precipitation processes (Kaufman et. al., 2002). Theoretical and experimental studies have shown that the effect of the mineral dust component in the atmospheric radiation budget is comparable to the greenhouse effect gases but opposite in sign. In fact, dust can have either a net positive or negative radiative effect depending on the surface albedo and the aerosol single scattering albedo (Liao and Seinfeld, 1998).

The Saharan Dust Experiment (SHADE) experiment (Tanré et al., 2003), held off the coast of West Africa in September 2000, shows that the net radiative impact of African dust, if extrapolated to all sources of the entire Earth, would be approximately $-0.4\ W\ m^{-2}$. However, because of the complexities of the competing solar and terrestrial radiative forcings, even the sign of the net effect is unknown (IPCC, 2001), suggesting a dust radiative forcing in the range $+0.4$ to $-0.6\ W\ m^{-2}$. Mineral dust may also exert an indirect radiative effect by modifying cloud properties and precipitation process (Rosenfeld et al., 2001; Sassen et al., 2003).

The quantification of the dust impact on climate change is, however, particularly uncertain because of the lack of knowledge about the natural variability of dust emissions and the temporal and spatial variability of transported dust. Another major uncertainty is due to the lack of reliable estimates of the anthropogenic fraction of mineral dust in the atmosphere (Haywood and Boucher, 2005). A recent estimate of this anthropogenic fraction using climate models is of about 10% of the global dust load (Tegen et al., 2004).

Several studies have shown that dust particles, by absorbing and scattering solar radiation, modify the atmospheric radiative budget (e.g., Tegen et al., 1996; Sokolik and Toon, 1996; Weaver et al., 2002). Nonetheless, dense dust clouds over the oceans reduce insolation at the ocean surface, thereby reducing the heating of ocean surface waters (Diaz et al., 2001) and sea-surface temperatures, which in turn affects the ocean-atmosphere transfer of water vapor and latent heat, which are important factors in climate (Lelieveld et al., 2002). Reduced heating over the tropical Atlantic could contribute to the interhemispheric, tropical Atlantic, sea-surface temperature anomaly patterns that have been associated with Sudan-Sahel drought (Lamb and Peppler, 1992; Ward, 1998). Thus, increased dust could conceivably lead to more intense or more prolonged drought. Dust could also affect climate through cloud microphysical processes, possibly suppressing rainfall and conceivably leading to the perpetuation and propagation of drought (Rosenfeld et al., 2001).

Over south Florida, clouds are observed to glaciate at relatively warm temperatures in the presence of African dust (Sassen et al., 2003), an effect that could alter cloud radiative processes, precipitation, and cloud lifetimes. The frequency and intensity of Atlantic hurricanes have been linked to East African rainfall (Landsea et al., 1992), showing decreased activity during dry phases. Although there is no evidence that exposure to dust across this region presents a health problem, it does demonstrate how climate processes can bring about changes in our environment that could have a wide range of consequences on intercontinental scales. It is thus important to understand the long-term variability of dust distribution, to determine which processes are controlling such variability.

Hurrell (1995) has shown that the circulation and precipitation over Europe and the North Atlantic is modulated by the North Atlantic Oscillation (NAO) with a period of about 8 years. Ginoux et al. (2004) found that in winter a large fraction of the North Atlantic and Africa dust loading is correlated with the North Atlantic Oscillation (NAO) index. They show that a controlling factor of such correlation can be attributed to dust emission from the Sahel. The Bodélé depression is the major dust source in winter and its interannual variability is highly correlated with the NAO. Studies based on Meteosat/visible light spectrometer (VIS) and Total Ozone Mapping Spectrometer (TOMS) observations (22 years from 1979 to 2000) have established the link between Sahel drought, dust emissions in Sahel, and summer dust export over the Atlantic (Moulin and Chiapello, 2004) and have shown the role of the North Atlantic Oscillation (NAO) on winter dust transport (Chiapello and Moulin, 2002). Dust export to the Atlantic (Moulin and Chiapello, 2004) and to Barbados (Prospero and Lamb, 2003) are most highly correlated with Sahel rainfall of the previous year (i.e., the rainy season preceding the dust occurrence).

The significant correlations obtained between interannual variability of summer surface dust concentrations at Barbados, and that of TOMS/DOT over both Sahel and northeastern tropical Atlantic suggest that Sahel sources significantly contribute to the dust transport over the western Atlantic. Thus the Sahelian region, if not of first importance in terms of intensity of dust emissions compared to the Saharan sources (Prospero et al., 2002), is probably critical in controlling the year-to-year variability of dust export, which should allow progress in the understanding of the mechanisms of influence of these climatic parameters, and provide more accurate estimates of the anthropogenic fraction of mineral dust. In spite of coverage limitations of TOMS and DOT, their agreement among them, and with different kinds of ground-based measurements (aerosol optical thickness) (Moulin and Chiapello, 2004) and mineral dust concentrations (Chiapello et al., 1999) gives a picture of dust dynamics with a reasonable confidence level.

Finally, it should be emphasized that the frequency and intensity of Atlantic hurricanes have been linked to east African rainfall (Landsea et al., 1992), showing decreased activity during dry phases.

5.5. RELATIONSHIP BETWEEN NORTH ATLANTIC HURRICANE ACTIVITY AND AFRICAN DUST OUTBREAKS

The recent increase since 1995 in Atlantic tropical cyclones (including both hurricanes and tropical storms) affecting North America has raised the awareness of their impact on society and the economy. Currently, there is a debate surrounding the cause of this observed increase in cyclone activity. Several recent studies have explored the relationship between long-term trends in tropical cyclone activity (either in terms of their number or intensity) and environmental factors that may or may not be influenced by global warming (Emanuel,

2005a,b; Landsea, 2005; Trenberth, 2005; Webster et al., 2005). Other studies, however, have concluded that different environmental factors—not necessarily related to global warming—control trends in cyclone activity (Goldenberg et al., 2001; Knutson and Tuleya, 2004).

The role of atmospheric dust as a possible contributor to changing North Atlantic tropical cyclone activity was suggested by Dunion and Velden (2004), who showed that tropical cyclone activity may be influenced by the presence of the Saharan Air Layer, which forms, as previously mentioned, when a warm, well-mixed, dry and dusty layer over West Africa is overimposed to the low-level moist air of the tropical North Atlantic (Carlson and Prospero, 1972; Prospero and Carlson, 1981). In fact, historical data indicates that Saharan dust may have a stronger influence than El Niño on hurricane statistics in the subtropical western Atlantic/Caribbean region, while El Niño influence may be stronger in the tropical eastern Atlantic.

It was also mentioned in the previous sections that Atlantic tropical cyclone activity varies strongly over time, and that summertime dust transport over the North Atlantic also varies from year to year, but any connection between tropical cyclone activity and atmospheric dust has been only recently examined. Evan et al. (2006) reported a strong relationship between interannual variations in North Atlantic tropical cyclone activity and atmospheric dust cover as measured by satellite, for the years 1982 to 2005 (Figure 5.33).

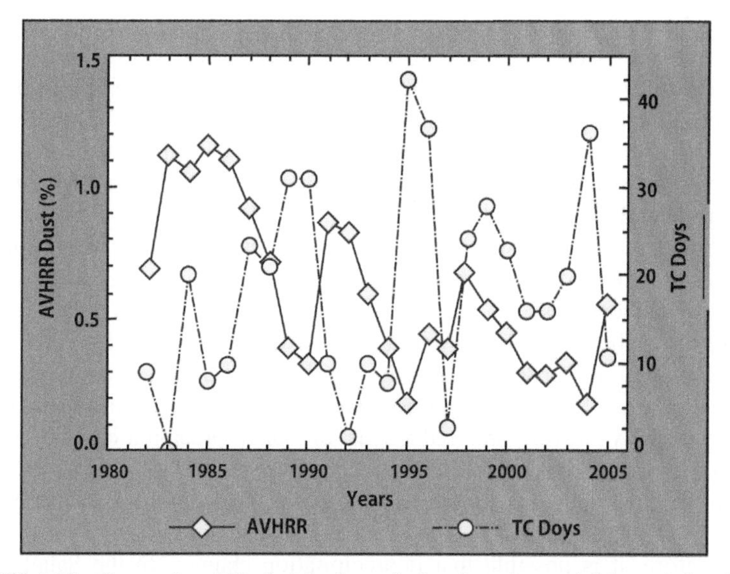

FIGURE 5.33 Time series of North Atlantic tropical cyclone days and Saharan-induced dust cover for 1982–2005. The black line and diamonds represent the detected dust cover and the gray line and circles represent the tropical cyclone days, over the region of 0–30 °N and 15–60 °W for the time period of August 20 through September 30. Source: *After Evan et al., 2006.*

Due to the fact that dust observations are a good proxy for the Saharan Air Layer, Evan et al. (2006) showed that the contrast between the presence of dust and the lack of tropical cyclone activity for 1983 and 1985, in particular the dust activity during the 1980s, was more intense than during any other period on record; the appearance of cyclone days where dust is lacking in the years 1995 and 2004, suggests an inverse correlation between dust and tropical cyclone activity, consistent with the hypothesis of Dunion and Velden (2004). Evan et al. (2006) suggest the possibility that the intense activity of the Saharan Air Layer indicates the presence of an environment less conducive to deep convection and tropical cyclogenesis, whereas the lack of Saharan Air Layer activity demonstrates the opposite situation. It can be speculated that the anomalous low hurricane activity during the 1970s and 1980s was due to a general high dust activity. In fact, that is true during the 1980s.

A correlation coefficient of ~0.51 is observed between tropical cyclone and dust activities time series, significant at ~99% during the last decade (Goldenberg et al., 2001). Sea surface temperature is important in shaping the interannual variability of North Atlantic tropical cyclones (Goldenberg et al., 2001; Landsea et al., 1999). However, over at least the last 26 years regional tropical cyclone activity and sea surface temperature exhibit an upward trend, while dust activity shows a downward one. The partial correlation coefficients of the detrended time series are both −0.50, significant at 98.5%, implying that Saharan dust activity can account for variance in the tropical cyclone record that cannot be attributed to ocean temperature.

The Accumulated Cyclone Energy (ACE) time series for the tropical Atlantic is also well correlated with dust cover series, with a correlation coefficient of 0.59, significant at 99.5%, possibly reflecting the effects of the Saharan Air Layer on cyclone intensity as well as genesis, as suggested by Dunion and Velden (2004). ACE index is defined as the sum of the squares of the maximum sustained surface wind speed (knots) measured every 6 hours for the Hurricane Best Track Files (Jarvinen et al., 1984).

Although the mean dust coverage and tropical cyclone activity are strongly (inversely) correlated over the tropical North Atlantic, this does not provide conclusive evidence that the dust itself is directly controlling tropical cyclone activity. It has been mentioned that a link exists between Sahel precipitation and North Atlantic hurricanes: increases in Sahel precipitation are thought to cause increases in North Atlantic hurricane activity through enhancement of African easterly waves, and reductions in Sahel precipitation and North Atlantic hurricane activity have been tied together through the associated changes in wind shear across the Atlantic basin (Gray, 1990; Landsea et al., 1992).

Therefore, it is possible that if precipitation changes in the Sahel alter West African dust outbreaks, then this variability in rainfall may be the cause of our observed correlations. However, it has been shown that, at least for the summertime months, interannual changes in dustiness over the North

Atlantic are related to changes in Sahel precipitation from the previous year and are not strongly correlated with same-year Sahel precipitation events (Moulin and Chiapello, 2004). Thus, Evan et al. (2006) suggest that because dust is a good tracer for the Saharan Air Layer, these observed correlations may result from the effect of the Saharan Air Layer acting as a control on cyclone activity in the tropical Atlantic, consistent with the hypotheses of Dunion and Velden (2004).

It is worth noting that the variability in the dust time series may not only reflect variations of the presence of the Saharan Air Layer, but it may also reflect changes in dust loadings within the Saharan Air Layer itself, which could also have important meteorological implications. It is interesting to note that a 5- to 8-year oscillating behavior is also seen in the dust record, superimposed over a downward trend in dustiness.

In contrast to the dynamic effects of the Saharan Air Layer (SAL) in suppressing cyclogenesis, Lau and Kim (2007a) suggest an extensive cooling over the subtropical North Atlantic may be related to the shielding of solar radiation (the so-called solar dimming effect) by dust. They exemplify with the particular behavior of the hurricane season in 2006, when it was expected a continuation of the trend of 9 preceding years of above- normal hurricane seasons, however, the 2006 hurricane season was near normal with 4 tropical storms and 5 hurricanes, but decidedly fewer as compared with the record numbers of 12 tropical storms and 15 hurricanes in 2005, including Katrina. Given the recent warming tendency in the Atlantic and the prevailing favorable preseason conditions, sea-surface temperature (SST) was above normal, vertical wind shear was low, and sea-level pressure was reduced over the tropical Atlantic at the beginning of the season, and all signs indicated that it would be more active than the 2005 season.

Though overall, the dust loading in 2006 was higher than in 2005, by the beginning of June 2006 there was a major increase in dust loading, and 2 weeks later a major episode of SST cooling occurred, concomitant with a long-term warming trend. Even if the 2006 SST in the Atlantic Ocean was above normal compared with the long-term climatology, there was an abrupt cooling of the Atlantic from 2005 to 2006 independently of the long-term variation (Figure 5.34). The most pronounced SST cooling began in mid-June 2006, reaching a maximum in late June and mid-July, until the end of September after which the SST returned to the 2005 level. The cooling was widespread, covering most of the subtropical and equatorial North Atlantic, with the strongest signal (~0.6–0.8 °C) over the WAC (western Atlantic and Caribbean).

The SST cooling of the North Atlantic appeared to be closely related to the variation of Sahara dust over the region: a substantial increase in atmospheric dust loading covering nearly all the northern tropical and subtropical Atlantic and western Africa with the oceanic maximum over the western Atlantic and Caribbean, where the negative SST anomaly was most

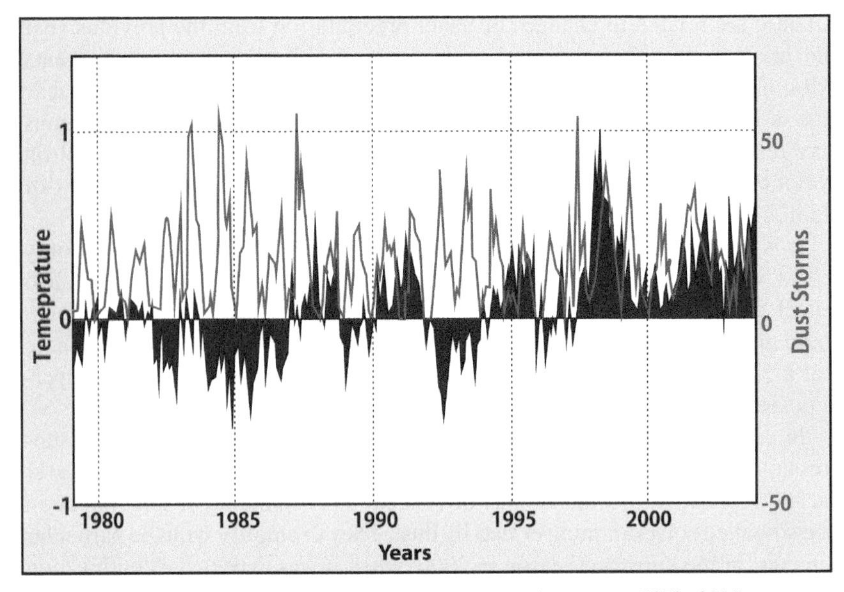

FIGURE 5.34 Superficial sea temperature versus dust storms (1979–2004).

pronounced. The major dust episode lasted for about a month, until the end of June. In 2006, no hurricanes were found over the WAC and the Gulf region. Therefore, it can be seen in Figure 5.34 that AD storms could be seen as a cooling factor of the Atlantic Ocean's superficial waters except during the periods 1990 to 1992, 1994 to 1996, and 1998 to 2004, when both signals AD and SST increased.

Evan et al. (2006) have contested the role of Sahara dust in triggering a series of rapid feedback processes in the ocean-atmosphere system resulting in unfavorable conditions for hurricane formation in the Atlantic in 2006. The main point is, how much initial dust radiative forcing is enough to trigger the feedback process? By an analysis of the aerosol optical thickness of the region in 2005 and 2006, Lau and Kim (2007b) estimated an increase of 28 to 30% in 2006 relative to 2005, which translated to a reduction of surface solar short-wave radiation flux of 4.3 to 8 watts m^{-2} Such values could explain the required forcing.

Whether there is a direct or indirect link remains elusive, since no direct causality has yet been established. Some authors suggest the variability in dust (and variability in the presence of the Saharan Air Layer), and others claim a solar dimming effect by Saharan dust as linking mechanisms for the changes in North Atlantic tropical cyclone activity. However, if up to the present there is not a conclusively direct causal relationship, there is conclusive evidence of robust link between tropical cyclone activity and dust transport over the tropical Atlantic.

5.6. SIGNAL THEORY AS A TOOL TO FIND COMMON CONNECTIONS AMONG DIFFERENT PHENOMENA[2]

We have mentioned in the previous sections some work that, by means of a correlational analysis, seems to indicate that certain extraterrestrial phenomena could have some kind of relation with the terrestrial phenomena, and particularly with the occurrence of hurricanes (Elsner and Kavlakov, 2001; Kavlakov, 2005; Kavlakov et al., 2008a,b,c). It is even speculated that these kinds of connections could seat the basis of deeper studies to use the results as indicators of hurricanes precursors. To give those results a higher meaning, it is convenient to carry out spectral studies of different times series to delimitate with more preciseness the existence of those potential relationships. With this goal in mind we use signal theory tools, searching to determine the most prominent signals between North Atlantic hurricanes of all categories, the sea-surface temperature (SST), the Atlantic Multidecadal Oscillation (AMO), African dust (AD) for one side, and for other side, solar activity (SS) and galactic cosmic rays (GCR).

Such analysis leads to establishing the evolution in frequency and time, as well as the phase between two time series of those phenomena, allowing us to infer the nature of any connection among them—that is, to find incident cosmophysical periodicities that may modulate terrestrial phenomena. As the role of the Sun in modulating these phenomena has not been clarified, it requires further assessment. In the next sections we describe the investigations done on the behavior of the main common periodicities among the AMO, SST, AD, solar activity phenomena (SS), galactic cosmic rays (GCR), and hurricanes.

Here, as before in the correlational analysis, it is assumed that if there is a good interconnection between the studied terrestrial phenomena and hurricanes, there is a good interconnection between these terrestrial phenomena and cosmophysical phenomena, therefore it should be also a good interconnection between hurricanes and cosmophysical phenomena.

5.6.1. Methods of Analysis

The simplest and most widely known technique to investigate common periodicities between two series of data is the Fourier transform, the fast Fourier transform, and regression analysis. However, while useful for stationary time series, these methods are not the best for time series that are not of stationary nature (Hudgins et al., 1993; Torrence and Compo, 1998; Torrence and Webster, 1999), as those analyzed in this work. A good description of the limitations and drawbacks of Fourier analysis (including the short-time-Garbor

[2] In collaboration with Dr. Victor Velasco Herrera, Instituto de Geofísica, Universidad Nacional Autónoma de México, México.

or the windowed Fourier transform, and the kind of wavelet transform was given by Polygiannakis et al. (2003).

In contrast to those methods, one of the most powerful tools to work with nonstationary series in signal theory is the so-called *wavelet spectral analysis*: within this context to find the time evolution of the main frequencies within a simple nonstationary series at multiple periodicities the *Morlet wavelet technique* is a useful technique for analyzing localized variations of power (Torrence and Compo, 1998; Grinsted et al., 2004). A way to analyze two nonstationary time series to discern whether there is a lineal or nonlinear relation is by means of the *coherence wavelet method* which furnishes valuable information about when and which periodicity coincides in time, and then about its nature, lineal or nonlinear relation between the given series (for instance, solar and terrestrial phenomena), provided there is not a noticeable diphase among them. The *wavelet coherence* is especially useful in highlighting the time and frequency intervals where two phenomena have a strong interaction.

5.6.1.1. The Morlet Wavelet Analysis

To analyze local variations of power within a single nonstationary time series at multiple periodicities, such as the dust or hurricanes series, we apply the wavelet (WT) using the Morlet wavelet (Grinsted et al., 2004). The Morlet wavelet consists of a complex exponential modulated by a Gaussian $e^{i\omega_o t/s}\ e^{-t^2/(2s^2)}$ where t is the time, $s = 1$/frequency is the wavelet scale, and ω_o is a nondimensional frequency. Here $\omega_o = 6$ is used to satisfy the admissibility condition (Farge, 1992). Torrence and Compo (1998) defined the wavelet power $|W_n^X|^2$, where W_n^X is the wavelet transform of a time series X and n is the time index. The power spectra for each one of the parameters described in the study was computed using a Morlet wavelet as a mother wave.

For the Morlet wavelet spectrum, the *significance level* is estimated for each scale, using only values inside the *cone of influence* (COI). The COI is the region of the wavelet spectrum where edge effects become important: it is defined as the e-folding time for the autocorrelation at each scale of the wavelet power. This e-folding time is chosen such that the wavelet power for a discontinuity at the edge drops by a factor e^{-2}, and ensures that the edge effects are negligible beyond that point (Torrence and Compo, 1998).

Wavelet power spectral density (WPSD) is calculated for each parameter; the contour (mask) of this cone defines the interval of 95% confidence, that is, within the COI. To determine significance levels of the global wavelet power spectrum, it is necessary to choose an appropriate background spectrum.

5.6.1.2. The Coherence-Cross Wavelet Analysis

For analysis of the covariance of two time series X and Y, such as the dust and hurricanes series, we used the cross-wavelet $W_k^{XY}(\psi)$ (*XWT*), which is a measure of the common power between the two series. The cross wavelet analysis was introduced by Hudgins et al. (1993). Torrence and Compo (1998) defined the cross wavelet spectrum of two time series X_1 and X_2, with wavelet transforms $(W_n^{X_1})$ and $(W_n^{X_2})$, as $W_n^{X_1,X_2} = W_n^{X_1} W_n^{X_2}$, where (*) denotes complex conjugation. The cross wavelet energy is defined as $|W_n^{X_1,X_2}|$. The complex argument is the local relative phase between X_1 and X_2 in time-frequency space. Torrence and Webster (1999) defined the cross-wavelet power as $|W_n^{X_1,X_2}|^2$. The *coherence* is then defined as the cross-spectrum normalized to an individual power spectrum. It is a number between 0 and 1, and gives a measurement of the cross-correlation between two time-series and a frequency function.

Statistical significance level of the wavelet coherence is estimated using Monte Carlo methods with red noise to determine the 5% significance level (Torrence and Webster, 1999). The coherence significance level scale appears at the bottom of the figures of the next sections.

5.6.1.3. The Wavelet-Squared Transformer Coherence Analysis

The wavelet-squared transform coherency $R_k^2(\psi)$ (*WTC*) is especially useful in highlighting the time and frequency intervals, when the two phenomena have a strong interaction (Torrence and Compo, 1998; Torrence and Webster, 1999). The wavelet square coherency $R_n^2(s)$ is defined as the absolute value squared of the smoothed cross-wavelet spectrum (*XWT*), normalized by the smoothed wavelet power spectra. Unlike the cross wavelet power, which is a measure of the common power, the wavelet square coherency is a measure of the intensity of the covariance of the two series in time-frequency space (Torrence and Compo, 1998).

The *WTC* measures the degree of similarity between the input (*X*) and the system output (*Y*), as well as the measure of the consistency of the output signal (*X*) due to the input (*Y*) for each frequency component. It is used to identify frequency bands within which two time series are covarying. By definition the condition $0 \leq R_n^2 \leq 1$ is fulfilled. The sub index (*n*) indicates the corresponding time interval for the evaluation. $R_k^2(\psi) = 1$ indicates that all frequency components of the output signal (*Y*) correspond to the input (*X*). If $R_k^2(\psi) << 1$ or $R_k^2(\psi) \approx 0$ then output *Y* is not related to input *X*, because of the presence of noise, nonlinearities, and time delays in the system.

5.6.1.4. The Wavelet Coherence Transformer Signal/Noise Analysis

The coherence of the system can be calculated through the relation signal/noise ($WTC_{s/n}$) given in Velasco et al. (2010) as

$$R^2_{s/n\ k}(\psi) = \frac{R_k^2(\psi)}{1 - R_k^2(\psi)} \tag{5.1}$$

Also, for definition of $WTC_{s/n}$,

$$0 \leq R_{s/n}^{2}{}_{k}(\psi) \leq 1$$

The $WTC_{s/n}$ is just what we are using in this work, Equation (5.1), because it allows us to find linear and nonlinear relationships, while verifying that the periodicities of cross-wavelet are not spurious, to minimize the effects of noise.

If the *XWT* and the $WTC_{s/n}$ of two series are high enough, the arrows in the *XWT* and $WTC_{s/n}$ spectra show the phase between the phenomena. The arrows in the coherence spectra show the phase between the phenomena: arrows at 0 ° (pointing to the right) indicate that both time series are correlated (in phase) and arrows at 180 ° (pointing to the left) indicate that they are anti-correlated (in anti-phase).

It is important to point out that these two cases imply a linear relation between the considered phenomena. Nonhorizontal arrows indicate an out-of-phase situation, meaning that the two studied phenomena do not have a linear relation but a more complex relationship. As in the Morlet wavelet, the 95% confidence level of the coherence is inside the black contour (the COI).

5.6.1.5. The Wavelet Global Spectra

On the right blocks of figures of the next sections, are shown the global spectra

$$GXWT(\psi) = \sum_{k} \langle W_{k}^{XY}(\psi) \rangle \tag{5.2}$$

and

$$GWTC_{s/n}(\psi) = \sum_{k} \left| \left\langle R_{s/n}^{2}{}_{k}(\psi) \right\rangle \right| \tag{5.3}$$

which is an average of the power of each periodicity in both the wavelet and the coherence spectra (*XWT* and *WTCs/n*). It is usually used to notice, at a glance, the global periodicities of either the time series, or of the coherence analysis results. The significance level of the global wavelet spectra is indicated by the dashed curves, which refer to the power of the red noise level at the 95% confidence, which increases with decreasing frequency (Grinsted et al., 2004). It is a way to show the power contribution of each periodicity inside the COI, delimiting the periodicities that are on, or above, the red noise level. The uncertainties of the periodicities of both global wavelet and coherence spectra are obtained at the half maximum of the full width peak.

5.7. COHERENCE ANALYSIS OF TERRESTRIAL AND COSMOPHYSICAL FORCINGS

To assess the long-term relations between space phenomena and indicators of the global climate it is often necessary to use reconstructions of galactic cosmic rays (CR), solar activity (SS), and climate phenomena. Direct measurements of solar activity based on sunspot numbers exist from 1749, but trustable CR data is only available since the 1950s when the neutron monitor stations began to operate.

Records of climatic phenomena exist from the end of the nineteenth century: for the AMO the annual time series between 1851 and 1985 and data from the World Data Center for Paleoclimatology is used (*http://www.appinsys.com/globalwarming/AMO.htm*). For solar activity we use the daily number of sunspots (SS): *http://www.ngdc.noaa.gov/nndc/struts/results?t=102827&s=5&d=8,430,9*. Data on sea-surface temperature in the Atlantic are given in the literature as temperature anomalies: for the period 1851 to 2005 (Figure 5.35); we use here data from *http://badc.nerc.ac.uk/view/badc.nerc.ac.uk__ATOM__dataent_11704341770613707.*

Concerning the ^{10}Be, there is a polemic about whether it can be considered as a *proxy* of galactic cosmic rays (see for instance Stozhkov et al., 2004; Stozhkov, 2007). Nevertheless, a good number of researchers still support the use of ^{10}Be as a proxy of CR (e.g., Wagner et al., 2000; McCracken, 2001; McCracken and MacDonald, 2001; Usokin et al., 2005; Usokin and Kovaltzov, 2007) and so on. Under this last context we have considered the ^{10}Be concentration in the Dye 3 ice core (65.2 °N, 43.8 °W, 2477 m altitude) from Beer et al. (1990), for which

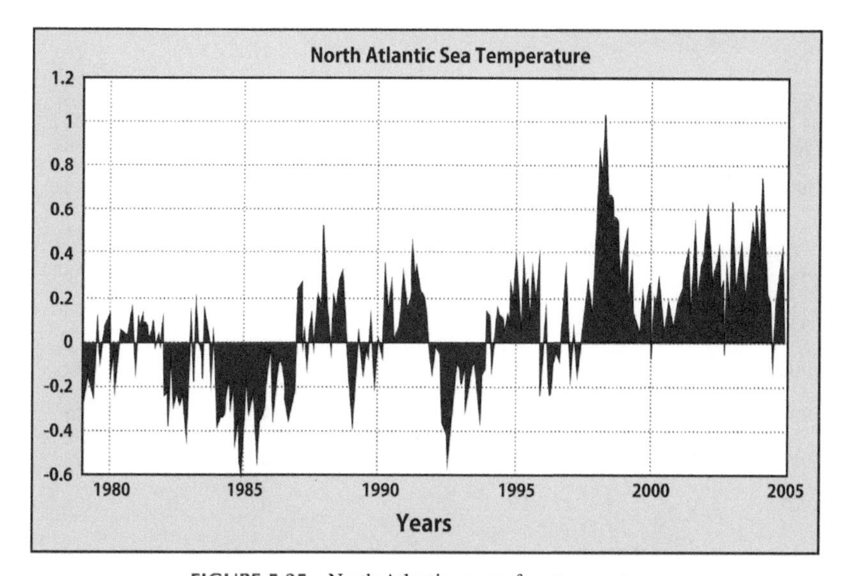

FIGURE 5.35 North Atlantic sea surface temperature.

TABLE 5.7 Data of Cyclones by Annual Number and per Category

Year	5	4	3	2	1	TS	ST	Total
1965	0	1	0	1	2	2	0	6
1966	0	1	2	0	4	4	0	11
1967	1	0	0	1	4	2	0	8
1968	0	0	0	0	4	3	0	7
1969	1	0	5	2	4	6	0	18
1970	0	0	2	1	2	5	0	10
1971	1	0	0	1	4	7	0	13
1972	0	0	0	1	2	1	3	7
1973	0	0	1	0	3	3	1	8
1974	0	1	1	1	1	3	4	11
1975	0	1	2	2	1	2	1	9
1976	0	0	2	2	2	2	2	10
1977	1	0	0	0	4	1	0	6
1978	0	2	0	1	2	6	0	11
1979	1	1	0	1	2	3	1	9
1980	1	0	1	3	4	2	0	11
1981	0	1	2	1	3	4	1	12
1982	0	1	0	0	1	3	1	6
1983	0	0	1	0	2	1	0	4
1984	0	1	0	1	3	7	1	13
1985	0	1	2	0	4	4	0	11
1986	0	0	0	1	3	2	0	6
1987	0	0	1	0	2	4	0	7
1988	1	2	0	0	2	7	0	12
1989	2	0	0	2	3	4	0	11
1990	0	0	1	2	5	6	0	14
1991	0	1	1	1	1	4	0	8

TABLE 5.7 Data of Cyclones by Annual Number and per Category—cont'd

Year	5	4	3	2	1	TS	ST	Total
1992	1	0	0	2	1	2	1	7
1993	0	0	1	1	2	4	0	8
1994	0	0	0	1	2	4	0	7
1995	0	3	2	2	4	8	0	19
1996	0	2	4	0	3	4	0	13
1997	0	0	1	0	2	4	1	8
1998	1	1	1	4	3	4	0	14
1999	0	5	0	3	0	4	0	12
2000	0	2	1	1	4	6	1	15
2001	0	2	2	1	4	6	0	15
2002	0	1	1	1	1	8	0	12
2003	1	1	1	1	3	9	0	16
2004	1	3	2	1	2	5	1	15
Total	13	34	40	43	105	166	19	

Category 5 = 5, Category 4 = 4, Category 3 = 3, Category 2 = 2, Category 1 = 1, tropical storm = TS, tropical depressions (subtropical storms) = ST

data from the period 1851 to 1985 were offered to us by the author. Regarding data of hurricanes, *http://weather.unisys.com/hurricane/* has been considered. Table 5.7 shows the distribution of hurricanes of different categories that are formed in the Atlantic every year, from 1965 to 2004.

As shown in Figure 5.36 the number of hurricanes of high category have been increasing since 1995. The years 1969, 1995, and 2004 have the highest number of hurricanes of high category (red bars) registered, but 1969 and 1995 have the highest amount of total number of hurricanes registered (12 and 11 respectively).

Results of the coherence analysis of terrestrial and cosmophysical phenomena are displayed through the next Figures 5.37–5.39. The upper block of each of these figures shows the time series of the data involved. The power level color code used throughout this chapter is indicated at the bottom of each figure. Areas inside black contours correspond to the 95% significance level. As we are working with two time series, the wavelet coherence and phase difference are obtained.

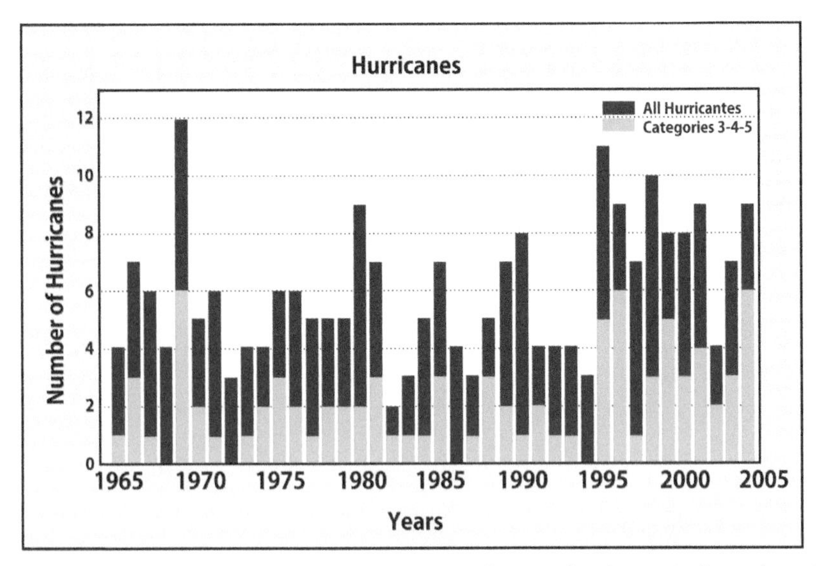

FIGURE 5.36 Number of North Atlantic Hurricanes: Category 3 and more (red), yearly total number of hurricanes (blue), including hurricanes Categories 1 to 2.

As indicated by the blocks of the wavelet coherence, the time and frequency intervals are where the two phenomena have a strong interaction.

The global spectra (on the right blocks of each figures 5.37 to 5.39) allows us to notice at a glance the global periodicities of either the time series or the coherence analysis. The *significance level of the global spectra* is indicated by the dashed curves. It refers to the power of the red noise level: peaks below the line imply a global periodicity with a confidence lower than 95% at the corresponding frequency, whereas a peak above the line indicates a confidence level higher than 95% at the given frequency. The spectral power (abscissa axis) is given in arbitrary units.

The main results that can be drawn from Figure 5.37 can be summarized as follows:

1. There is a coherence of 0.95 inside the COI between the AMO and SST anomalies through the band of 15 to 32 years, in the time interval 1900 to 1980 (Figure 5.37B–C). The oscillation in the 30 years frequency is completely in phase, indicating a lineal relation among both phenomena, which is not surprising because it is something very well known by climate specialists.
2. There is a coherence of 0.6 between the SST anomalies and SS, also limited to short intervals from 1895 to 1910 and 1945 to 1960 at the frequency of 11 years, with a tendency to be in anti-phase, and 1940 to 1980 at the 22 years frequency, with tendency to be in phase (Figures 5.37E–F). It can be seen from these figures that no frequency at the 30 years periodicity was found

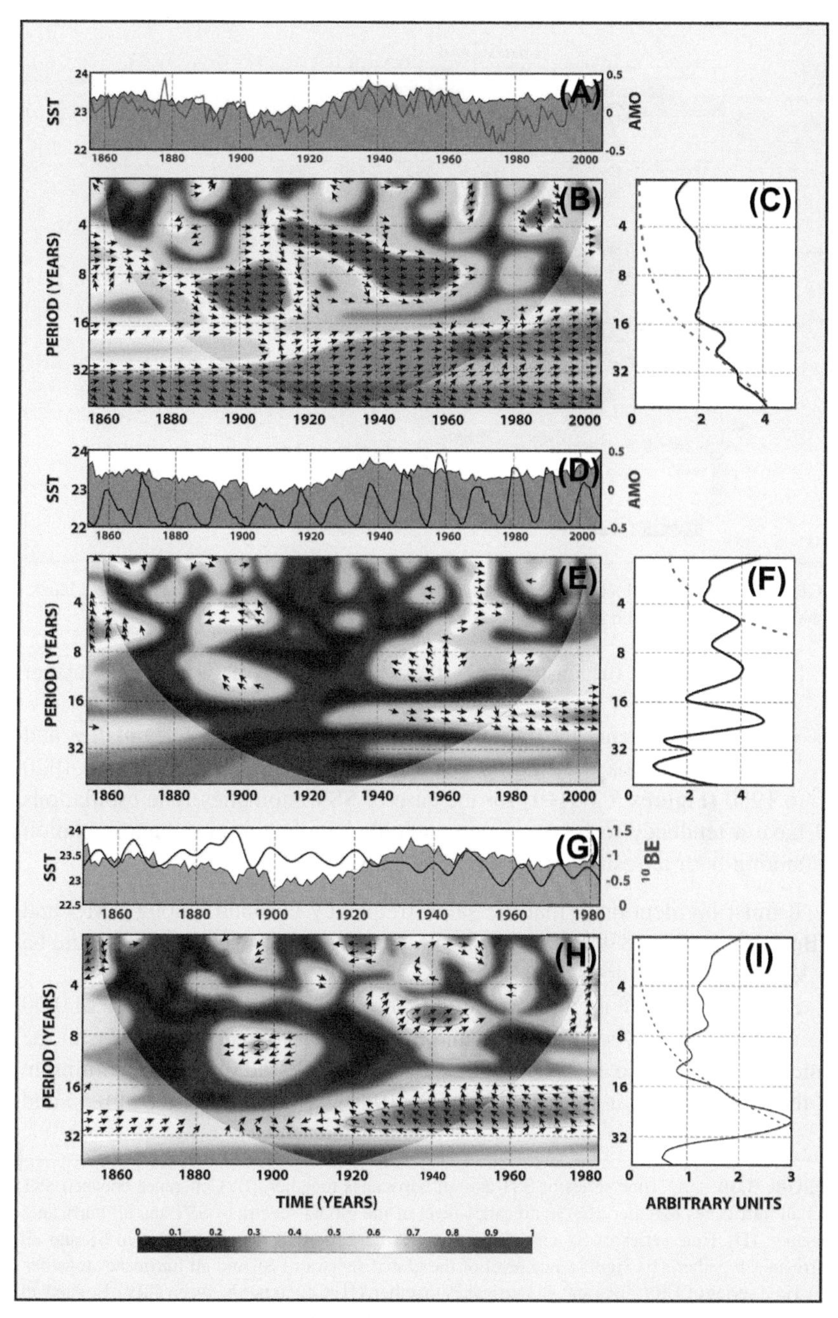

FIGURE 5.37 (A) Time series of AMO (blue line) and SST (gray area). (B) Coherence between SST and AMO. (C) Significance level of the global spectra of SST and AMO. (D) Time series (blue line) of SS and SST. (E) Coherence between SST and SS. (F) Significance level of the global spectra of SST and SS. (G) Time series of SST and CR (^{10}Be) (blue line). (H) Coherence between SST and CR (^{10}Be). (I) Significance level of the global spectra of SST and CR (^{10}Be).

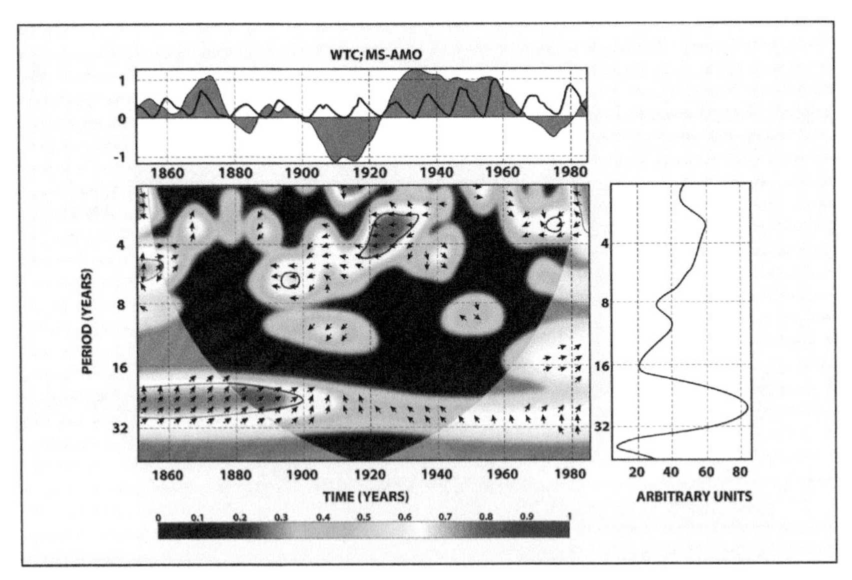

FIGURE 5.38 Upper block: Time series of AMO (gray area) and SS (black line). Lower block: Coherence between SS and AMO. Right block: Global spectrum.

for solar activity (at least through the use of SS), by means of the wavelet spectral analysis.

3. There is a coherence of 0.90 inside the COI between SST anomalies and ^{10}Be (the CR proxy) at the 30 years frequency, in the time interval 1920 to 1950 (Figures 5.37H–I) for the case of SST anomalies. The oscillations have a tendency to be quasi-perpendicular, indicating a complex relation among both terrestrial phenomena and CR.

It must be mentioned that the same frequency is found among AMO and ^{10}Be in the period 1870 to 1950, but with a lower coherence of ~0.75 (figure 6a in Velasco and Mendoza, 2008)

From Figure 5.38 it can be seen that there is a nonlinear coherence of 0.90 inside the COI between the AMO and SS near the 30 years frequency, in the time interval 1875 to 1895. This is gradually attenuated during the minimum of the modern secular solar cycle (1890–1940), with a coherence of 0.55 and

FIGURE 5.39 (A) Time series of SST and all hurricanes together. (B) Coherence between SST and all hurricanes together. (C) Significance level of the global spectra of SST and all hurricanes together. (D) Time series of SS and all hurricanes together. (E) Coherence between SS and all hurricanes together. (F) Significance level of the global spectra of SS and all hurricanes together. (G) Time series of CR (^{10}Be) and all hurricanes together. (H) Coherence between CR (^{10}Be) and all hurricanes together. (I) Significance level of the global spectra of CR (^{10}Be) and all hurricanes together. (J) Time series of CR (^{10}Be) and hurricanes of magnitude-4. (K) Coherence between CR (^{10}Be) and hurricanes of magnitude-4. (L) Significance level of the global spectra of CR (^{10}Be) and hurricanes of magnitude-4.

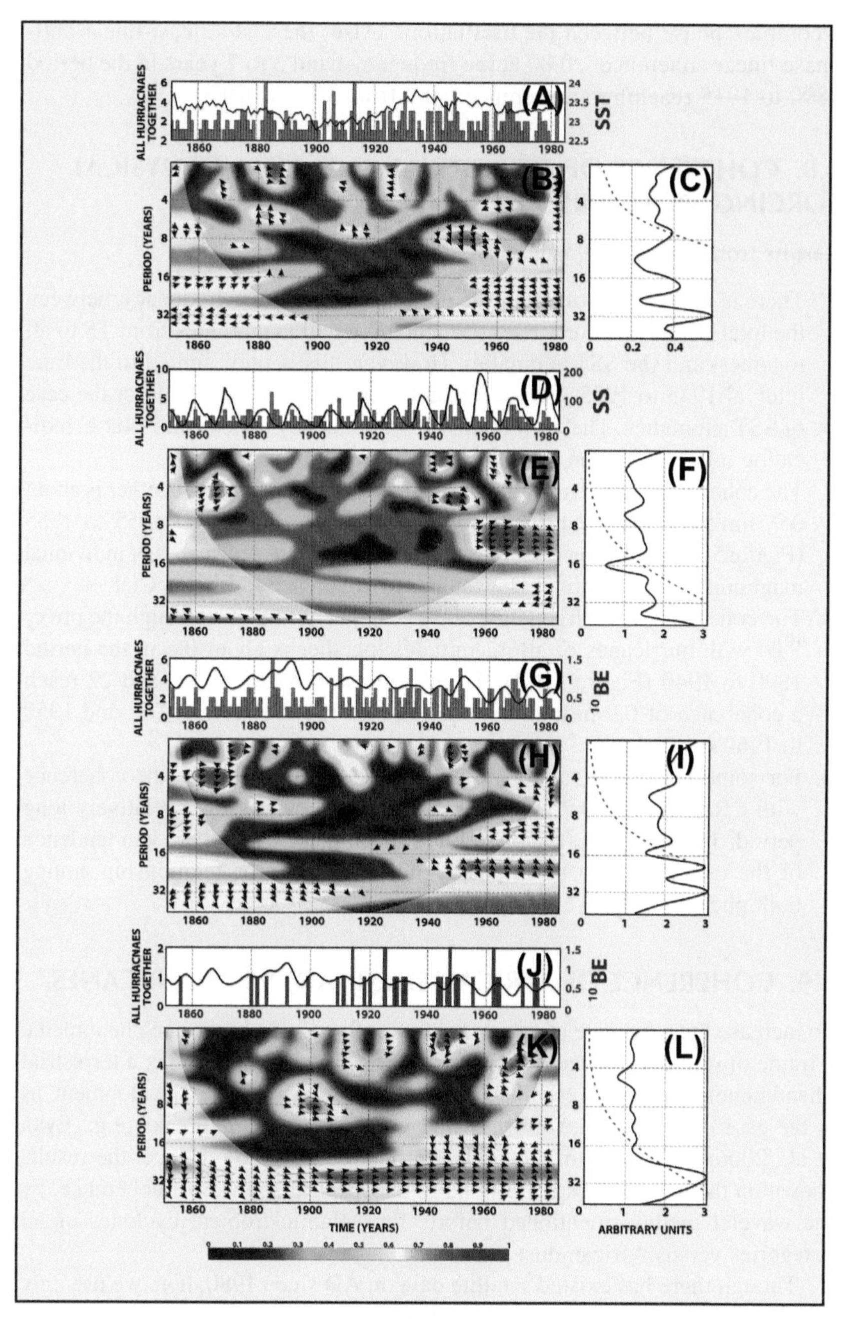
ALL HURRACNAES TOGETHER
(A)
SST
PERIOD (YEARS)
(B)
(C)
(D)
SS
(E)
(F)
(G)
^{10}BE
(H)
(I)
(J)
(K)
(L)
TIME (YEARS)
ARBITRARY UNITS

FIGURE 5.39

a complex phase between the oscillations. Also, there is a quasi-linear anti-phase linear coherence <0.80 at the frequency band 3 to 7 years in the period 1890 to 1915 reaching a coherence >0.9 from 1915 to 1930.

5.8. COHERENCE OF TERRESTRIAL AND COSMOPHYSICAL FORCING VS. HURRICANES

Results from Figure 5.39 can be summarized as follows:

1. There is a coherence of 0.9 inside the COI at the 7 years frequency, between the total number of hurricanes (i.e., including all magnitudes from TS to all together) and the SST anomalies. However, this is only limited to the time interval 1945 to 1955. This is illustrated in Figure 5.39(B−C), for the case of SST anomalies. The oscillations have a tendency to be in anti-phase, indicating a lineal relation of both phenomena with hurricanes.
2. The coherence between SS and hurricanes of all magnitudes together is about 0.9, limited at the frequency of 11 years during the period 1955 to 1965 (Figure 5.39E−F). However, the analysis of SS versus hurricanes of individual magnitudes gives relatively low values of coherence inside the COI.
3. The coherence at the frequency of 30 years, between CR (through the proxy ^{10}B) with hurricanes of all magnitudes together is about 0.6 in the period 1890 to 1940 (Figure 5.39H−I). In contrast, those of 5, 11, and 22 reach a coherence of 0.9 in the intervals 1860 to 1870, 1960 to 1970, and 1950 to 1960 respectively.
4. For some hurricanes, as for instance those of magnitude-4, the coherence with CR (^{10}B) is >0.9 at the 30 years frequency, during a relatively long period, 1890 to 1950 (Figure 5.39K−L). In these cases there is a tendency of the oscillations to be in-phase, indicating a linear relationship among both phenomena.

5.9. COHERENCE OF AFRICAN DUST (AD) VS. HURRICANES

To increase the relevance of extraterrestrial influence on hurricane phenomena, a frame of reference is needed. We consider here, as such a frame, a terrestrial phenomenon that is well established to be related with cyclone development, as is the case of the dust cover originated in African dust outbreaks (e.g., Evan et al., 2006; Lau and Kim, 2007a,b; Chronis et al., 2007). Hence, the results shown on the next figures correspond to the spectral analysis of coherence, by the wavelet method mentioned before, for Atlantic tropical cyclones of all categories versus African dust outbreaks.

Though there has existed satellite data on AD since 1980, here we use only multidecadal continuous *in-situ* monthly data available from Barbados from 1966 to the present (Prospero, 1999, 2006). Monthly data of hurricanes was taken from the National Weather Service and transformed into a date series of

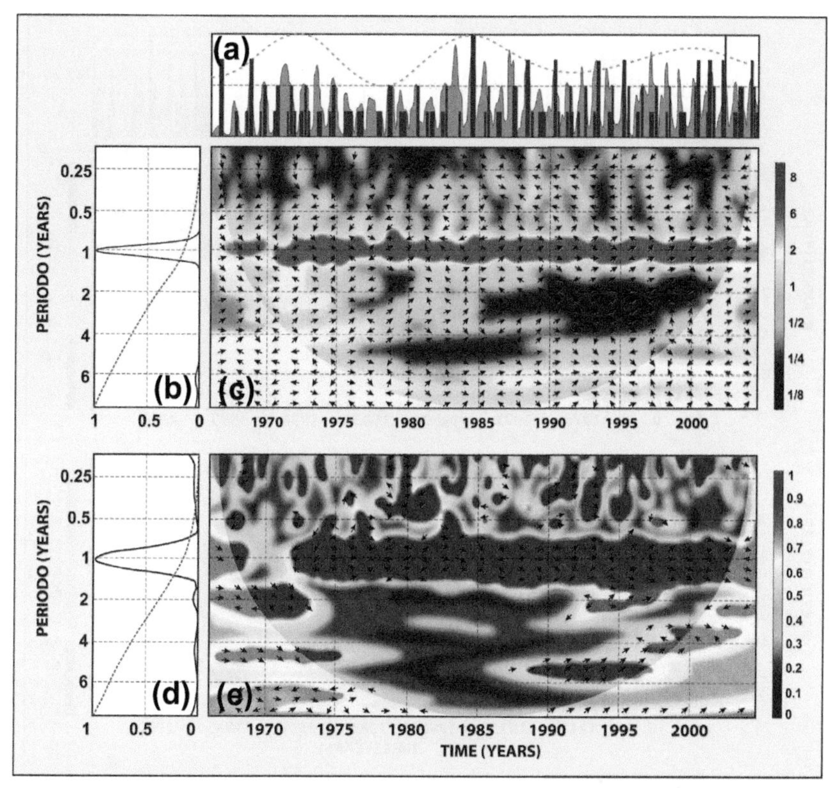

FIGURE 5.40 (a) Time series of African dust (shaded area): tropical storms (black bars), decadal tendency of dust (dotted line). (b) Global cross-wavelet coherence, *GXWT*. (c) Cross-wavelet coherence *XWT*. (d) Global wavelet-squared transform coherency (signal/noise), $GWTC_{s/n}$. (e) Wavelet-squared transform coherency (signal/noise), $WTC_{s/n}$.

pulses with the technique of pulse width modulation (PWM) as: n = number of hurricanes, where 0 = no hurricane (Holmes, 2003). Both data series are shown in blocks (a) of Figures 5.40 through 5.45. On these figures the global spectra are shown on the left side blocks.

Figure 5.40 shows the *GXWT* (panel b) between dust and tropical storms where it can be seen that the most prominent periodicity is that of 1 year. These annual variations do not have the same intensity throughout the period studied, as can be observed in the *XWT* (panel c) when the periodicity was low and parsimonious from the second half of the 1960s up to the first half of the 1970s, due to an increase in precipitation in North Africa. At this point, the coherence becomes very intense due to the very severe drought in West Africa, known as the Sahel drought, that began in the middle of the 1970s and lasted for several decades. The in-phase behavior of this annual periodicity with a linear tendency seems to indicate

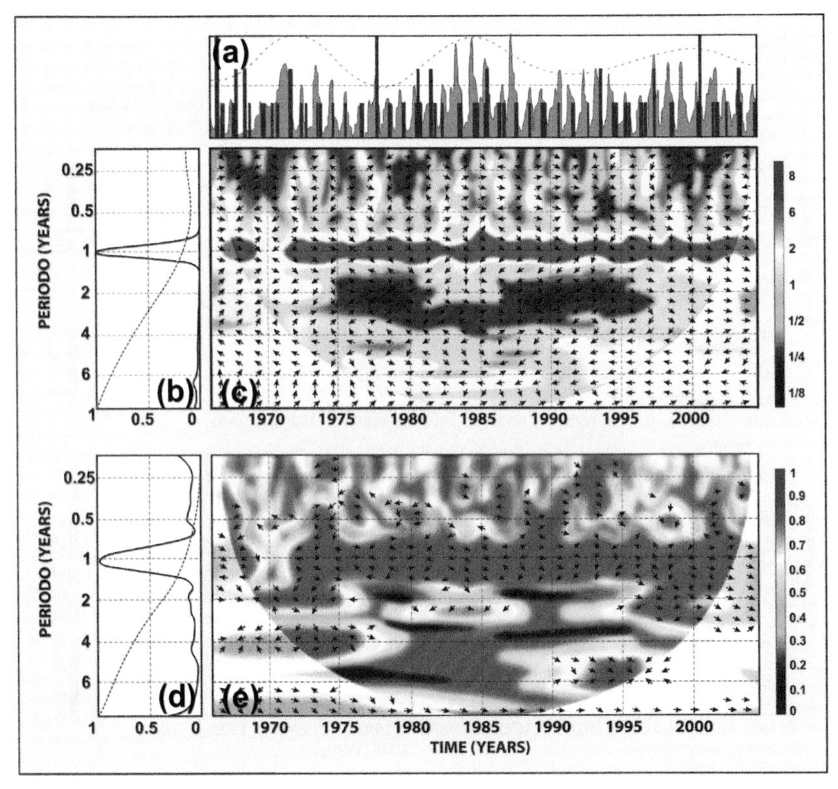

FIGURE 5.41 (a) Time series of African dust (shaded area): Category 1 hurricanes (black bars), decadal tendency of dust (dotted line). (b) *GXWT*. (c) *XWT*. (d) $GWTC_{s/n}$. (e) $WTC_{s/n}$.

that variability in dust has a quasi-immediate effect on the genesis and evolution of tropical storms.

There are other periodicities lesser and greater than 1 year but with inconsistent patterns. This may be interpreted as that dust concentration and the evolution of tropical storms are the result of many external and internal factors occurring on different time scales.

The *XWT* and $WTC_{s/n}$ show that the multiannual periodicities, including the decadal periodicity, are not essential factors in the formation of tropical storms.

The global *GXWT* and $WTC_{s/n}$ (panels b and d), as well as the cross-wavelet and coherence (panels c and e) between dust and Category 1 hurricanes (Figure 5.41) show a great deal of similarity with tropical storms, though with different intensities. These differences may be due to the increase of precipitation in North Africa, as indicated by the Sahel rainfall index (*http://jisao.washington.edu/data/sahel/022208*). Decadal periodicity is absent for Category 1.

It can be observed in Figure 5.42 that the *GXWT* and $GWTC_{s/n}$ (panels b and d), as well as the *XWT* and $WTC_{s/n}$ (panels c and e) between dust and hurricanes for Category 2 show an annual periodicity that has a less continuous time

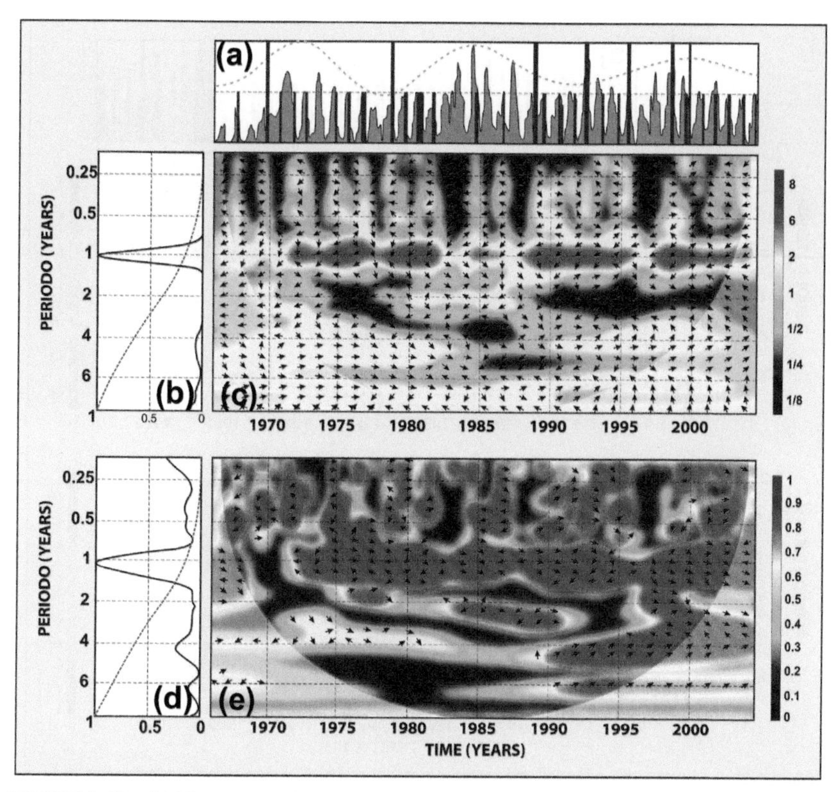

FIGURE 5.42 (a) Time series of African dust (shaded area): Category 2 hurricanes (black bars), decadal tendency of dust (dotted line). (b) *GXWT*. (c) *XWT*. (d) $GWTC_{s/n}$. (e) $WTC_{s/n}$.

interval as compared to Category 1 hurricanes (Figure 5.42) and tropical storms (Figure 5.40), becoming more intense from 1972 to 1982 and 1988 to 2004. Additionally, there exists less prominent periodicities of 3.5 to 5.5 years that are associated with the El Niño Southern Oscillation (ENSO) (*www.cpc.noaa.gov/products/analysis_monitoring/ensostuff/ensoyears.shtml*). Decadal periodicity is absent for Category 2 hurricanes.

The *GXWT* and $GWTC_{s/n}$ (panels b and d), as well as the *XWT* and *WTC* (panels c and e) in Figures 5.43 and 5.44 between dust and Category 3 and 4 hurricanes, respectively, show that the annual periodicity is also not continuous over the period studied. It can also be observed in panel (e) that the periodicity of 3.5 years for Category 3 hurricanes becomes more intense during the interval 1980 to 1992. For Category 4 hurricanes the periodicities of 3.5 to 5.5 years (panel e) are anti-correlated during 1988 to 2002. Decadal periodicities for Category 3 and 4 hurricanes are practically absent.

The *GXWT* (panel b of Figure 5.45) between dust and Category 5 hurricanes shows a very prominent annual periodicity with confidence higher than 95%. As can be seen in the *XWT* (panel c), this periodicity presents a high variability;

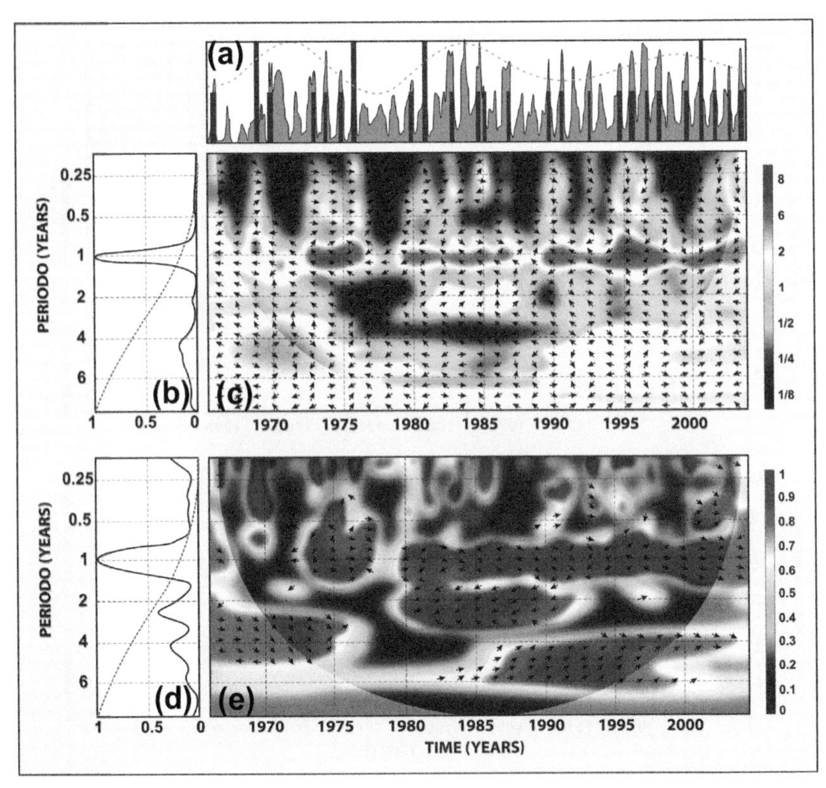

FIGURE 5.43 (a) Time series of African dust (shaded area): Category 3 hurricanes (black bars), decadal tendency of dust (dotted line). (b) *GXWT.* (c) *XWT.* (d) $GWTC_{s/n}$. (e) $WTC_{s/n}$.

it is not continuous and doesn't have the same intensity throughout the 1966 to 2004 period, becoming more intense during the 1970s and at the beginning of the 1980s, just when Category 5 hurricanes were, on average, the most intense. This annual periodicity is presumably related to the dust cycle in North Africa and to seasonal changes in atmospheric circulation (Husar et al., 1997).

Additionally, there are periodicities (11–13 years) that are present throughout the entire time interval with an anti-correlation tendency (panel c) (Velasco Herrera et al., 2010): its temporal tendency can be observed in panel (a) of Figure 5.45 with a dotted line obtained by means of a *Daubechies* type modified wavelet filter. This decadal periodicity is presumably related to the Atlantic trade wind variations and the dominant meridional mode of SST variability in the tropical Atlantic (Shanahan et al., 2009), as well as with solar activity (Hodges and Elsner, 2010) and cosmic rays (Pérez-Peraza et al., 2008b). The interaction of solar activity and cosmic rays with hurricanes is probably accomplished through the modulation of the Atlantic multidecadal oscillation (Pérez-Peraza et al., 2008a).

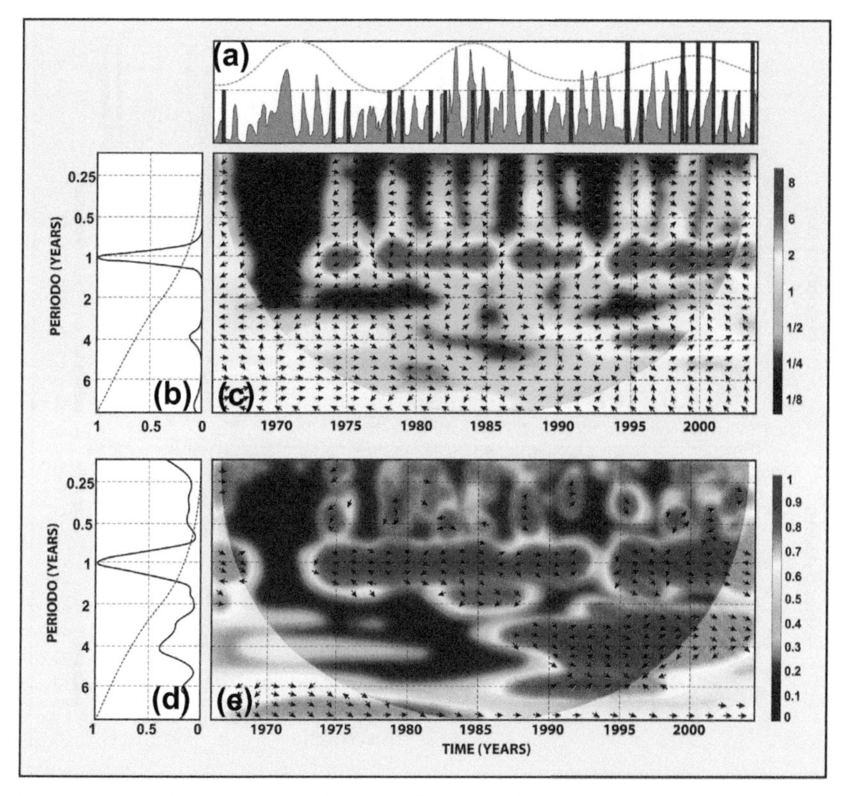

FIGURE 5.44 (a) Time series of African dust (shaded area): Category 4 hurricanes (black bars), decadal tendency of dust (dotted line). (b) *GXWT*. (c) *XWT*. (d) $GWTC_{s/n}$. (e) $WTC_{s/n}$.

This decadal variation shows that Category 5 hurricanes occur around the decadal minimum (panel a), because the local vertical wind shear ($V_z > 8$ m/s) is unfavorable for the genesis of tropical cyclones (Shapiro, 1987; DeMaria et al., 1993). This would explain why from 2008 up to the present, there have not been any Category 5 hurricanes, since it was precisely during this time that African dust in the atmosphere has been increasing. This would imply that if such a tendency continues, the next group of tropical cyclones will not evolve to Category 5 until the next decadal minimum of African dust occurs. To confirm if the annual and decadal periodicities obtained with the cross-wavelet are intrinsically related to the modulation of African dust on Category 5 hurricanes, we also obtained the modified wavelet coherence ($WTC_{s/n}$) and found, in addition to these two periodicities, two others, of 125 days and 1.8 years with a confidence level higher than 95% (panels d and e of Figure 5.45).

Figure 5.46 shows the *GXWT* (panel b) between dust and all hurricane categories together: it can be seen that the most prominent periodicity is again that of 1 year, mainly after 1972, with an anti-phase tendency and confidence

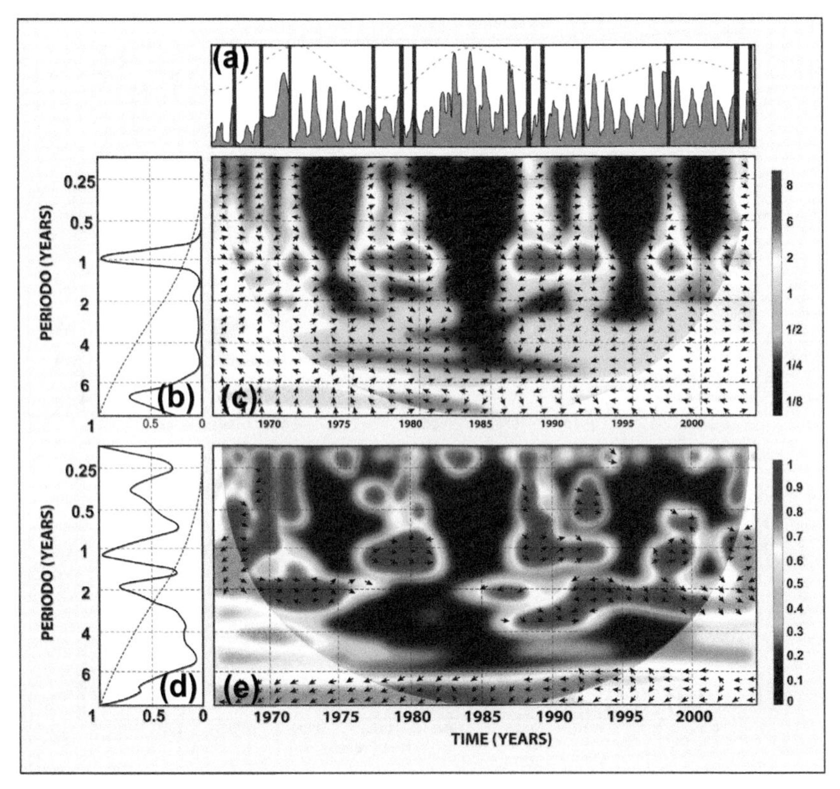

FIGURE 5.45 (a) Time series of African dust (shaded area): Category 5 hurricanes (black bars), decadal tendency of dust (dotted line). (b) *GXWT.* (c) *XWT.* (d) $GWTC_{s/n}$. (e) $WTC_{s/n}$.

higher than 98%. The decadal frequency is practically absent. The 4-year periodicity is also present with a gap between 1975 and 1983, and with intermittent periods in phase and anti-phase.

Finally, it should be emphasized that the observed correlations show not only a direct effect of African dust on hurricane activity but also reflect an indirect relationship between the wind, SST, AMO, the Modoki cycle, El Niño, la Niña, precipitation, solar activity, and cosmic rays that to a greater or lesser degree modulate the evolution of Atlantic hurricanes.

5.10. COHERENCE BETWEEN COSMIC RAYS AND HURRICANES

For comparison of the influence of CR on cyclones with the influence of AD on cyclones (for which the period time is relatively short, 1966–2005), it is not necessary to use a proxy for CR, since data of NM stations is quite reliable since the late 1950s. Therefore we use data from the worldwide NM station network,

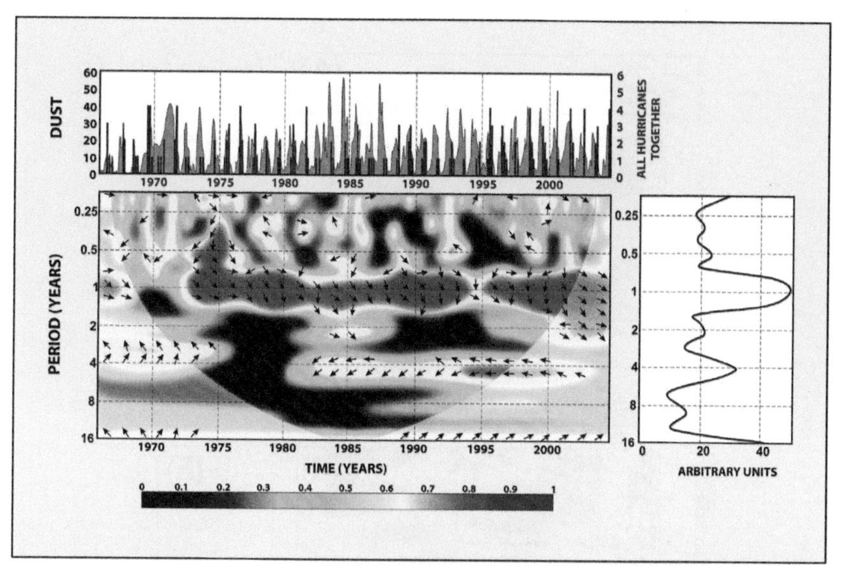

FIGURE 5.46 Time series of African dust (shaded area): all hurricane categories together (black bars). The lower panel is wavelet-squared transform coherency (*WTC*) and the right panel is the global spectrum *GXWT*.

in units of counts/min that we transform to monthly and annual data. We present results corresponding to annual data.

The results obtained in Section 5.10 as compared with those of Section 5.9 can be summarized as follows:

1. It can be seen on Figures 5.47(a–c) and 5.40 for tropical storms, that the coherence is higher and more continuous in time with AD than with CR.
2. It can be appreciated from Figures 5.47(d–f) for hurricanes of magnitude-1, that the coherence with dust is very high, of the order of 1, almost during all the studied period. However, Figure 5.41 shows that with CR a coherence higher than 0.9 occurs at the 2 years frequency only in the short time interval from 1996 to 1999.
3. For hurricanes of magnitude-2 the coherence with CR is in the period 0.7 years in short periods with a coherence near 0.8, again stronger than with dust (Figure 5.47G–H). Figure 5.42 shows that the coherence with AD is >0.9 at the periodicity of 1 year in anti-phase over all the studied period.
4. For hurricanes of magnitude-3, it can be seen from Figure 5.47(J–K) the coherence with CR is of 0.6 at the frequency of 0.6 years for very short time periods. In contrast from Figure 5.43 it can be seen that for AD the coherence is >0.9 at the 1-year periodicity in anti-phase all the time interval.

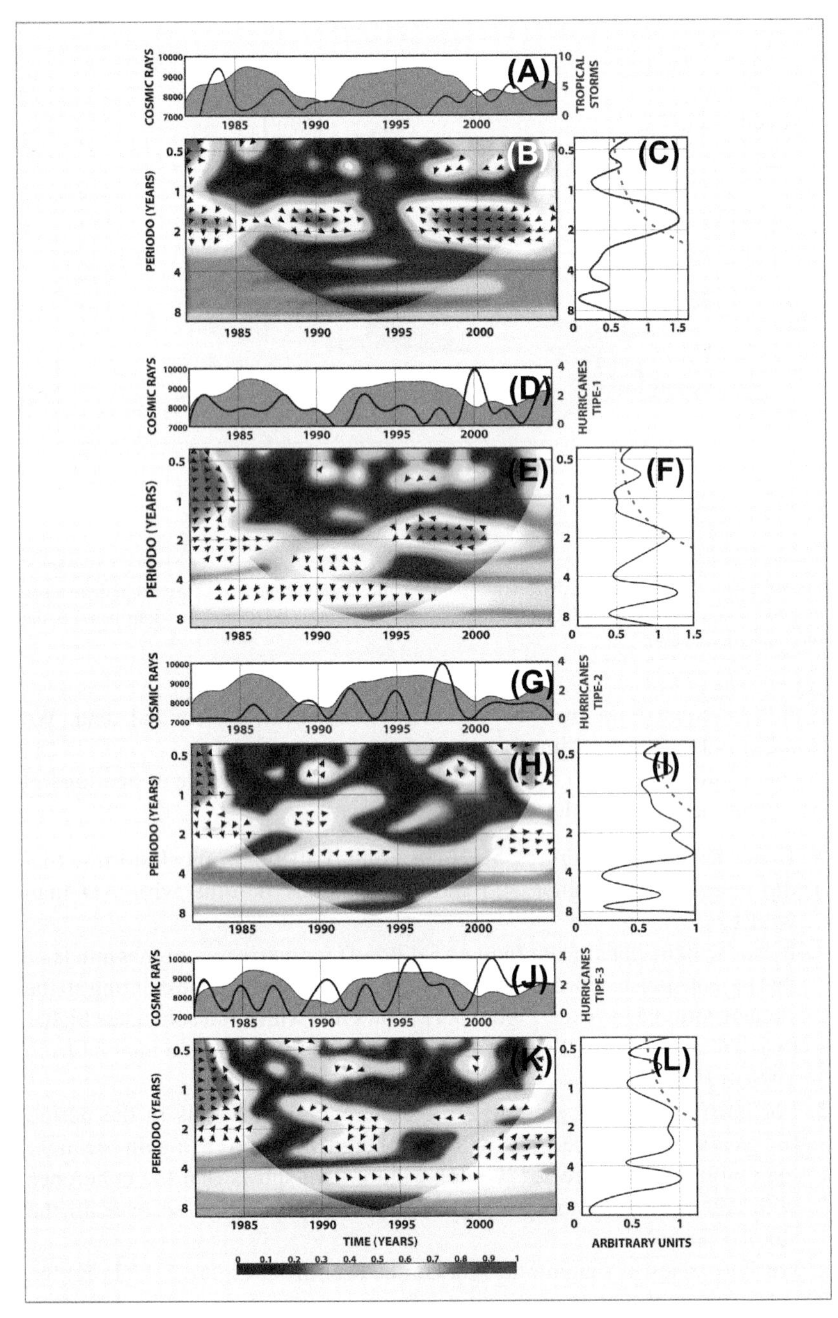
COSMIC RAYS
TROPICAL STORMS
HURRICANES TIPE-1
HURRICANES TIPE-2
HURRICANES TIPE-3
PERIODO (YEARS)
TIME (YEARS)
ARBITRARY UNITS
(A)
(B)
(C)
(D)
(E)
(F)
(G)
(H)
(I)
(J)
(K)
(L)

FIGURE 5.47

5. It can be seen on Figure 5.47 (a–c) and 5.40 for tropical storms, that the coherence is higher and more continuous in time with AD than with CR.
6. It can be appreciated from 5.48(d–f) and 5.42 for hurricanes of magnitude-1, that the coherence with dust is very high, of the order of 1, almost during all the studied period, but a short period 1996 to 1999 is only observed with CR, with a coherence higher than 0.9 at the 2 years frequency.
7. It can be seen on Figure 5.47(A–C) and 5.40 for tropical storms, that the coherence is higher and more continuous in time with AD than with CR.
8. It can be appreciated from Figures 5.47(D–F) and 5.41 for hurricanes of magnitude-1, that the coherence with dust is very high, of the order of 1, almost during all the studied period, but a short period 1996–1999 is only observed with CR, with a coherence higher than 0.9 at the 2 years frequency.
9. For hurricanes of magnitude-2 the coherence with CR is in the period 0.7 years in short periods with a coherence near 0.8, again stronger than with dust (Figure 5.47G–H). Figure 5.42 shows that the coherence with AD is >0.9 at the periodicity of 1 year in anti-phase over all the studied period.
10. For hurricanes of magnitude-3, it can be seen from Figure 5.47(J–K) that for CR the coherence is of 0.6 at the frequency of 0.6 years, in short time periods, whereas for AD it can be seen from Figure 5.43 that the coherence is >0.9 at the 1 year periodicity in anti-phase all the time.
11. For hurricanes of magnitude-4, Figure 5.47(L–O) shows the coherence with CR of the order of 0.95 at the periodicity of 1.7 years in limited time intervals in anti-phase. Coherence with AD is in anti-phase at the frequency of 1 year, with a coherence >0.9 after 1973 (Figure 5.44).
12. For the more dangerous hurricanes, those of magnitude-5, we can see from Figure 5.47(P–R) that the coherence is of complex nature in the 0.7 and 1.7 years periodicities, in short time intervals, with a coherence around 0.9. From Figure 5.45 we can see that coherence is of the order of 1 at the 1 and 10 to 11 years periodicity, always in anti-phase. The 1-year periodicity occurs mainly in limited periods occurring every 10 years.
13. For hurricanes of magnitude-2, the coherence with CR is in the period 0.7 years in short periods with a coherence near 0.8, again stronger than with dust (Figure 5.47G–H). Figure 5.42 shows that the coherence with AD is >0.9 at the periodicity of 1 year in anti-phase over all the studied period.

FIGURE 5.47 (A) Time series of CR and tropical storms. (B) Coherence between CR and tropical storms. (C) Significance level of the global spectra of CR and tropical Storms. (D) Time series of CR and hurricanes of magnitude-1. (E) Coherence between CR and hurricanes of magnitude-1. (F) Significance level of the global spectra of CR and hurricanes of magnitude-1. (G) Time series of CR and hurricanes of magnitude-2. (H) Coherence between CR and hurricanes of magnitude-2. (I) Significance level of the global spectra of CR and hurricanes of magnitude-2. (J) Time series of CR and hurricanes of magnitude-3. (K) Coherence between CR and hurricanes of magnitude-3. (L) Significance level of the global spectra of CR and hurricanes of magnitude-3.

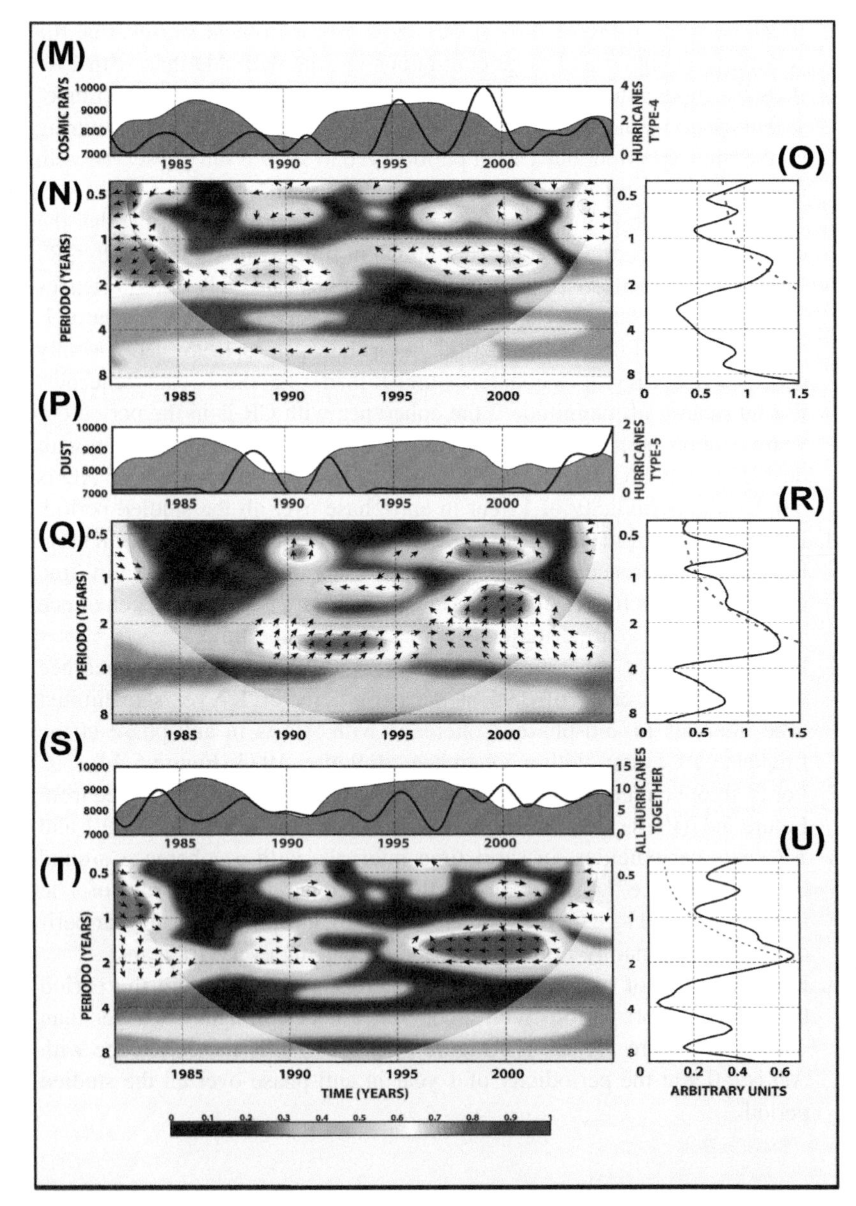

FIGURE 5.47 *(Continued)* (M) Time series of CR and hurricanes of magnitude-4. (N) Coherence between CR and hurricanes of magnitude-4. (O) Significance level of the global spectra of CR and hurricanes of magnitude-4. (P) Time series of CR and hurricanes of magnitude-5. (Q) Coherence between CR and hurricanes of magnitude-5. (R) Significance level of the global spectra of CR and hurricanes of magnitude-5. (S) Time series of CR and hurricanes of all magnitudes together. (T) Coherence between CR and hurricanes of all magnitudes together. (U) Significance level of the global spectra of CR and hurricanes of all magnitudes together.

14. For hurricanes of magnitude-3, it can be seen from Figure 5.47(J–K) that the coherence for CR is of 0.6 in the frequeny of 0.6 years in short time intervals, whereas for AD it can be seen from Figure 5.43 that the coherence is >0.9 at the 1-year periodicity in anti-phase all the time.
15. For hurricanes of magnitude-4, Figure 5.47(L–O) shows the coherence with CR of the order of 0.95 at the periodicity of 1.7 years in limited time intervals in anti-phase. Coherence with AD is in anti-phase at the frequency of 1 year, with a coherence >0.9 after 1973 (Figure 5.44).
16. For the more dangerous hurricanes, those of magnitude-5, we can see from Figure 5.47(P–R) that the coherence is of a complex nature, in the 0.7 and 1.7 years periodicities, in short time intervals, with a coherence around 0.9. From Figure 5.45 we can see that coherence is of the order of 1 at the 1 and 10 to 11 years periodicity always in anti-phase. The 1-year periodicity occurs mainly in limited periods, occurring every 10 years.
17. It can be seen from Figure 5.47(I–T) and (S–U) that the coherence between all kinds of Atlantic hurricanes together (from tropical storms

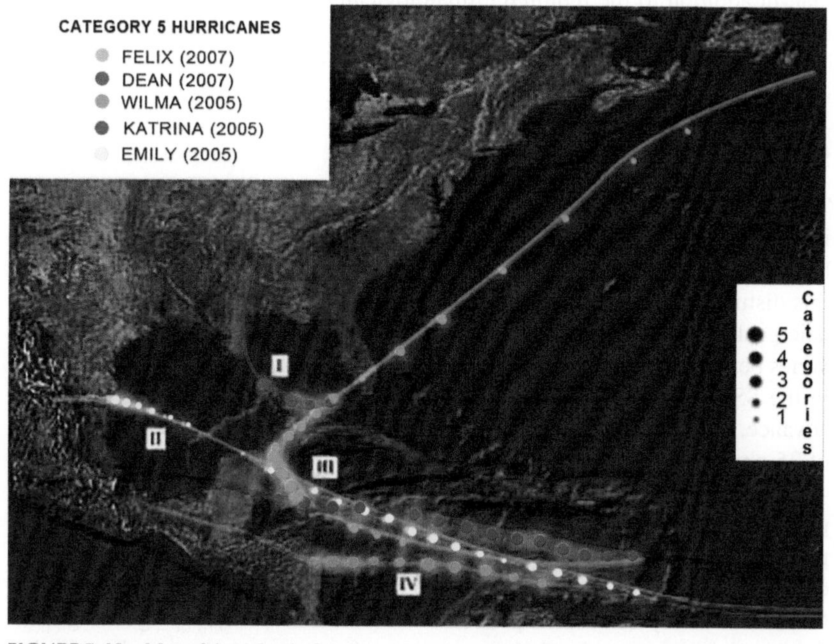

FIGURE 5.48 Map of the Atlantic Ocean centers of cyclogenesis of Category 5 hurricanes: (I) the east coast of the United States, (II) the Gulf of Mexico, (III) the Caribbean Sea, and (IV) the Central American coast. The 5 color paths illustrate the barometric distribution of the eyes of Category 5 hurricanes illustrated for six of these kind of hurricanes, between 2005 and 2007: Wilma (2005), Rita (2005), Katrina (2005), Emily (2005), Dean (2007), and Felix (2007). The black circles at the right side column indicate the evolution of each hurricane, from Category 1 to 5, during their trajectory. Source: *Velasco Herrera et al., 2010.*

to magnitude-5) and cosmic rays is at the 1.3 and 1.7 years frequencies. Variations seem to be in phase from 1987 to 1991 with coherence of 0.6, and suddenly they switch to anti-phase during 1996 to 2002, with coherence higher than 0.9.

18. In contrast to (7), on hurricanes versus CR, in Figure 5.46 it is shown that the coherence between all kinds of Atlantic hurricanes together and the African dust outbreaks is more continuous than with CR, with coherence higher than 0.9, and it is concentrated around the 1.3 and 4 years periodicities, for long time periods, with complex nonlinear phases. Moreover, it should be mentioned that the red noise of the confidence level of the global spectrum in Figure 5.47 is not shown, because it is far above the frequency picks.

By tracing the trajectories of all hurricanes, we have delimited the existence of four areas of deep water in the Atlantic Ocean where the eye of the hurricane has the lowest pressure (<920mb) as can be seen in Figure 5.48: (I) the east coast of the United States, (II) the Gulf of Mexico, (III) the Caribbean Sea, and (IV) the Central American coast. This must be pointed out because it implies that, in addition to the required climatologically conditions for the genesis of this kind of Category 5 hurricanes to take place, the geography of the marine bottom also plays an important role, and that these hurricanes do not originate in hazardous places (figure 1 in Velasco Herrera et al., 2010).

5.11. DISCUSSION

Trying to generalize the large amount of information described in Section 5.3, it can be said that CR, SS, AP, and K_p showed much more intensive disturbances in the periods preceding and following the hurricane appearance. For SS this disturbance gradually increases with the hurricane strength. A characteristic peak in the CR intensity appears before the hurricane start. But its place varies between 5 and 20 days before that start. Specific changes were observed in the SS. For major hurricanes they begin sometimes more than 20 days in advance. The AP and the K_p show series of bursts, spread over the whole period of 35 preceding days.

The chosen long preceding period of days reveals the behavior of these parameters long before the cyclone appearance. Specific precursors exist persistently before the cyclone start. During the time of major cyclone development, specific changes are also noticeable: a considerable change in the solar activity and the dependence on it, CR intensity, and geomagnetic field activities, precede the appearance of intense cyclones, though the preceding time fluctuates considerably from event to event. The obtained interconnections show that these parameters should be taken in consideration, when complicated processes in the upper atmosphere are used to determine the hurricane formation that potentially will contribute to hurricane development forecasts.

On the other hand, in relation to results displayed through Sections 5.7 to 5.10 we must keep in mind that the quasi anti-correlation between SST and dust storms, as reported in the literature, is not a perfect one and that a linear correlation between SST and North Atlantic cyclones (Figure 5.39B–C) is also not perfect. However, the anti-phase situation is systematically found between AD and hurricanes, then we assume here, in the first instance, that the more abundant the AD storms the fewer the number of North Atlantic cyclones; conversely, a linear correlation would mean that the more abundant the AD storms the higher the occurrence of cyclones, at least that were completely independent phenomena, which seems not to be the case.

In this context, the spectral analysis carried out here for the study of common periodicities among cosmic rays and phenomena that are presumably associated with hurricanes (AMO and SST anomalies) is in agreement with previous results (Pérez-Peraza et al., 2008a,b). We have found that there are common periodicity between some extraterrestrial and terrestrial phenomena, namely 3.5, 5.5, 7, 11, and the more prominent one, that of 30 ± 2 years is often present (as can be seen in most of Figures 5.37–5.38) with the exception of SS. It should be mentioned that no figure with dust (Figures 5.40–5.46) shows the 30-year periodicity, because data of dust only cover 40 years. Such a periodicity is also present in CR versus hurricanes of magnitude-4 as was shown in Pérez-Peraza et al. (2008b), where the applied Morlet-wavelet technique allowed us to put in evidence, for the first time the existence of a periodicity of 30 ± 2 years in cosmic ray fluctuations. The most dangerous hurricanes do not show that periodicity, since confident data on Category 5 exist only since 1980. It should be mentioned that this frequency is also found in other indexes of hurricane activity, as has been previously mentioned. Preliminarily, we speculate that this 30 years cycle may be associated with a semi-phase (either of the maximum or the minimum) of the secular cycle of 120 years of solar activity, that is, half of the so called Yoshimura–Gleissberg cycles (Yoshimura, 1979; Velasco and Mendoza, 2008).

If the coherences found in this work among the studied phenomena may be interpreted as a modulator factor, then from the analysis of the previous results, it could be speculated that the modulator agent of terrestrial phenomena is the open solar magnetic field, translated in GCR (via the ^{10}Be). This modulation seems to be more important in the period 1880 to 1960. That does not mean there is no modulation after and before, but according to the coherence wavelet technique the significance level is lower than 95%. It seems then that GCR are modulating in some way both the SST (Figure 5.37H–I) and AMO (figure 6a in Velasco and Mendoza, 2008), and these in turn modulate in some way hurricanes, as can be seen from the coherence wavelet analysis (Figure 5.37B–C), which confirms the conventional statement of hurricanes to be linked to warmer oceans. We then assume here that the action of GCR on the clouds is an additional warming factor of the SST and AMO, as are the greenhouse gases on the Earth's surface temperature (Ram et al., 2009), which in some way is to be translated in the development of hurricanes. In contrast, the indicator of closed solar magnetic field (via SS) presents, within the COI, a lower and attenuated coherence with the terrestrial phenomena.

It is obvious from the correlational analysis carried out in Section 5.3 that the parameters SS, CR, AP, and K_p are not the basic driving factors for hurricane appearance and development. That is, at present, an accurate forecast of cyclone activity cannot be claimed only on the basis of preceding K_p and CR, or Forbush effects data drastic changes. However, the results confirmed the preliminary hypothesis suggesting that there is some kind of interconnection between these parameters and the appearance of tropical cyclones, especially with the most powerful of them. It is not pretended by the moment to forecast the creation of a dangerous vortex on the basis of peculiar behavior preceding of those data changes. However, looking at the presented results it could be strongly asserted that if a package of large SS and a sharp rise in K_p index is recorded together with Forbush decreases appearing during the summer (that is, by investigating all the parallel atmospheric data in coincidence with a trend of the deeper minimum in CR intensity and the higher maximum in SS, as shown in Figure 5.30 for higher hurricane categories), we could be closer to a most probable prediction.

Very interesting results are obtained with correlational studies, however, it should be noted that this kind of analysis is only the necessary first step to be done when it is to be determined whether or not there is a connection between two different physical phenomena. In fact, correlational analysis only provides global information about the degree of linear dependence between two time series but does not gives information when the correlation dependence is of nonlinear nature. Even if the global correlation coefficient is low, that does not mean there is no physical relation. In fact, there is the possibility that such a relation could be of a nonlinear nature, or that there is a strong phase shift between the cosmophysical phenomenon and the plausible associated terrestrial effect, or there is a time delay between one series time (input) and the system reaction (output). Nevertheless, the obtained results are exciting enough for motivating us to jump to a next step, that is, to reinforce such analysis with collateral methods, by means of a more precise statistical analysis technique.

In this regard, let us emphasize again that one of the most powerful tools of signal theory is the so called *wavelet-coherence transformer spectral analysis* (Section 5.6), which furnishes not only global but also local information in time, and per frequency band (Hudgins et al., 1993; Torrence and Compo, 1998; Torrence and Webster, 1999); that is, it provides the coherence between two series, by means of the evolution of the relative phase between both series, determining whether their correlation is linear or not in different band widths. It may appear that the global correlation coefficient is low, but in some periods of the studied time interval the coherence can be relatively high, indicating a nonlinear correlation (a complex one) in those periods. The important feature of wavelet spectral analysis is that it gives the evolution of the synchronization in time–frequency space. In contrast, the so-called *Pearson correlation coefficient* does not provide the evolution of the common synchronized periodicities, nor the evolution of the relative phase between two time series.

Therefore, using such a tool we presented in Sections 5.7 through 5.10 the most prominent signals between North Atlantic hurricanes of all categories, and terrestrial phenomena SST, AMO, AD (African dust), geoexternal phenomena such as solar activity (SS), geomagnetic activities (K_p), and GCR (galactic cosmic rays). Such periodicities among cosmic rays and phenomena that are presumably associated with hurricanes (AMO and SST-anomalies) is in agreement with previous results (Pérez-Peraza et al., 2008a,b). Results indicate that among the most prominent common signals for hurricanes, SST, AMO, and GCR are those of period of 30 $\pm$ 2 years, with exception of the SS graphs. The high level of coherence between CR and hurricanes indicates a stronger modulation effect by GCR, probably on the basis of the mechanisms described by Ram et al. (2009).

Though, it is well known that cosmic rays and solar activity phenomena are inversely related in time, such a relation is not directly translated to their influence on hurricane development. The temporal scale of their influence is certainly different: cosmic rays influence is a relatively prompt effect, whereas solar activity seems to act as a result of a slower buildup effect (Pérez-Peraza, 1990). It should be appreciated the good coherence between CR and hurricanes of all magnitude, particularly with those of magnitude-4 (Figures 5.39k–l and 5.45). In contrast, the indicator of closed solar magnetic field (via SS) does not present the 30-year periodicity, and only presents within the COI, a very low and attenuated coherence with terrestrial phenomena, at the frequencies of 3.5, 5, 7, 11, and 22 years (Figures 5.37E–F and 5.38). Nevertheless, the analysis must be extended with data of other solar indexes (as for instance, radio in 10.3 cm and coronal holes) versus specific parameters of the hurricanes (vorticity, linear velocity, duration, energy, power destruction index, PDI, accumulated cyclone energy, ACE, and storm intensity). There is also the possibility that the periodicity of 30 years could be associated in a nonlinear way to the solar Hale cycle (Raspopov et al., 2005) with a certain phase shift.

Also, we would like to state that though we cannot say in a conclusive way that CR modulates the AMO and SST, we must keep in mind that the AMO has intrinsic periodicities (at least since 1572) at 30, 60, and 100 years (Velasco and Mendoza, 2008) and the AMO is in turn a modulator of the SST (Sutton and Hudson, 2005). Because the only other phenomena we know that present such periodicities are SS and CR, we infer that such a modulation of AMO and SST may be related to one or both cosmophysical phenomena.

To estimate the relevance of cosmophysical influence on hurricane activity we have compared it with African dust outbreaks (Sections 5.7 through 5.10). For AD, the time interval data is relatively short (since 1966), nevertheless, several well-defined signals are found with hurricanes of different categories in the range 0.25 to 13 years. A clear anti-correlation is found between Categories 5, 4, and 3 hurricanes with AD. While the most prominent periodicity for hurricanes of Category 4 or less is 1 year, for Category 5 hurricanes it is a well-defined decadal periodicity. It is found that the coherence at similar frequencies is in many cases higher with AD than with CR, though the influence of CR is not at all negligible. However, cosmophysical influences cannot be disregarded, and may eventually become of the same order of importance of some terrestrial effects, thought it seems not to be the case for the dust cover originated in African dust outbreaks.

Furthermore, the fact that two data series (cosmophysical and climatic) have similar periodicities does not necessarily imply that one is the cause and the other the effect: a physical mechanism must be behind it, able to give an explanation of such coincident signals. Since such interconnections are still in the stage for establishment of trustable evidences, it is perhaps still not time to look for physical mechanisms.

Special mention must be made of the results of Section 5.9, from where we can conclude that African dust influence on the genesis and evolution of Atlantic hurricanes varies, as we said above, in two main ways: (a) annually for tropical storms and hurricanes of all categories and (b) decadally for Category 5 hurricanes. Category 5 hurricanes develop during the minimums of decadal cycles of African dust, when winds are lower than 8 m/s, and ocean temperatures are around 26 °C. In addition to peculiar climatological conditions, the geography of the ocean bottom is an important factor in the development and evolution of Category 5 tropical cyclones, since in at least four Atlantic deep-water regions, hurricane eyes have the lowest pressure. Velasco et al. (2010) determined that every Category 4 hurricane crosses in some segment of its trajectory one or more of these four regions if it evolves to a Category 5 hurricane. Velasco et al. (2010) determined for the first time which hurricanes of Category 5 are able to evolve, or not, in Category 5, depending on whether or not their trajectory cross the geographic areas of deep waters (10 °W–20 °W latitude).

Finally, it is worth mentioning that if climatological tendencies continue as they have in recent decades, then according to our wavelet analysis, a prominent periodicity of 8 years exists for hurricanes of the higher categories and that must be deeply studied for prognosis goals. Furthermore, according to the coherence analysis between Category 5 hurricanes and African dust, a periodicity of around 10 years is quite suggestive (see Figure 5.45a). In a preliminary form we argue that future hurricanes will not be able to develop into Category 5 until the next decadal minimum, between 2012 and 2017.

References

Abramov, M.A., 2001. Structural and cyclical patterns in nature, society, art. Saratovsky Gosudarstvenny Tekhnichesky Universitet, Saratov, Russia.

Abrosov, V.N., 1973. Lake Balkhash. Nauka, Leningrad, Russia.

Abuzyarov, Z.K., 2003. Technology forecasting trends in the Caspian Sea to the prospect of 6 and 18 years. In: Hydrometeorological aspects of the Caspian Sea basin. Hydrometeoizdat, St. Petersburg, Russia, pp. 351–363.

Adler, Yu. P., Khunzidi, Yu. P., Shpeer, V.L., 2005. Methods of continuous improvement through the prism of Shewhart-Deming cycle. Methods of Quality Management 3, 29–36.

Afonin, S.V., Belov, V.V., Gridnev, Yu, V., Protasov, K.T., 2009. Passive satellite remote sensing the Earth's surface in visible light wavelengths. Atmospheric and Ocean Optics 22 (10), 945–949.

Akasofu, S., Chepman, S., 1975. Solar-terrestrial physics. Mir, Moscow, Russia.

Alexeev, G.V., 2006. Arctic climate change in the twentieth century. The possibility of preventing climate change and its adverse effects. Nauka, Moscow, Russia.

Alvarez-Madrigal, M., 2010. Comment on "Correlation between cosmic rays and ozone depletion." Physical Review Letters 105 (16), 169801.

Alvarez-Madrigal, M., Pérez-Peraza, J., 2005. Analysis of the evolution of the Antarctic ozone hole size. Journal of Geophysical Research 110 D02107. doi:10.1029/2004JD004944.

Alvarez-Madrigal, M., Pérez-Peraza, J., 2007. Differences in the monthly evolution of the Antarctic ozone whole size. Atmósfera 20–22, 215–221.

Alvarez-Madrigal, M., Pérez-Peraza, J., Velasco, V.M., 2007. A plausible relation between cosmic rays and the Antarctic ozone hole size. In: Proceedings of the 30th international cosmic ray conference, vol. 1. Universidad Nacional Autónoma de México, Mexico City, Mexico, pp. 789–792.

Alvarez-Madrigal, M., Pérez-Peraza, J., Velasco, V.M., 2009. Influence of cosmic rays on the size of the Antarctic ozone hole. Current Development in Theory and Applications of Wavelets 3 (3), 233–248.

Andersen, N., 1976. Statistical analysis of time series. Mir, Moscow, Russia.

Anderson, R.Y., 1992. Long-term changes in the frequency of occurrence of El Niño events. In: Diaz, H.F., Markgraf, V. (Eds.), El Niño. Historical and paleoclimatic aspects of the southern oscillation. Cambridge University Press, Cambridge, U.K, pp. 193–200.

Andreasen, G.K., 1993. Solar irradiance variations a candidate for climate change (Scientific report No. 93–5). Danish Meteorological Institute, Copenhagen.

Anikeev, V.V., Chyasnavichyus, Yu. K., 2008. Environmental security as a condition for sustainable development in the northern regions of Russia. In: Proceedings of the conference, ensuring integrated safety of the northern regions of the Russian Federation, April 22, 2008. NCUKS MCS Russia, Moscow, Russia, pp. 75–96.

Anisimov, O.A., Belolutskaya, M.A., 2002. Assessing the impact of climate change and degradation of permafrost on infrastructure in the northern regions of Russia. Meteorology and Hydrology 6, 15–22.

Anisimov, O.A., Nelson, E.F., 1997. Impact of climate change on permafrost in the North Hemisphere. Meteorology and Hydrology 5, 71–80.

Anisimov, O.A., Lavrov, S., 2008. Global warming and the melting of permafrost: risk assessment for industrial facilities TEC Russia Retrieved from: www.bestreferat.ru/referat-3089.html (in Russian).

Ariel, N.E., Shakhmeyster, V.A., Murashova, A.V., 1986. Spectral analysis of the characteristics of ocean-atmosphere energy exchange. Meteorology and Hydrology 2, 49–53.

Artekha, S.N., Golbraykh, E., Erokhin, N.S., 2003. The role of electromagnetic interactions in the dynamics of powerful atmospheric vortices. Problems of Atomic Science and Technology 4, 94–99.

Artekha, S.N., Erokhin, N.S., 2004. Electromagnetic force and eddy processes in the atmosphere. In: Proceedings of the international conference of the MCC-04, transformation of waves, coherent structures and turbulence. Rokhos, Moscow, Russia, pp. 326–332.

Artekha, S.N., Erokhin, N.S., 2005. On the relationship between large-scale atmospheric processes in the vortex of electromagnetic phenomena. Electromagnetic Phenomena 5 (1), 14, 3–20.

Attolini, M.R., Ceccini, S., Galli, M., 1983. A search for heliosphere pulsation in the range 1 c/yr to 1c/10yr. In: Proceedings of the 18th international cosmic ray conference, vol. 10, pp. 174–177.

Attolini, M.R., Ceccini, S., Galli, M., 1984. The power spectrum of cosmic ray. The low-frequency range. Nuovo Cimento C7 (4), 413–426.

Attolini, M.R., Galli, M., Cini Castagnoli, G., 1985. On the R-sunspot relative number variations. Solar Physics 95 (2), 391–395.

Avdiushin, S.I., Danilov, A.D., 2000. The Sun, weather and climate: present view of the problem (a review). Geomagn 40 (5), 3–14.

Ayrapetian, M.S., 2003. Actual problems of the theory of economic cycles. TEK (Russian trade magazine) 1, 147–151.

Babayev, E., Allahverdiyeva, A., 2007. Effects of geomagnetic activity variations on the physiological and psychological state of functionally healthy humans: some results of Azerbaijani studies. Advances in Space Research 40, 1941–1951.

Babcock, H.D., 1959. The Sun's polar magnetic field. The Astrophysical Journal 130, 364–366.

Babkin, V.I., 2005. Moisturizing areas of internal drainage of Eurasia (for example, the Aral Sea, Caspian Sea and Lake Balkhash) (Doctoral Dissertation). St. Petersburg State University, Department of Geography, St. Petersburg, Russia.

Babkin, V.I., 2008. Long-term fluctuations in flow of the Volga, Oka, Don, Dnieper and its methods of forecasting. Izvestiya Akademii Nauk SSSR - Seriya Geograficheskaya 3, 92–98.

Babkin, V.I., 2009. Assessment of periodicity and long-term prognosis of changes in water levels of lakes (on the example of Lake Ladoga and Lake Vättern. Memoirs of the Russian State Hydrometeorological University 220 (9), 5–11.

Babkin, V.I., Vorobiev, V.I., Smirnov, N.P., 2004. Fluctuations in runoff of the Ob, Yenisei and Lena and the dynamics of atmospheric circulation in the Northern Hemisphere. Russian Meteorology and Hydrology 1, 74–80.

Baer, K., Helmersen, G., 1886. Beitrage zur Kenntniss des Russischen Reiches und der angranzenden Lander Asiens (B. 1. V. 29, Hrsg, in German). Moscow: Academia Nauk USSR, Russia. 1995, p. 272.

Baker, R.G.V., 2008. Exploratory Analysis of Similarities in Solar Cycle Magnetic Phases with Southern Oscillation Index Fluctuations in Eastern Australia. Geographical Research 46 (4), 380–398.

Baker, R.G.V., Haworth, J., Flood, P.G., 2005. An oscillating Holocene sea-level? Revisiting Rottnest Island, Western Australia, and the Fairbridge Eustatic Hypothesis. Journal of Coastal Research SI42, 3–14.

Barashkova, N.K., 2007. Dynamic meteorology. Ed. TGU, Tomsk, Russia.

Bashkirtsev, V.S., Mashnich, G.P., 2008. Solar activity effects on the Earth's climate. 12th European Solar Physics Meeting ESPM-12, 8–12 September, 2008, Freiburg, Germany, p. 80.

Bashkirtsev, V.S., Mashnich, G.P., 2003. Awaits us if global warming in the coming years? Geomagnetism and Aeronomy 43 (1), 131–135.

Bashkirtsev, V.S., Mashnich, G.P., 2004a. The variability of the Sun and the Earth's climate. Solar-Terrestrial Physics 6 (119), 135–137.

Bashkirtsev, V.S., Mashnich, G.P., 2004b. Solar activity and Earth's climate prediction. In: Solar-terrestrial physics, abstracts of the international conference, Irkutsk, pp. 66–67.

Bashmakov, I.A., 2009. Low carbon Russia: 2050. Center for Energy Efficiency, Moscow, Russia.

Belov, A.V., Gaidash, S.P., Ivanov, K.G., Kanonidi, Kh. D, 2004. Unusually high geomagnetic activity in 2003. Cosmic Research 42 (6), 1–10.

Belov, A.V., Gaidash, S.P., Kanonidi, Kh. D, Kanonidi, K. Kh., Kuznetsov, V.D., Kuznetsov, V.D., Eroshenko, E.A., 2005a. Operative center of the geophysical prognosis in IZMIRAN. Annales Geophysicae 23, 3163–3170.

Belov, A.V., Guschina, R.T., Obridko, V.N., Shelting, B.D., Yanke, R.T., 2005b. On the possibility of prediction of long-term variations of cosmic rays on the basis of various indices of solar activity. In: Proceedings of the all-Russian conference, experimental and theoretical studies of the foundations of forecasting heliogeophysical activity, October 10–15, Troitsk. IZMIRAN Retrieved from: http://helios.izmiran.rssi.ru/Solter/prog2005/prog/abstracts.htm (in Russian).

Belov, A.V., Berkova, M., Voronova, I.V., Guschina, R.T., Dorman, L.I., Eroshenko, E.A., Zhelezniy, V.B., Kazansky, S.S., Libin, I. Ya., Pérez-Peraza, J., Treyger, K.F., Yanke, V.G., 2011a. History of the muon telescope and the hodoscope. In: Current socio-economic problems of the modern world: science and practice. MAOK, Moscow, Russia, pp. 423–433.

Belov, A.V., Berkova, M., Voronova, I.V., Eroshenko, E.A., Zhelezniy, V.B., Kazansky, S.S., Leyva-Contreras, A., Libin, I. Ya., Mikalajunas, M.M., Pérez-Peraza, J., Treyger, E.M., Jaani, A., Yanke, V.G., 2011b. Accounting for the temperature effect of the muon component of cosmic rays, used for diagnostics of the interplanetary medium in the vicinity of the Earth and the analysis of global atmospheric processes (analysis of modern models of temperature). In: Current socio-economic problems of the modern world: science and practice. MAOK, Moscow, Russia, pp. 434–467.

Belov, A.V., Berkova, M., Eroshenko, E.A., Zjelezniy, V.B., Korotkov, V., Leyva-Contreras, A., Libin, I. Ya., Mikalajunas, M.,M., Pérez-Peraza, J., Treyger, E.M., Jaani, A., Yanke, V.G., 2011c. Accounting for the effects of atmospheric processes in the observations of cosmic radiation on Earth (the effect of snow). In: Current socio-economic problems of the modern world: science and practice. MAOK, Moscow, Russia, pp. 468–474.

Belov, A.V., Berkova, M., Eroshenko, E.A., Zjelezniy, V.B., Leyva-Contreras, A., Libin, I. Ya, Mikalajunas, M.M., Pérez-Peraza, J., Smirnov, D., Treyger, E.M., Jaani, A., Yanke, V.G., 2011d. Accounting for the effects of atmospheric processes in the observations of cosmic radiation on Earth (the effect of temperature). In: Current socio-economic problems of the modern world: science and practice. MAOK, Moscow, Russia, pp. 475–484.

Bendat, G., Pirsol, A., 2008. Applications of correlation and spectral analysis. Nauka, Moscow, Russia.

Benestad, R.E., 2006. Solar activity and Earth's climate. Springer, Praxis Publishing Ltd, London, U.K.

Beer, J., Blinov, A., Bonani, G., Hofmann, H.J., Finkel, R.C., Lehmann, B., Oeschger, H., Sigg, A., Schwander, J., Staffelbach, T., Stauffer, B., Suter, M., Wötfli, W., 1990. Use of ^{10}Be in polar ice to trace the 11-year cycle of solar activity. Nature 347 (6289), 164–166.

Berg, L.S., 1938. Fundamentals of climatology. Russia, Leningrad: Public education and teacher publishing RSFSR People's Commissariat, USSR, p. 220.

Birman, B.A., 2007. The basic weather and climatic features of the Northern Hemisphere of the Earth: an analytical review. Hydrometeorological Research Center of the Russian Federation, Moscow, Russia.

Bjørn Helland-Hansen, Fridtjof Nansen, T., 1909. The Norwegian Sea. Its Physical Oceanography based upon the Norwegian Researches 1900-1904. Report on Norwegian Fishery and Marine-Investigations Vol. 11, No. 2. Kristiania, Det Mallingske Ogtrykkeri.

Blackman, R.B., Tukey, J.W., 1959. The measurement of power spectra, from the point of view of communications engineering. Dover Publications, p. 190. ISBN 0-486-60507-8.

Bogachev, S.A., Bugaenko, O.I., Ignatiev, A.P., Kuzin, S.V., Mitrofanov, A.V., Pertsov, A.A., Shestov, S.V., Slemzin, V.A., Suhodrev, N.K., Zhitnik, I.A., 2009a. The TESIS experiment on EUV imaging spectroscopy of the Sun. Advances in Space Research 43 (6), 1001–1006.

Bogachev, S.A., Kuzin, S., Shestov, S.V., Urnov, A.M., Zhitnik, I.A., 2009b. Diagnostics of plasma electron density structures along the lines of the solar corona ions FeXI–FeXIII range 176–207 Å in the experiment SPIRIT/CORONAS-F. Astronomy Letters 35 (1), 45–56.

Bochkareva, E.G., Romanov, L.N., 2007. The forecast of dangerous winds and precipitation using the method of plane rotations to the territory of Western Siberia. Meteorology and Hydrology 8, 5–16.

Borrero Francisco, Frances Scelsi Hess, Juno Hsu, Gerhard Kunze, Stephen A. Leslie, Stephen Letro, Michael Manga, Len Sharp, Theodore Snow, 2008. Earth science: geology, the environment, and the universe. Glencoe/McGraw-Hill, New York, p. 348.

Brasseur, G.P., Orlando, J.J., Tyndall, G.S., 1999. Atmospheric chemistry and global change. Oxford University Press, New York.

Braulov, K.A., Goliyandina, N.E., Nekrutkin, V.V., Solntsev, V.N., 1999. The method of "Caterpillar" and its possibilities, and some research directions. Multivariate statistical analysis and probabilistic modeling of real processes. In: Ayvazian, S.A. (Ed.), Abstracts of the International Jubilee session of the Scientific Seminar. CEMI RAS, Moscow, Russia, pp. 45–47.

Brand, S., Dethlof, K., Handorf, D., 2007. Atmospheric variability in a coupled atmosphere-ocean-sea ice model with interactive stratospheric ozone chemistry. The Open Atmospheric Science Journal 2, 6–19.

Brooks, C.E.P., Mirrlees, S.T.A., 1932. A study of the atmospheric circulation over tropical Africa. Geophysical Memoirs 55, 3–109.

Bruevich a, Yakunina, G.V., 2011. Solar Activity Indices in 21,22 and 23 Cycles, Los Alamos Laboratory: arXiv:1102.5502v1[astro-ph.SR].

Bucha, V., 1988. Influence of solar activity on atmospheric circulation types. Annales Geophysicae 6, 513–524.

Bushuev, V.V., Golubev, S.V., Plujnikov, V.B., 2002. Oil and cycles of solar activity. Energy Policy 1, 53–56.

Carlson, T.N., Prospero, J.M., 1972. The large-scale movement of Sahara air outbreaks over the northern equatorial Atlantic. Journal of Applied Meteorology 11, 283–297.

Carslaw, K.S., Harrison, R.G., Kirkby, J., 2002. Atmospheric science: cosmic rays, clouds, and climate. Science 298 (5599), 1732–1737. doi:10.1126/science.1076964.

Chan, J.C.L., 2006. Comment on "Changes in Tropical Cyclone number, Duration and Intensity in a Warming Environment". Science 311, 1713a.

Chernavskaya, M.M., Kononova, N.K., Val'chuk, T.E., 2006. Correlation between atmospheric circulation processes over the Northern Hemisphere and parameter of solar variability during 1899–2003. Advances in Space Research 37 (8), 1640–1645.

Cherry, N.J., 2002. Schumann Resonances, a plausible biophysical mechanism for the human health effects of solar/geomagnetic activity. Natural Hazards 26, 279–331.

Cheshihin-Vetrinsky, E.V., 1879. Livonian Chronicle, Herman Warthberg, Volume II. Latvia, Riga.

Chiapello, I., Prospero, J.M., Herman, J.R., Hsu, N.C., 1999. Detection of mineral dust over the North Atlantic Ocean and Africa with the Nimbus 7 TOMS. Journal of Geophysical Research 104, 9277–9291.

Chiapello, I., Moulin, C., 2002. TOMS and METEOSAT satellite records of the variability of Saharan dust transport over the Atlantic during the last two decades (1979–1997). Geophysical Research Letters 29 (8), 1176. doi:10.1029/2001GL013767.

Charakhchian, T.N., 1979. 11-year modulation of galactic cosmic rays in interplanetary space and solar activity. Studies on Geomagnetism and Aeronomy 34, 12–27.

Chertkov, A.D., 1985. The solar wind and the internal structure of the Sun. Nauka, Moscow, Russia.

Chijzhevsky, A.L., 1976. Terrestrial echo of solar storms. Mysl, Moscow, Russia.

Chijzhevsky, A.L., 1924. Physical factors of the historical process. In: Kaluga, Abbrev. (Ed.), Chemistry and Life (1990) Nos. 1–3, 22–32, 82–90, 22–33.

Chijzhevsky, A.L., 1995. Cosmic pulse of life: Earth in the embrace of the Sun. Geliotaraksiya. M.: Misl, Moscow, Russia, p. 179.

Chirkov, N.P., 1978. On the cyclicality of geomagnetic activity and solar wind velocity. Izvestiya Akademii Nauk SSSR—Seriya Fizicheskaya 42 (5), 1016–1017.

Chistiakov, V.F., 2000. Oscillations of the solar radius at the epoch of minimum Maunder and Dalton. In: Solar activity and its influence on the Earth. Vladivostok: Dal'nauka, Russia, pp. 84–107.

Chronis, T., Williams, E., Emmanouil, A., Petersen, W., 2007. African lightning: indicator of tropical Atlantic cyclone formation. EOS 88 (40), 397.

Clayton, H.H., 1933. Sunspots and the weather (Abstract of papers by Schostakovich & Memery). Bulletin of the American Meteorological Society 14 (3), 65–69.

Climate Change, 2007. Synthesis Report. Contribution of Working Groups I, II and III to the Fourth Assessment Report of the Intergovernmental Panel on Climate Change, 2007. In: Pachauri, R.K. and Reisinger, A. (Eds.), Core Writing Team, IPCC, Geneva, Switzerland. p. 104.

Cook, E.R., Meko, D.M., Stockton, C.W., 1997. A new assessment of possible solar and lunar forcing of the bi-decadal drought rhythm in the western United States. Journal of Climate 10, 1343.

Danilov, D.L., Solntsev, V.N., 1997. On the method of "Caterpillar." The basic ideas and methods of implementation. In: Danilov, D.L., Zhiglyavskogo, A.A. (Eds.), The main components of time series: the method of "Caterpillar". St. Peterburg, Univ. house PRESSKOM, pp. 48–72.

Demaria, M., Kaplan, J., 1999. An updated statistical hurricane intensity prediction scheme (SHIPS) for the Atlantic and eastern North Pacific Basins. Wea. Forecasting 14, 326–337.

DeMaria, M., Balk, J.-J., Kaplan, J., 1993. Upper level eddy angular momentum fluxes and tropical cyclone intensity change. Journal of the Atmospheric Sciences 50, 1133–1147.

Denkmayr, K., 1993. On long-term predictions of solar activity (Diploma thesis). University of Linz, Linz, Austria.

Dentener, F.J., Carmichael, G.R., Zhang, Y., Lelieveld, J., Crutzen, P.J., 1996. Role of mineral aerosol as a reactive surface in the global atmosphere. Journal of Geophysical Research 101, 22.

Dergachev, V.A., 2006. The impact of solar activity on climate. Izvestiya Akademii Nauk—Rossiyskaya Akademiya Nauk, Seriya Fizicheskaya 70 (10), 1544–1548.

Dergachev, V.A., 2009. Cosmogenic radionuclides 14C and 10Ve: solar activity and climate. Izvestiya Akademii Nau—Rossiyskaya Akademiya Nauk, Seriya Fizicheskaya 73 (3), 399–401.

Dergachev, V.A., Raspopov, O.M., 2000. Long-term processes on the Sun, determine trends in solar radiation and surface temperature of the Earth. Geomagnetism and Aeronomy 40 (3), 9–14.

Dergachev, V.A., Dmitriev, P.B., 2005. Periodic variations of cosmic rays on ground-based monitors from 1953 to 2004. In: Proceedings of the all-Russian conference, experimental and theoretical studies of the foundations of forecasting heliogeophysical activity, October 10–15, 2005, Troitsk, IZMIRAN Retrieved from: http://helios.izmiran.rssi.ru/Solter/prog2005/prog/abstracts.htm (in Russian).

Dhanju, M.S., Sarabhai, V.A., 1967. Short-period variations of cosmic-ray intensity. Physical Review Letters 3, 252–255.

Dhanju, M.S., Sarabhai, V.A., 1970. Fluctuations of Cosmic Ray Intensity at the Geomagnetic Equator and Their Solar and Terrestrial Relationship. Journal of Geophysical Research 5, 1795–1801.

Diaz, J.P., Expósito, F.J., Torres, C.J., Herrera, F., Prospero, J.M., Romero, M.C., 2001. Radiative properties of aerosols in Saharan dust outbreaks using ground-based and satellite date: applications to radiative forcing. Journal of Geophysical Research 106, 18403.

Dickerson, R.R., Kondragunat, S., Stenchikov, G., Civerolo, K.L., Doddridge, B.G., Holben, B.N., 1997. The impact of aerosols on solar ultraviolet radiation and photochemical smog. Science 278, 827–830.

Djenkins, G., Watts, D., 1972. Spectral analysis and its applications, vols. 1–2. Mir, Moscow, Russia.

Dmitrieva, I.V., Khabarova, O.V., Ragoulskaia, M.V., 1999. Influence of natural variations of electromagnetic terrestrial field on electrical conductivity of acupunctural points. In: Proceedings of the second international conference, electromagnetic fields and human health, September 20–24, Moscow, pp. 363–364.

Dmitrieva, I.V., Khabarova, O.V., Obridko, V.N., Ragoulskaia, M.V., Reznikov, A.E., 2000. Experimental confirmations of the bioeffective influence of magnetic storms. Astronomical and Astrophysical Transactions 19 (1), 54–59.

Doganovsky, A.M., 1982. Cyclical fluctuations of lake levels in the last century. Geography and Natural Resources 3, 152–156 (in Russian).

Doganovsky, A.M., 1990. Patterns of vibration levels of lakes and their influence on the basic elements of the regime of water bodies. In: Proceedings of V All-Union Hydrological Congress, vol. 8. Gidrometeoizdat, Leningrad, Russia.

Domysheva, V.M., Sakirko, M.V., Panchenko, M.V., Pestunov, D.A., 2006. The interaction of CO_2 between the atmosphere and surface waters of Lake Baikal and the influence of water composition. In: Advances in the geological storage of carbon dioxide, international approaches to reduce anthropogenic greenhouse gas emissions, NATO science series, IV: Earth and environmental sciences, vol. 65. Springer, The Netherlands, pp. 35–45. Part 1.

Domysheva, V.M., Panchenko, M.V., Pestunov, D.A., Sakirko, M.V., 2007. Influence of atmospheric precipitation on the CO_2 exchange with the water surface of Lake Baikal. Doklady Academy of Sciences 414 (5), 1–4.

Dorman, L.I., 1967. Modulation of cosmic rays in interplanetary space. In: Cosmic rays. Nauka, Moscow, Russia, pp. 305–320 (No. 8).

Dorman, L.I., 1975. Experimental and theoretical basis of cosmic ray astrophysics. Nauka, Moscow, Russia.

Dorman, L.I., 1982. Sun and galactic cosmic rays. Nauka, Moscow, Russia.

Dorman, L.I., Libin, I. Ya., 1984. Cosmic ray scintillations. Space Science Reviews 27, 91.

Dorman L.I., 1989. Solar activity and Geophisical processes. Preprint IZMIRAN, M.: IZMIRAN, p. 32.

Dorman, L.I., 1991. Cosmic ray modulation. Nuclear Physics 22B (Suppl.), 21–45.

Dorman, L.I., 1998. On the prediction of great flare energetic particle events to save electronics on spacecrafts. In: Meeting of Israel Physics Society. Weizmann Institute of Science, Israel, pp. 18–19.

Dorman, L.I., 2005. Variations of cosmic rays and solar-terrestrial relations. Tel-Aviv: Gaavaht. p. 192.

Dorman, L.I., 2006. Long-term cosmic ray intensity variation and part of global climate change, controlled by solar activity through cosmic rays. Advances in Space Research 37 (8), 1621–1628.

Dorman, L.I., 2009. The role of space weather and cosmic ray effects in climate change. In: Letcher, T.M. (Ed.), Climate change: observed impacts on planet Earth. Elsevier, St. Louis, pp. 43–76.

Dorman, L.I., Kozin, I.D., Sacuk, V.V., Seregina, N.G., Churunova, L.F., 1978. The study of hysteresis phenomena, fluctuations in barometric and ionospheric effects in cosmic rays. Izvestiya Akademii Nauk SSSR - Seriya Fizicheskaya 42 (7), 1501–1506.

Dorman, L.I., Libin, I. Ya., Blokh, Ya. L., 1979. Scintillation method in the study of cosmic ray variations. Nauka, Moscow, Russia.

Dorman, L.I., Libin, Ya. I., 1984. Cosmic ray scintillations and dynamic processes in space. Space Science Reviews 39, 91–102.

Dorman, L.I., Libin, Ya. I., 1985. Short-period variations in cosmic ray intensity. Successes of Physical Sciences 145 (3), 403–440.

Dorman, L.I., Libin, I. Ya., Yudakhin, K.F., 1986a. The role of the energy spectrum anisotropy in variations of the cosmic ray fluctuations power-spectrum before interplanetary medium disturbances. Astrophysics and Space Science 123, 53–58.

Dorman, L.I., Libin, I. Ya., Mikalajunas, M.M., 1986b. On possibility of the influence of cosmic factors on the weather: cosmic factors and shtormistost (tornado). In: Regionne Hidrometeorologia, vol. 10. Academy of Scince of the Lithuanian SSR, Vilnius, Lithuania Hidrology and Agricultural Chemistry.

Dorman, L.I., Libin, I. Ya., Mikalajunas, M.M., Yudakhin, K.F., 1987a. Variations of space physics and geophysical parameters in 18–21 cycles of solar activity. Geomagnetism and Aeronomy 27 (3), 483–485.

Dorman, L.I., Libin, I. Ya., Mikalajunas, M.M., Yudakhin, K.F., 1987b. Relationship of space physics and geophysical parameters in 19–20 cycles of solar activity. Geomagnetism and Aeronomy 27 (2), 303–305.

Dorman L.I., Gulinsky, O.V., Libin I. Ya., Sitnov, A.M., Starkov, F.A., Khamirzov, Kh., Chechenov, A., Schoya, L., Yudakhin, K.F., 1983. Joint analysis of the data register of common muon and neutron component of cosmic ray intensity May 7, 1978. In book "Cosmic Ray", M.: Radio and communication. pp. 20–26.

Dorn, W., Dethlof, K., Rinke, A., Kurgansky, M., 2008. The recent decline of the arctic summer sea-ice cover in the context of internal climate variability. The Open Atmospheric Science Journal 2, 91–100.

Dragan, Ya. P., Rojkov, V.A., Yavorsky, N.N., 1984. Application of the theory of periodically correlated random processes to the probabilistic analysis of oceanographic time series. In: Probabilistic modeling and analysis of oceanographic time series. Gidrometeoizdat, Leningrad, Russia, pp. 4–23.

Drummond, C.N., Wilkinson, B.H., 2006. Interannual variability in millennial climate proxy data. Journal of Geology 114, 325–339.

Duce, R.A., 1995. Distributions and fluxes of mineral aerosol. In: Charlson, R.J., Heintzenberger, J. (Eds.), Aerosol forcing of climate. Wiley, Chichester, U.K, pp. 43–72.

Dunion, J.P., Velden, C.S., 2004. The impact of the Saharan Air Layer on Atlantic tropical cyclone activity. Bulletin of the American Meteorological Society 85 (3), 353–365.

Durand, G., 1968. La imaginación simbólica (The symbolic imagination). Amorrortu Editores, Buenos Aires.

Eddy, J.A., 1984. The Maunder minimum. Science 192 1189–1902.

Efroimson, V.P., 2002. Genetics of genius. Russia, Moscow: Nauka (pp. 65–66).

Efron, B., 1979. Bootstrap methods: another look at the jackknife. Annals of Statistics 7 (1), 1–26. doi:10.1214/aos/1176344552.

Elansky, N.F., 2008. Environmental monitoring: validation of a system for observing the interaction space means of terrestrial ecosystems and the atmosphere. Environmental Engineering 4, 4–23.

Elsner, J.B., Kara, A.B., 1999. Hurricanes of the North Atlantic climate and society. Oxford University Press, New York.

Elsner, J.B., Kavlakov, S.P., 2001. Hurricane intensity changes associated with geomagnetic variation. Atmospheric Science Letters 2, 86–93.

Emanuel, K., 2005a. Meteorology: Emanuel replies. Nature 438 (7071), E13.

Emanuel, K., 2005b. Increasing destructiveness of tropical cyclones over the past 30 years. Nature 436 (7051), 686–688.

Erlykin, A., 2010, February 24. Cosmic rays, climate and the origin of life. CERN Courier.

Eselevich, V.G., 2002a. Relationship of the ray structure of the coronal streamer belt and abrupt large increases in solar wind plasma density at the Earth's orbit. In: Eselevich, V.G., Eselevich, M.V. (Eds.), Third Russian-Chinese conference on space weather, Irkutsk: SibIZMIR, 19–21 June, 2002, pp. 12–13.

Eselevich, V.G., 2002b. Large and fast solar wind ion flux (density) pulses and their possible solar source. In: Riazantseva, M.O., et al. (Eds.), The 27th General Assembly of the European Geophysical Society (EGS), Nice, France, 21–26 April 2002, Geophysical Research Abstracts, 4.

Eselevich, V.G., Eselevich, M.V., 2002c. Study of the nonradial directional property of the rays of the streamer belt and chains in the solar corona. Solar physics 208 (1), 5–16.

Evan, A.T., Dunion, J., Foley, J.A., Heidinger, A.K., Velden, C.S., 2006. New evidence for a relationship between Atlantic tropical cyclone activity and African dust outbreaks. Geophysical Research Letters 33, L19813. doi:10.1029/2006GL026408.

Falkowski, P.G., Barber, R.T., Smetacek, V., 1998. Biogeochemical controls and feedbacks on ocean primary production. Science 281, 200.

Farge, M., 1992. Wavelets and turbulence. In: Legras, B. (Ed.), Geophysical fluid dynamics. publication CNRS-INSU, pp. 25–32.

Fastrup, B., Pedersen, E., Lillestol, E., Thorn, E., Bosteels, M., Gonidec, A., Harigel, G., Kirkby, J., Mele, S., Minginette, P., Nicquevert, B., Schinzel, D., Seidl, W., Grundsøe, P., Marsh, N., Polny, J., Svensmark, H., Viisanen, Y., Kurvinen, K., Orava, R., Hämeri, K., Kulmala, M., Laakso, L., Mäkelä, J.M., O'Dowd, C.D., Afrosimov, V., Basalaev, A., Panov, M., Laaksonen, A., Joutsensaari, J., Ermakov, V., Makhmutov, V., Maksumov, O., Pokrevsky, P., Stozhkov, Y., Svirzhevsky, N., Yin, Y., Trautmann, T., Arnold, F., Wohlfrom, K.-H., Hagen, D., Schmitt, J., Whitefield, P., Aplin, K., Harrison, R.G., Bingham, R., Close, F., Gibbins, C., Irving, A., Kellett, B., Lockwood, M., Petersen, D., Szymanski, W.W., Wagner, P.E., Vrtala, A., 2001. A study of the link between cosmic rays and clouds with a cloud chamber at the CERN PS, Los Alamos National Laboratory. Retrieved from: http://arxiv.org/abs/physics/0104048v1.

Filatova, T.N., Kvon, V.I., Solntsev, V.N., Nechaev, M.V., Avinsky, V.A., Arshanitsa, N.M., 2003. Trends in the elements of the hydrological regime of Peipsi-Pskov Lake in relation to the assessment of its ecological status (A collection of papers on hydrology, No. 26) pp. 172–199. Gidrometeoizdat, St. Petersburg, Russia.

Forbush, S.E., 1946. Three unusual cosmic-ray increases possibly due to charged particles from the Sun. Physical Review 70, 771–772.

Forbush, S.E., 1954. World-wide cosmic-ray variations, 1937–1952. Journal of Geophysical Research 59 (4), 525–542.

Forbush, S.E., 1958. Cosmic-ray intensity variations during two solar cycles. Journal of Geophysical Research 63 (4), 651–669.

Freeman, C., Louca, F., 2001. As time goes by: from the industrial revolutions to the information revolution. Oxford University Press, Oxford, New York.

Friis-Christensen, E., Lassen, K., 1991. Length of the solar cycle: an indicator of solar activity closely associated with climate. Science 254 (5032), 698–700. doi:10.1126/science.254.5032.

Friis-Christensen, E., Lassen, K., 1992. Global temperature variations and a possible association with solar activity variations (Scientific Report No 92–3). Danish Meteorological Institute, Copenhagen.

Friis-Christensen, E., Svensmark, H., 2003. What do we really know about the Sun-climate connection pp. 913–921. Retrieved from: www-ssc.igpp.ucla.edu/IASTP/43.

Gallegos-Cruz, A., Pérez-Peraza, J., 1995. Derivation of analytical particle spectra from the solution of the transport equation by the WKBJ method. The Astrophysical Journal 446 (1), 400–420.

Gazina, E.A., Klimenko, V.V., 2008. Analysis of climate change in Eastern Europe in the last 250 years of instrumental data. Vestnik Moskovskogo Universiteta 1, 60–66.

Gierenes, K., Ponater, M., 1999. Variation of cosmic ray flux and global cloud coverage—a missing link in solar-climate relationship. Journal of Atmospheric and Terrestrial Physics 61 (11), 795–797.

Gilbert, E., Cotterell, M., 2000. The Mayan prophecies: unlocking the secrets of a lost civilization. Veche, Moscow, Russia.

Gillman, M., Erenler, H., 2008. The galactic cycle of extinction. International Journal of Astrobiology 7 (1), 17–26. doi:10.1017/S1473550408004047.

Ginoux, P., Chin, M., Tegen, I., Prospero, J., Holben, B., Dubovik, O., Lin, S.-J., 2001. Sources and global distributions of dust aerosols simulated with the GOCART model. Journal of Geophysical Research—Atmospheres 106 (D17), 20255–20273.

Ginoux, P., Prospero, J.M., Torres, O., Chin, M., 2004. Long-term simulation of global dust distribution with the GOCART model: correlation with North Atlantic Oscillation. Environmental Modelling and Software 19 (21), 113–128.

Ginzburg, V.L., 1947. Radio emission from the Sun and the galaxy. Success in Physical Science 32, 26.

Ginzburg, V.L., 1948. New data on the radio emission on the Sun and the galaxy. Success in Physical Science 34, 1.

Glokova, E.S., 1952. Some data on the effect of variations of cosmic rays on solar activity cycle Proc. NIIZM, M.: NIIZM 8, 59–70.

Gnevyshev, M.N., 1963. The crown and the 11-year cycle of solar activity. Astronomer Magazine 40 (3), 401–412.

Goldenberg, S.B., Landsea, C.W., Mestas-Nuez, A.M., Gray, W.M., 2001. The recent increase in Atlantic hurricane activity: causes and implication. Science 293 (5529), 474–479.

Goliyandina, N.E., Solntsev, V.N., Filatova, T.N., Jaani, A., 1997. The study of periodic component in the dynamics of hydrological indicators. In: Danilov, D.L., Zhiglyavskogo, A.A. (Eds.), The main components of time series: the method of "Caterpillar." St. Petersburg State University, St. Petersburg, Russia.

Goliyandina, N.E., Nekrutkin, V., Zhglyavivsky, A., 2001. Analysis of time series structure: SSA and related techniques. Chapman & Hall, London, England.

Gorbatenko, V.P., Ippolitov, I.I., Podnebesnikh, N.V., 2007. The atmospheric circulation over West Siberia in 1976–2004 years. Meteorology and Hydrology 5, 28–36.

Gray, W.M., 1990. Strong association between West African rainfall and U.S. landfall of intense hurricanes. Science 249 (4974), 1251–1256.

Gray, L.J., Haigh, J.D., Harrison, R.G., 2005. The influence of solar changes on the Earth's climate. Hadley Centre Technical Note Series 62.

Grinsted, A., Moore, J., Jevrejera, S., 2004. Application of the cross wavelet transform and wavelet coherence to geophysical time series. Nonlinear Processes in Geophysics 11, 561–566.

Gritsevitch, I.G., Garnak, A., Kokorin, A.O., Safonov, G.V., 2008. Economic development and tackling climate change. Danish Energy Agency, Moscow, Russia.

Gro Harlem, (Ed.), 1987. "Our Common Future" Brundtland, the Prime Minister of Norway Oxford University Press, ISBN 0-19-282080-X.

Gulinsky, O.V., Guschina, R.T., Dorman, L.I., Libin, I. Ya., Mikalajunas, M.M., Yudakhin, K.F., 1992. Modeling the mechanism of action heliophysical parameters on atmospheric processes. In: Cosmic Rays. Nauka, Moscow, Russia, pp. 98–106 (No. 26).

Gulinsky, O.V., Glushkov, V., Leyva-Contreras, A., Libin, I. Ya., Pérez-Peraza, J., Yudakhin, K.F., 2002. Mathematical and statistical explorations of the regional climate variabilities. Publishing Moscow State Government.

Gushchina, R.T., Dorman, I.V., Dorman, L.I., Pimenov, I.A., 1968. The impact of solar activity on the electromagnetic conditions in interplanetary space from the data on the effects of modulation of cosmic rays. Izv. Academy of Science USSR, ser.fiz., t.32. Number 11. pp. 1924–1928.

Guschina, R.T., Dorman, L.I., 1970. Helioshirotny index of solar activity and HL 11 year variations of cosmic rays, Izvetia Akad. 34-11, 2426–2433.

Guschina, R.T., Dorman, L.I., Ilgach, S.F., Kaminer, N.S., Pimenov, I.A., 1970. UV index HL and annual variations of cosmic rays, Izvetia Akad Ser. Phys., 34-11, 2434–2438.

Guschina, R.T., 1983. Relationships of various of solar activity to long-term changes in cosmic rays. Geomagnetism & Aeronomy 23-3, 378–381.

Haigh, J.D., 2001. Climate variability and the influence of the Sun. Science 294 (5549), 2109–2111.

Haigh, J.D., Lockwood, M., Giampapa, M.S., 2005. The Sun, solar analogs and the climate (Saas-Fee Advanced Course 34, Swiss Society for Astrophysics and Astronomy). Springer-Verlag, Berlin, Germany.

Halberg, F., Cornélissen, G., Otsuka, K., Watanabe, Y., Katinas, G.S., Burioka, N., Delyukov, A., Gorgo, Y., Zhao, Z.Y., Weydahl, A., Sothern, R.B., Siegelova, J., Fiser, B., Dusek, J., Syutkina, E.V., Perfetto, F., Tarquini, R., Singh, R.B., Rhees, B., Lofstrom, D., Lofstrom, P., Johnson, P.W.C., Schwartzkopff, O., 2000. International BIOCOS Study Group. Cross-spectrally coherent ~10.5- and 21-year biological and physical cycles, magnetic storms and myocardial infarctions. Neuroendocrinol Lett 21, 233–258.

Hale, G.E., Ellerman, F., Nicholson, S.B., Joy, A.H., 1919. The Magnetic Polarity of Sun-Spots. Astrophys. J. 49, 153–178.

Hamilton, T.S., Evans, M.E., 1983. A magnetostratigraphic and secular variation study of Level Mountain, northern British Columbia. Geophys. J. Roy. Astron. Soc. 73, 39–49.

Hansen, 2009. James Storms of My Grandchildren. London: Bloomsbury Publishing. p. 242. ISBN: 1408807459.

Haywood, J., Boucher, O., 2005. Estimates of the direct and indirect radiative forcing due to tropospheric aerosols: a review. Reviews of Geophysics 38 (4), 513–543.

Hodges, R.E., Elsner, J.B., 2011. Evidence linking solar variability with U.S. hurricanes. International Journal of Climatology 31 (13), 1–41. doi:10.1002/joc.2196. Also available in (2010) 29th Conference on Hurricanes and Tropical Meteorology, Tucson, AZ.

Hoerling, M., Hurrel, J., Eischeid, J., Phillips, A., 2006. Detection and attribution of twentieth-century northern and southern African rainfall change. Journal of Climate 19, 3990.

Holmes, D.G., 2003. Pulse width modulation for power converters: principles and practice. John Wiley & Sons, New York.

Huber, P.J., 1981. Robust statistics. John Wiley & Sons, New York.

Hudgins, L., Friehe, C., Mayer, M.E., 1993. Wavelet transforms and atmospheric turbulence. Physical Review Letters 71, 3279–3282.

Hurrell, J.W., 1995. Decadal trends in the North Atlantic oscillation: regional temperatures and precipitation. Science 269 (5224), 676–679.

Husar, R.B., Prospero, J.M., Stowe, L.L., 1997. Characterization of tropospheric aerosols over the oceans with the NOAA advanced very high resolution radiometer optical thickness operational product. Journal of Geophysical Research 102 (D14), 16889.

http://photojournal.jpl.nasa.gov/targetFamily/Sun?sort=ASC&start=700

http://klimatkomfort.ru/modules/pictures/viewcat.php?id=24&cid=4&min=0&orderby=titreA&show=20 (in Russian).

Intergovernmental Panel on Climate Change (IPCC), 2001. Climate change: the scientific basis. In: Houghton, J.T., et al. (Eds.), Working Group I contribution to the Third Assessment Report (TAR) of the Intergovernmental Panel on Climate. Cambridge University Press, New York.

Intergovernmental Panel on Climate Change (IPCC), 2007. Climate change: the physical science basis. In: Working Group I contribution to the Intergovernmental Panel on Climate Change Fourth Assessment Report (FAR): impacts, adaptation and vulnerability. Cambridge University Press, New York.

Ishkov, V.N., Shibaev, I.G., 2005. Cycles of solar activity: general characteristics and boundaries of modern forecasting. In: Proceedings of the all-Russian conference, "Experimental and theoretical studies of the foundations of forecasting heliogeophysical activity," 10–15, October 2005, Troitsk, IZMIRAN Retrieved from: http://helios.izmiran.rssi.ru/Solter/prog2005/prog/abstracts.htm (in Russian).

Ishkov, V.N., 2007. Sun in the current 23rd solar activity cycle. Retrieved from: http://crydee.sai.msu.ru/ Universe_and_us/4num (in Russian).

Ishkov, V.N., 2008. POP-UP magnetic fluxes and flare phenomena on the Sun (Dissertation for the degree of candidate of physical and mathematical). Nauka, IZMIRAN, Russia.

Ivanov, V.V., 2002. Periodic weather and climate variations. Soviet Uspekhi Fizicheskikh Nauk 172 (7), 777–811 (in Russian).

Ivanov-Kholodny, G.S., Chertoprud, V.E., 2005. Relationship between the quasi-biennial variations in the processes of the Sun and Earth. In: Proceedings of the all-Russian conference "Experimental and theoretical studies of the foundations of forecasting heliogeophysical activity, " 10–15, October 2005, Troitsk, IZMIRAN Retrieved from: http://helios.izmiran.rssi.ru/Solter/prog2005/prog/abstracts.htm (in Russian).

Ivaschenko, Yu. V., 2001. Cyclicality of global crises in Russia's development and their nature. In: The cycles of nature and society: Mater IX International Conference, Stavropol, Sept. 25–28. Stavropol University V. D. Chursin, Stavropol, Russia, pp. 148–149.

Jaani, A., 1973. Cyclic changes in the abundance of water. Eesti Loodus 12, 758–764 (in Estonian).

Jaani, A., 1987. How to call Tchudskoe Lake? Izvestiya Akademii Nauk SSSR—Seriya Biologicheskaya 36 (2), 169–172.

James Hansen, Reto Ruedy, Makiko Sato, Ken Lo, If It's That Warm, How Come It's So Darned Cold? An Essay on Regional Cold Anomalies within Near-Record Global Temperature http://www.columbia.edu/~jeh1/mailings/2010/20100127_TemperatureFinal.pdf, http://www.realclimate.org/index.php/archives/2010/01/2009-temperatures-by-jim-hansen/

Jarvinen, B.R., Neumann, C.J., Davis, M.A., 1984. A tropical cyclone data tape for the North Atlantic basin, 1886–1983: contents, limitations, and uses. NOAA Technical Memorandom: National Weather Service/National Hurricane Center (NWS/NHC) 22.

Jickells, T.D., An, Z.S., Andersen, K.K., , Baker, A.R., Bergametti, G., Brooks, N., Cao, J.J., Boyd, P.W., Duce, R.A., Hunter, K.A., Kawahata, H., Kubilay, N., IaRoche, J., Liss, P.S., Mahowald, N., Prospero, J.M., Ridgwell, A.J., Tegen, I., Torres, R., 2005. Global iron connections between desert dust, ocean biogeochemistry, and climate. Science 308 (5718), 67–71.

Jones, P.D., 1986. Northern Hemisphere surface air temperature variations: 1851–1984. Journal of Climate and Applied Meteorology 25, 161–179.

Juglar, C., 1889. Des crises commerciales et de lear retour périodique, second ed. France (in French), Paris: Piccard, pp. 153–163.

Kachanov, S.A., Kozlov, K.A., 2008. The problems of monitoring and forecasting of emergencies in the Arctic and the Far North of the Russian Federation. In: Proceedings of the conference "Ensuring integrated safety of the northern regions of the Russian Federation", April 22, 2008. NCUKS MCS Russia, Moscow, Russia, pp. 140–154.

Kalashnikov, B.G., Moroshkin, Yu. V., Skopintsev, V.A., 2002. Assessment of the seasonal cycles of accidents in electric power systems. Electricity 7, 2–8.

Kanipe, J., 2006. Climate change: a cosmic connection. Nature 443, 141–143.

Kattsov, V.M., Malevsky-Malevich, S.P., Mokhov, I.I., Nadezhdina, E.D., Semenov, V.A., Sporyshev, P.V., Khon, Ch. V., 2003. Anthropogenic climate changes in Russia in the 21st century: an ensemble of climate model projections. Russian Meteorology and Hydrology 4, 38.

Kattsov, V.M., Alekeev, G.A., Pavlova, T.V., Sporyshev, P.V., Bekryaev, R.V., Govorkova, V.A., 2007a. Modeling the evolution of the ice cover of the world ocean in the 20th and 21st centuries. Izvestiya RAN: Physics of the Atmosphere and Ocean 43 (2), 165–181.

Kattsov, V.M., Meleshko, V.P., Chicherin, S.S., 2007b. Climate change and the national security of the Russian Federation. Journal of Law and Security 1–2, 22–23.

Kaufman, Y.K., Tanré, D., Boucher, O., 2002. A satellite view of aerosols in the climate system. Nature 419, 215.

Kavlakov, S.P., 2005. Global cosmic ray intensity changes, solar activity variations and geomagnetic as North Atlantic hurricane precursors. International Journal of Modern Physics 20 (29), 6699–6701.

Kavlakov, S.P., Elsner, J.B., Pérez-Peraza, J., 2008a. Atlantic hurricanes, geomagnetic changes and cosmic ray variations, part I. geomagnetic disturbances and hurricane intensifications. In: International Union for Pure and Applied Physics (Ed.), Proceedings of the 30th ICRC, Mérida, Yucatán, Mexico, vol. 1, pp. 693–696 (SH).

Kavlakov, S.P., Elsner, J.B., 2008b. Atlantic hurricanes, geomagnetic changes and cosmic ray variations. Part II, Forbush decreases and hurricane velocity changes. In: International Union for Pure and Applied Physics (Ed.), Proceedings of the 30th ICRC, Mérida, Yucatán, Mexico, vol. 1, pp. 697–700 (SH).

Kavlakov, S.P., Elsner, J.B., Pérez-Peraza, J., 2008c. A statistical link between tropical cyclone intensification and major geomagnetic disturbances. Geofisica Internacional 47, 207–213.

Kazansky, S., Libin, I. Ya., Mikalajunas, M.M., 2011. On the possible impact of solar activity on long-term changes in precipitation. In: Current socio-economic problems of the modern world: science and practice. Ed. by MAOK, Moscow, Russia, pp. 485–488.

Keldysh, M.V., 1980. The creative legacy of Sergei Pavlovich Korolev. Selected works and documents. Nauka, Moscow, Russia.

Kerry, E., 2006. Hurricanes: tempests in a greenhouse. Physics Today 59 (8), 74–75.

Key, S., Marple, S.L., 1981. Spectrum analysis—a modern perspective. In: Proceedings of the IEEE 69 (11), 5–48.

Khorozov, S.V., Budovy, V.I., Martin, I.M., Medvedev, V.A., Belogolov, V.S., 2006. The influence of solar activity and cosmic rays on precipitation budget in different regions of the planet. In: The 36th COSPAR scientific assembly, 16–23 July, 2006, Beijing, China (TCI- 0232; C4.2-0056-06).

Khramova, M.N., Krasotkin, S.A., Kononovich, E.V., 2002. In: Sawaya-Lacoste, H. (Ed.), Proceedings of the second solar cycle and space weather euroconference, 24–29 September 2001, Vico Equense, Italy. ESA Publications Division, Noordwijk, The Netherlands, pp. 229–232 (ESA SP-477).

Kitchin, J., 1923. Cycles and trends in economic factors. Review of Economics and Statistics 5 (1), 10–16. doi:10.2307/1927031.

Klimenko, V.V., 2007a. The impact of climate change on the level of heat consumption in Russia. Energy 2, 2–8.

Klimenko, V.V., 2007b. Complete reconstruction of climate in the Russian Arctic XV–XX centuries. Vestnik Moskovskogo Universiteta, Seriya 6, Biologiya, 16–24 (in Russian).

Klimenko, V.V., 2008. Reconstruction of the climate of the Russian Arctic over the past 600 years on the basis of documentary evidence. Reports of the Academy of Sciences 418 (1), 110–113.

Klimenko, V.V., 2009a. A composite reconstruction of the Russian Arctic climate back to A.D. 1435. In: Przybylak, R., Majorowicz, J., Brázdil, R., Kejna, M. (Eds.), The Polish climate in the European context: an historical overview. Springer Verlag, Berlin, Germany, pp. 295–326.

Klimenko, V.V., 2009b. Climate: unread chapter of history. Publishing House MEI, Moscow.

Klimenko, V.V., Khrustalev, L.N., Mikushina, O.V., Emelianova, L.V., Ershov, E.D., Parmuzin, S. Yu., Tereshin, A.G., 2007. Climate change and the dynamics of permafrost in the north-west Russia in the next 300 years. M.: Earth's Cryosphere 11 (3C), 3–13.

Klimenko, V.V., Solomina, O.N., 2009. Climatic variations in the East European plain during the last millennium: state of the art. In: Przybylak, R., Majorowicz, J., Brázdil, R., Kejna, M. (Eds.), The Polish climate in the European context: an historical overview. Springer Verlag, Berlin, Germany, pp. 71–101.

Klimenko, V.V., Khrustalev, L.N., Emelianova, L.V., Ershov, E.D., Parmuzin, S. Yu., Mikushina, O.V., Tereshin, A.G., 2009. Reconstruction of the climate of the Russian Arctic over the last 600 years based on documented evidence. In: Battle on the ice, Arctic shelf in world politics and economy of the XXI century. M.: Tribune, pp. 232–237.

Klinov, V.G., 2002. U.S. impact on long business cycles formation. USA v Canada: Economics, Politics, Culture 2 (386), 33–49. Ed. by RAS (Russian Academy of Sciences).

Klinov, V.G., 2003. Scientific and technological progress and the larger cycles of the world market. M.: Forecasting Problems 1, 118–135.

Klotzbach, Philip J, 1986–2005. Trends in global tropical cyclone activity over the past twenty years. Geophysical Research Letters Vol. 33, L10805, p. 4, 2006 doi:10.1029/2006GL025881.

Knutson, T.R., Tuleya, R.E., 2004. Impact of CO_2-induced warming on simulated hurricane intensity and precipitation: sensitivity to the choice of climate model and convective parameterization. Journal of Climate 17 (18), 3477–3495.

Kononovich, E., 1967. Course of general astronomy. Nauka, Moscow, Russia.

Kocharov, G.E., Ogurtsov, M.G., 1999. The generation of solar protons in the last 415 years based on data on concentrations of nitrate in polar ice. Izvestiya Akademii Nauk SSSR—Seriya Fizicheskaya 63 (8), 16–19.

Kondratyev, Ya. K., 2004a. Priorities of global climatology. In: Proceedings of the Russian Geographical Society 136 (2), 1–25 (in Russian).

Kondratyev, Ya. K., 2004b. Global climate change: observations and numerical simulation results. Study of Earth from Space 2, 61–96.

Kondratyev, K. Ya., Nikolsky, G.A., 1982. Stratospheric mechanism of solar and anthropogenic influence on climate. In: Mir, (Ed.), Solar-terrestrial communications, weather and climate. Moscow, Russia, pp. 354–360.

Kondratyev, K. Ya., Nikolsky, G.A., 1983. The solar constant and climate. Solar Physics 89, 215–222.

Kondratyev, K. Ya., Nikolsky, G.A., 1995a. Solar activity and climate. 1. Observational data. Condensation and ozone hypothesis. Investigation of the Earth from Space 5, 3–17 (in Russian).

Kondratyev, K. Ya., Nikolsky, G.A., 1995b. Solar activity and climate. 2. The direct impact of changes in extra-atmospheric spectral distribution of solar radiation. Investigation of the Earth from Space 6, 3–17.

Kondratyev, K. Ya., Nikolsky, G.A., 2005. The impact of solar activity on the structural components of the Earth. 1. Weather conditions. A Study of Earth from Space 3, 1–10.

Kondratyev, K. Ya., Nikolsky, G.A., 2006. Impact of Solar Activity on Structure Component of the Earth. I. Meteorological Conditions // Il Nuovo Cimento. Geophysics and Space physics Vol. 29 C, (NO 2), 253–268.

Kondratiev, N.D., 1935. The long waves in economic life. Review of Economic Statistics 17 (6) 105–115.

Kondratiev, N.D., 1984. The long wave cycle. Richardson & Snyder, New York.

Kondratiev, N.D., Yakovets, J., Abalkin, L.I., 2002. Larger cycles and conditions theory prediction. M.: Orphus, p. 764.

Kononova, N.K., 2008. The growth of diurnal amplitude of air temperature in the Arctic region in the late XX–early XXI century as a risk factor for emergency situations. In: Proceedings of the conference “Ensuring integrated safety of the northern regions of the Russian Federation,” April 22, 2008. NCUKS MCS, Moscow, Russia, pp. 115–121.

Konstantinovskaya, L.V., 2001a. The classification scale solar cycles and global catastrophes. Bulletin of the Peoples’ Friendship University of Russia: Ecology and Life Safety 5, 86–88.

Konstantinovskaya, L.V., 2001b. Modern solar cycle and global catastrophes. In: Actual problems of ecology and environmental management: Sat scientific papers, vol. 2. RUDN, Moscow, Russia, pp. 51–55.

Kotlyakov, V., 1996. Variations of snow and ice in the past and at present on a global and regional scale. In: International Hydrological Programme. UNESCO, Paris, France (Technical documents in hydrology, 1).

Kotlyakov, V., 1997. Science. Society. Environment. Science, Moscow, Russia.

Kovalenko, V.A., 1983. The Solar Wind, M.: Nauka, p. 272.

Kovalishina, G., 2009. The invisible dangers of the climate change. Institute for Financial Studies, Vjscow, Russia.

Krainev, M.B., Webber, W.R., 2005a. The medium energy galactic cosmic rays according to the spacecraft and stratospheric data (Preprint No. 11). Lebedev Physical Institute RAS, Moscow, Russia.

Krainev, M.B., Webber, W.R., 2005b. The development of the maximum phase of solar cycle 23 in the galactic cosmic ray intensity. International Journal of Geomagnetism and Aeronomy 5, GI3008. doi:10.1029/2004GI000067.

Krainev, M.B., Kalinin, M.S., The models of the infinitely thin global heliospheric current sheet. Proc. 12th Intern. Solar Wind Conf., Saint-Malo, France, 2009. AIP Conf. Proc., 2010, v. 1216, 371–374.

Kristjánsson, J.E., Staple, A., Kristiansen, J., Kaas, E., 2002. A new look at possible connections between solar activity, clouds and climate. Geophysical Research Letters 29 (23), 2107. doi:10.1029/2002GL015646.

Kudela, K., Storini, M., Hofer, M.Y., Belov, A., 2000. Cosmic rays in relation to space weather. Space Science Reviews 93, 153–174.

Kudela, K., Storini, M., 2005. Cosmic ray variability and geomagnetic activity: A statistical study. Journal of Atmospheric and Terrestrial Physics 67, 907–912.

Kuzhevskaya, I.V., Dubrovskaya, L.I., 2007. Analysis and forecasting of meteorological data. IDO TGU, Tomsk, Russia.

Kuzhevsky, B.M., 2002. Science 4, 4–11 (In Russian).

Kuzhevsky, B.M., Nechaev, O. Yu., Panasyuk, M.I., Sigaeva, E.A., 2002. Studies of neutrons distributions near the Earth surface in order to predict space weather. In: Safrankova, J. (Ed.), Proceedings of the Week of Doctoral Students, 11–14 July, 2002, Czech, Prague. Czech Republic: Matfyzpress, Prague, pp. 83–87.

Kuznetsov, V.D., Oraevsky, V.N., 2002a. Polar-ecliptic patrol (PEP) for solar research and monitoring of space weather. Space and Rocket 1 (26), 51–59.

Kuznetsov, V.D., Oraevsky, V.N., 2002b. Polar-ecliptic patrol (PEP) for solar studies and monitoring of space weather. Journal of the British Interplanetary Society 55 (11–12), 398–403.

Kuzhevsky, B.M., 2005. The neutron field of the Earth. Geophysical Processes and Biosphere 4 (1–2), 18–26.

Kuzin, S.V., Bogachev, S.A., Zhitnik, I.A., Pertsov, A.A., Ignat’ev, A.P., Mitrofanov, A.V., Slemzin, V.A., Shestov, S.V., Suhodrev, N.K., Bugaenko, O.I., (2009) TESIS experiment on EUV imaging spectroscopy of the Sun. Advances in Space Research, 43 (6), 1001–1006.

Kuznetsov, V.D., Zeleny, L.M., 2008. Space projects on solar-terrestrial physics. Solar-Terrestrial Physics 12 (1), 83–92.

Kuznets, S., 1930. Secular movements in production and prices: their nature and their bearing upon cyclical fluctuations. Houghton Mifflin, Boston, MA.

Lamb, P.J., Peppler, R.A., 1992. Further case studies of tropical Atlantic surface atmospheric and oceanic patterns associated with sub-Saharan drought. Journal of Climate 5 (5), 476–488.

Landsea, C.W., 2005. Meteorology—hurricanes and global warming. Nature 438 (7071), E11–E13.

Landsea, C.W., Gray, W.M., Mielke, P.W., Berry, K.J., 1992. Long-term variations of Western Sahelian monsoon rainfall and intense U.S. landfalling hurricanes. Journal of Climate 5 (5), 1528–1534.

Landsea, C.W., Pielke, R.A., Mestas-Nunez, A., Knaff, J.A., 1999. Atlantic basin hurricanes: indices of climatic changes. Climatic Change 42 (1), 89–129.

Latif, M., 2009. Advancing climate prediction science—decadal prediction. In: World Climate Conference 3 (WCC-3), Geneva, Switzerland, 31 Aug–4 Sep 2009. World Meteorological Organization, Geneva, Switzerland.

Lau, W.K.M., Kim, K.-M., 2007a. How nature foiled the 2006 hurricane forecasts. EOS 88 (9), 105–107.

Lau, W.K.M., Kim, K.-M., 2007b. Reply to comment on "How nature foiled the 2006 hurricane forecasts." EOS 88 (26), 271.

Laut, P., 2003. Solar activity and terrestrial climate: an analysis of some purported correlations. Journal of Atmospheric and Solar-Terrestrial Physics 65, 801–812.

Lelieveld, J., et al., 2002. Global air pollution crossroads over the Mediterranean. Science 298 (5594), 794. doi:10.1126/science.1075457.

Leyva-Contreras, A., Libin, I. Ya., Pérez-Peraza, J., Jaani, A., 1996a. Temperature variations of the Baja California and a possible association with solar activity variations. In: The solar cycle: recent progress and future research meeting April 1–4, 1996, Hermosillo Sonora, Mexico. Universidad de Sonora, Sonora, Mexico, p. 24.

Leyva-Contreras, A., Libin, I. Ya., Pérez-Peraza, J., Jaani, A., 1996b. The solar radiation on the Earth and its possible communication with changes of the solar activity. In: The solar cycle: recent progress and future research meeting April 1–4, 1996, Hermosillo Sonora, Mexico. Universidad de Sonora, Sonora, Mexico, p. 25.

Levitin, A.E., 2006. Interaction between the solar wind with the magnetosphere. Institute of Terrestrial Magnetism, Ionosphere and Radio Wave Propagation, Troitsk, Russia.

Leighton, R.B., 1964. Transport of magnetic fields on the Sun. The Astrophysical Journal 140, 1547.

Liao, H., Seinfeld, J.H., 1998. Radiative forcing by mineral dust aerosol: sensitivity to key variables. Journal of Geophysical Research 103 (D24), 31637–31646.

Liao, H., Yung, Y.L., Seinfeld, J.H., 1999. Atmospheric chemistry-climate feedbacks. Journal of Geophysical Research 104, 23.

Libin, I., Perez-Peraza, J., Jaani, A., Mikalaiunas, M., 1999. Long fluctuations of the ice cover of the Baltic sea. Geografijos Metrastis. v.XXXII, 39–46. Vilnius, Mokslas.

Libin, Ya. I., 1983a. The study of fluctuations of cosmic rays during Forbush-decreases. Cosmic Rays (Kosmicheskie Luchi) 22, 21–43 (in Russian).

Libin, Ya. I., 1983b. The spectral characteristics of fluctuations of cosmic rays. Cosmic Rays (Kosmicheskie Luchi) 22, 14–20 (in Russian).

Libin, I. Ya., 2005. Methods for analysis of autoregressive helioclimatologic research. In: Dangers caused by global climate change. Gidrometeoizdat, Russia, pp. 34–67.

Libin, I. Ya., Dorman, L.I., 1985. Short-period cosmic ray variations. Soviet Physics—Uspekhi 145 (3), 403–440.

Libin, I. Ya., Gulinsky, O.V., 1979. Fluctuation phenomena in cosmic rays according to the multi-directional scintillation supertelescopes. Moscow: IZMIRAN, Preprint IZMIRAN 30 (258), p. 32 (in Russian).

Libin, I. Ya., Jaani, A., 1987. Influence of variations of solar activity on geophisical and hydrological processes. Izvestiya Akademii Nauk Estonskoi SSR—Seriya Biologicheskaya 38 (2), 97–100 (in Russian).

Libin, I. Ya., Jaani, A., 1989. The impact of changes in solar activity on geophysical and hydrological processes. I. The spectral characteristics of oscillations characterized conductivity of Lake Peipsi. Izvestiya Akademii Nauk Estonskoi SSR—Seriya Biologicheskaya 38 (2), 97–106.

Libin, I. Ya., Jaani, A., 1990. The impact of changes in solar activity on geophysical and hydrological processes. II. Short-period fluctuations in water volume of Lake Peipsi. Izvestiya Akademii Nauk Estonskoi SSR—Seriya Biologicheskaya 39 (3), 98–107.

Libin, I. Ya., Gulinsky, O.V., Guschina, R.T., Dorman, L.I., Mikalajunas, M.M., Yudakhin, K.F., 1992. Modeling the mechanism of action heliophysical parameters on atmospheric processes. Cosmic Rays (Kosmicheskie Luchi) 26, 22–56.

Libin, I. Ya., Guschina, R.T., Pérez-Peraza, J., Leyva-Contreras, A., Jaani, A., Fomichev, V.V., Yudakhin, K.F., 1994. The impact of solar activity on hydrological processes (autoregressive analysis of solar activity and lake levels). M.: Radio and Sviaz, Cosmic Rays (Kosmicheskie Luchi) 27, 115–127.

Libin, I. Ya., Guschina, R.T., Pérez-Peraza, J., Leyva-Contreras, A., Jaani, A., 1996a. Influence of solar activity variations on hydrological processes (autoregressive analysis of solar activity and levels of lakes). Geomagnetism & Aeronomy 36 (1), 79–83.

Libin, I. Ya., Guschina, R.T., Pérez-Peraza, J., Leyva-Contreras, A., Jaani, A., 1996b. The influence of solar activity on atmospherical processes (cyclic variations of precipitation). Geomagnetism & Aeronomy 36 (1), 83–86.

Libin, I. Ya., Guschina, R.T., Pérez-Peraza, J., Leyva-Contreras, A., Jaani, A., Mikalayunene, U., 1996c. The modulation effect of solar activity on the solar radiation. Geomagnetism & Aeronomy 36 (5), 109–114.

Libin, I. Ya., Guschina, R.T., Leyva-Contreras, A., Pérez-Peraza, J., Jaani, A., 1996d. The changes of solar activity and their influence to large- scale variations of surface-air temperature. Geomagnetism & Aeronomy 36 (5), 115–119.

Libin, I. Ya., Guschina, R.T., Pérez-Peraza, J., Leyva-Contreras, A., Jaani, A., 1996e. The impact of solar activity on atmospheric processes. Autoregressive analysis of cyclic changes in precipitation. Geomagnetism and Aeronomy 36 (5), 83–86.

Libin, I. Ya., Guschina, R.T., Pérez-Peraza, J., Leyva-Contreras, A., Jaani, A., 1996f. Changes in solar activity and their possible impact on long-term variations in surface temperature. Geomagnetism and Aeronomy 36 (5), 115–119.

Libin, I. Ya., Mikalajunas, M.M., Yudakhin, K.F., 1987. Variations of cosmophysical and hydrological parametrs in 18–21 cycles of the solar activity. Geomagnetism and Aeronomy 27 (3), 483–486.

Libin, I. Ya., Pérez-Peraza, J., Jaani, A., 1998. Effects of geomagnetic storms on atmospheric processes. In: Proceedings of the XXIII general assembly of the European Geophysical Society, Nice, France, April 1998.

Libin, I. Ya., Pérez-Peraza, J., 2007. Global warming: myths and realities. In: The conceptual basis for the development of socio-economic space in the context of globalization. MAOK, Moscow, Russia, pp. 8–16. Proceedings of the conference of the applied MAOK.

Libin, I. Ya., Pérez-Peraza, J., 2009. Helioclimatology. MAOK, Moscow, Russia (in Russian). p. 252. ISBN 978-5-89513-161-9.

Libin, I. Ya., Pérez-Peraza, J., Dorman, L.I., Mikalajunas, M.M., Jaani, A., 2011. A possible source of long-term climate change. In: Current socio-economic problems of the modern world: science and practice. Ed. by MAOK, Moscow, Russia, pp. 400–422.

Lisetskii, F.N., 2008. Agrogenic transformation of soils in the dry steppe zone under the impact of antique and recent land management practices. Eurasian Soil Science 41 (8), 805–817.

Loginov, V.F., 1978. The reaction of atmospheric circulation on the conditions in outer space. In: Proceedings of VNII GMI-ICD, vol. 37, pp. 117–130.

Loginov, V.F., Visotsky, A.M., Sherstiukov, B.G., 1975. IMF sector structure and atmospheric circulation. In: Proceedings of VNII-MCD, vol. 23, pp. 43–49.

Loginov, V.F., Rakirova, L.B., Sukhova, G.I., 1980. The effects of solar activity in the stratosphere. Leningrad (Now San Peterbourg): Gidrometeoizdat, Russia.

Lovelius, N.V., 1979. Variability of tree growth. In: Dendroindikatsiya natural processes and human impacts. Ed. by Nauka, Leningrad, Russia.

Lukin, V.P., Iliasov, S.P., Nosov, V.V., Odintsov, S.L., Tillaev, Yu.A., 2009. Study astroclimate region of southern Siberia and Central Asia. Atmospheric and Ocean Optics 22 (10), 973–980.

Lychak, M.M., 2002. Elements of the theory of chaos and its applications. Cybernetics and Informatics 5, 52–56.

Lychak, M.M., 2004. Study and forecasting of solar activity, April 6–9, 2004, Pushchino-on-Oka. Rotaprint IKI RAS, Moscow, Russia, p. 401.

Lychak, M.M, 2006. Analysis of the cyclical processes of solar activity, 03/10 September 2006 NCSVCT, Evpatoria, Proceedings. IKI NANU-NKAU, Kiev, Russia.

Maddison, A., 1962, June. Growth and fluctuation in the world economy 1870–1960. Banca Nazionale del Lavoro Quarterly Review, 3–71.

Maddison, A., 1964. Economic Growth in the West. Allen and Unwin, London, Twentieth Century Fund, and Norton, New York (also in Japanese, Russian and Spanish). Reprinted, 2006 by Routledge. p. 264.

Maddison, A., 2006. Economic growth in the west. Allen & Unwin, London, England (Original book published 1964).

Maddison, A., 2007. Contours of the world economy 1–2030 AD. Oxford University Press, Oxford, England.

Makarenko, N.G., 2005. Modern methods of nonlinear time series prediction. In: Proceedings of the All-Russian Conference "Experimental and theoretical studies of the foundations of forecasting heliogeophysical activity." Troitsk, IZMIRAN, October, 2005 Retrieved from: http://helios.izmiran.rssi.ru/Solter/prog2005/prog/abstracts.htm.

Makhov, S.A., Posashkov, S.A., 2007. Analysis of strategic risks on the basis of mathematical modeling. In: IPM, M.V., (Ed.), Keldish Russian Academy of Sciences, Moscow, Russia, p. 24.

Malkov, A.S., Korotaev, A.V., Khalturina, D.A., 2005. A mathematical model of population growth, economics, technology and education (No. 13). Preprint, IPM Keldish RAS. Moscow, Russia, p. 24.

Majorowicz, J.A., Skinner, W.R., 1997. Anomalous ground warming versus surface air warming in the Canadian Prairie provinces. Climatic Change 4, 485–500.

Marsh, N.D., Svensmark, H., 2003. Solar influence on Earth's climate. Space Science Reviews 107, 317–325.

Marsh, N.D., Svensmark, H., 2000. Low cloud properties influenced by cosmic rays. Physical Review Letters 85 (23), 5004–5007.

Martin, J.H., Gordon, R.M., 1988. Northeast Pacific iron distributions in relation to phytoplankton productivity. Deep Sea Research 35, 177–196.

Martin, R.V., Jacob, D.J., Logan, J.A., Bey, I., Yantosca, R.M., Staudt, A.C., Li, Q., Fiore, A.M., Duncan, B.N., Liu, H., Ginoux, P., Thouret, V., 2002. Interpretation of TOMS observations of tropical tropospheric ozone with a global model and in-situ observations. Journal of Geophysical Research 107 (D18), 4351. doi:10.1029/2001JD001480.

Mason, S.J., Tyson, P.D., 1992. The modulation of sea surface temperature and rainfall associations over southern Africa with solar activity and the quasi-biennial oscillation. Journal of Geophysical Research D5, 5847.

Mashnich, G.P., 2004a. 2-D velocity field structure and oscillations nearby quiescent filaments. In: Mashnich, G.P., Bashkirtsev, V.S., et al. (Eds.), Multi-wavelength investigations of solar activity: book of abstracts of the 223rd IAU International Symposium, Russia. St.-Petersburg-Pulkovo, June 14–19, 2004, p. 141.

Mashnich, G.P., 2004b. Solar activity and Earth's climate prediction. In: Bashkirtsev, V.S., Mashnich, G.P. (Eds.), Solar-terrestrial physics: a program mes. Abstracts. Intern. Conference, Irkutsk, Russia, pp. 66–67.

Mashnich, G.P., 2007. Sun and the Earth's climate prediction. In: Solar activity and its influence on the Earth—Vladivostok, Proceedings of the Ussuriysk Astrophysical Observatory/FEB RAS, pp. 13–19 (Issue 10).

Mavromichalaki, H., Souvatzoglou, G., Sarlanis, C., Mariatos, G., Plainaki, C., Gerontidou, M., Belov, A., Eroshenko, E., Yanke, V., 2006. Space weather prediction by cosmic rays. Journal of Advances in Space Research 37 (6), 1141–1147, (Elsevier).

Max, G., 1983. Methods and techniques of signal processing, vol. 1. Ed. by Mir, Moscow. p. 352.

Mazzarella, A., Palumbo, F., 1992. Rainfall fluctuations over Italy and their association with solar activity. Theoretical and Applied Climatology 45 (3), 201–207.

McCracken, K.G., 2001. Variations in the production of ^{10}Be due to the 11 year modulation of the cosmic radiation, and variations in the vector geomagnetic dipole. In: Proceedings of 27th ICRC, Hamburg, Germany, Copernicus Gesellschaft, pp. 4129–4132.

McCracken, K.G., McDonald, F.B., 2001. The long term modulation of the galactic cosmic radiation, 1500–2000. In: Proceedings of 27th ICRC, Hamburg, Germany, Copernicus Gesellschaft, pp. 3753–3756.

McCracken, K.G., Beer, J., McDonald, F.B., 2004. Variations in the cosmic radiation, 1890–1986, and the solar and terrestrial implications. Advances in Space Research 34 (2), 397–406.

McMichael, A.J., Campbell-Lendrum, D.H., Corvalan, C.F., Ebi, K.L., Githeko, A.K., Scheraga, J.D., Woodward, A. (Eds.), 2003. Climate change and human health. Risks and responses. World Health Organization, Geneva, Switzerland.

Meleshko, V.P., Kattsov, V.M., Govorkova, V.A., Sporyshev, P.V., Shkol'nik, M., Shneerov, B.E., 2004. Man-made climate change in the 21st century in northern Eurasia. Meteorology and Hydrology 7, 5–26.

Mendoza, B., Perez-Enriquez, R., 1993. Association of coronal mass ejections with the heliomagnetic current sheet. Journal of Geophysical Research 98, 9365.

Michalenco, T.D., Leonova, E.Y., 2007. El Niño and La Nina. Retrieved from: http://primpogoda.ru/articles (in Russian).

Mikalajunas, M.M., 1973a. Application of the modified coefficient of variation for the characteristics of seasonal storms in the North Sea. In: Mokslas, Vilnius, (Ed.), Articles on Hydrometeorology, vol. 6. Lithuania, pp. 185–199.

Mikalajunas, M.M., 1973b. The main characteristics of the regime of gales in the North Sea. In: Mokslas, Vilnius, (Ed.), Articles on Hydrometeorology, vol. 6. Lithuania, pp. 177–183.

Mikalajunas, M.M., 1985. Estimated parameters of tornadoes. U.S.S.R. Academy of Science, Moscow, Russia.

Mikalajunas, M.M., 2010. Predictions of heavy storms and precipitations in Europe. Vilnius Pedagogical University, Vilnius, Lithuania.

Mikisha, A.M., Smirnov, M.A., 1999. Terrestrial catastrophe caused by the fall of celestial bodies. Herald of the Russian Academy of Science 69 (4), 327.

Monin, A.S., Shishkov, Yu. A., 2000. Climate as a problem of physics. Physics-Uspekhi 170 (4), 419–445.

Moulin, C., Chiapello, I., 2004. Evidence of the control of summer atmospheric transport of African dust over the Atlantic by Sahel sources from TOMS satellites (1979–2000). Geophysical Research Letters 31 (2) L02107. doi:10.1029/2003GL018931.

Mulokwa, W.N., Mak, M., 1980. On the momentum flux by the interseasonal fluctuation of the circulation, monthly weather review. American Meteorological Society 108 (10), 1533–1537.

Mustel, E.R., Migulin, V.V., Medvedev, V.I., Parygin, V.N., 1981. Fundamentals of vibration theory, second ed. Nauka, Moscow, Russia.

Neff, U.S., Burns, J., Mangini, A., Mudelsee, M., Fleitmann, D., Matter, A., 2001. Strong coherence between solar variability and the monsoon in Oman between 9 and 6 kyr ago. Nature 411 (6835), 290–293.

Nelson, F.E., Anisimov, O.A., Shiklomanov, N.I., 2002. Climate change and hazard zonation in the circum-Arctic permafrost regions. Natural Hazards 3, 203–225.

Neumann, C.J., Jarvinen, B.R., McAdie, C.J., Hammer, G.R., 1999. Tropical Cyclones of the North Atlantic Ocean, 1871–1998. In: National Oceanic and Atmospheric Administration, (Ed.), p. 206.

Ney, E.P., 1959. Cosmic radiation and the weather. Nature 183 (4659), 451–452.

Ney, E.P., Winckler, J.R., Freier, P.S., 1959. Protons from the Sun on May 12, 1959. Physical Review Letter 3, 183–185.

Nikolsky, G.A., 2009. Effects and mechanisms of action of solar radiation at the vortex of the spiral structure of matter. In: Proceedings of the VI International Conference, "Natural and anthropogenic aerosol—2008" Institute of Physics, St. Petersburg State University. St. Petersburg State University Publishing House, St. Petersburg, Russia, pp. 187–194.

Nikolsky, G.A., 2010. Hidden solar emissions and Earth's radiation budget. In: Collection of articles on Inter-regional scientific seminar "Ecology and Space," dedicated to the 90th anniversary of Academician Kondratyev, February 8–9, pp. 230–240.

Nikolsky, G.A., 2011. The up trend of the Earth radiation balance. Variations of Astronomical and Meteorological Constant. Proc. in the Conf. Radiative climatology and algorithms in models for weather and climate forecasting. SPb, 156. Retrieved from:. www.rrc.phys.spbu.ru/msard11/session5.html.

Nikolsky, G.A., The sun, ocean and climate. http://vd2-777.narod.ru/files/SUN_AND_CLIMATE.pdf

Nikonov, A., 1996. Measuring instruments with computer equipment. Textbook, Moscow. p. 123.

Novikov, L.S., Panasiuk, M., 2007. Model of Space, M. University Book House, p. 2016.

Obridko, V.N., Shelting, B., 2008. Temporal variations of the heliospheric equator. The Astronomical Journal 85 (8), 750–754.

Obridko, V.N., 2011. Cycles of solar activity and especially the 23rd cycle. Earth and Universe 1, January 2011, 23–26.

Obridko, V.N., Ragulskaya, M.V., 2005. Orderliness and stochasticity of biological systems under the influence of external natural fields. Biophysics 6, 342. http://www.pribory-magic.narod.ru/MARY.htm.

Odintsov, S.D., Ivanov-Kholodny, G.S., Georgieva, K., 2005. Solar activity and global seismicity of the Earth. In: Proceedings of the all-Russian conference "Experimental and theoretical studies of the foundations of forecasting heliogeophysical activity." October 10–15, 2005, Troitsk, IZMIRAN Retrieved from: http://helios.izmiran.rssi.ru/Solter/prog2005/prog/abstracts.htm.

Odintsov, S.L., Fedorov, V.A., 2007. Study variations in wind speed scale of mesometeorological sodar observations. Atmospheric and Ocean Optics 20 (11), 986–993.

Ol', A.I., 1969. The 22-year cycle of solar activity in the Earth's climate. Reports of Arctic and Antarctic Research Institute (Trudy Arkticheskogo I Antarkticheskog Nauchno-Issledowatel'skogo Instituta) 289, 116–128 (in Russian).

Ol', A.I., 1973. Rhythmic processes in the Earth's atmosphere. Nauka, Leningrad, Russia.

Oraevsky, V.N., Breus, T.K., Baevsky, R.M., Rapoport, Z.C., Petrov, V.M., Barsukova, J.V., Gurfinkel, Yu. I., Rogoza, A.T., 1998. The influence of geomagnetic activity on the functional state of the organism. Biophysics 43 (5), 819–826.

Oraevsky, V.N., Kuleshova, V.P., Gurfinkel, Iu. F., Guseva, A.V., Rapoport, S.I., 1998. Medico-biological effect of natural electromagnetic variations. Biophysics 43 (5), 844–888.

Oraevsky, V.N., Sobelman, I.I., Jitnik, I.A., Kuznetsov, V.D. 2002a. 172, 949.

Oraevsky, V.N., Kanonidi, Kh. D., Belov, A.V., Gaidash, S.P., 2002b. Operations Centre Heliophysical forecast IZMIRAN/Proc. Conference on the physics of Solar-Terrestrial relationsships, Irkutsk, 24–29 September, 2001. Solar-Terrestrial Physics 2 (115), 114–116.

Osterkamp, T.E., Romanovsky, V.E., 1999. Evidence for warming and thawing of discontinuous permafrost in Alaska. Permafrost and Periglacial Processes 10, 17–37.

Pai, G.L., Sarabhai, V.A., 1963. Intensity of green coronal emission and the velocity of plasma wind. Proc. 8th ICRC, Jaipur, 1, pp. 190–197.

Palmer, S.J., Rycroft, M.J., Cermack, M., 2006. Solar and geomagnetic activity, extremely low frequency magnetic and electric fields and human health at the Earth's surface. Surveys in Geophysics 27, 557–595.

Panchenko, M.V., Domisheva, V.M., Pestunov, D.A., Sakirko, M.V., Zavoruev, V.V., Novitsky, A.L., 2007. Experimental study of CO_2 gas exchange in the system "atmosphere-surface water" Lake Baikal (experimental setup). Atmospheric and Ocean Optics 20 (5), 448–452.

Paris, J.-d., Arshinov, M. Yu., Ciais, P., Belan, B.D., Nédélec, P., 2009. Large-scale aircraft observations of ultra-fine and fine particle concentrations in the remote Siberian troposphere: new particle formation studies. Atmospheric Environment 43 (6), 1302–1309.

Parker, E., 1958. Dynamics of the Interplanetary gas and magnetic fields. Astrophys. J. 128, 664–676.

Parker, E., 1965. Dynamic processes in the interplanetary medium. Mir, Moscow, Russia.

Parker, E., 1982. Cosmic magnetic fields. Mir, Moscow, Russia.

Pavlishkina, E.S., Mikhaylov, V.N., 2002. Joint vibration analysis of runoff of the Volga and the Caspian Sea level for the 100-year period. In: Janus-K, (Ed.), Atlas of temporal variations of natural, human and social processes, vol. 3. Moscow, Russia, pp. 390–395.

Pavlov, A., 2010. The emergence of the biosphere and the evolution of geo-biological systems. In: X-th conference EANA (European Network of Associations astrobiological), Pushchino, Russia.

Pérez-Peraza, J., 1990. Space plasma physics (invited talk). In: Space conference of the Americas: perspectives of cooperation for the development, PNUD, San José de Costarica, 1, pp. 96–113.

Pérez-Peraza, J., 1998. Acceleration of solar particles during cycle 22 (invited talk). In: 16th European cosmic ray symposium, Alcala de Henares, Madrid, pp. 97–112.

Pérez-Peraza, J., Gallegos-Cruz, A., 1994. Weightiness of the dispersive rate in stochastic acceleration process. The Astrophysical Journal (Suppl.) 90 (2), 669–682.

Pérez-Peraza, J., Gallegos-Cruz, A., 1998. Diagnostics of solar particle acceleration processes. Advances in Space Research 21 (4), 629–632.

Pérez-Peraza, J., Leyva-Contreras, A., Valdés-Barrón, M., Libin, I., Yudakhin, K., Jaani, A, 1995. Influence of solar activity on hydrological processes: spectral and autoregressive analysis of solar activity and levels of lakes Patzcuaro and Tchudskoye (Reportes tecnicos 95–3). Instituto de Geografia, Universidad Nacional Autónoma de México, Mexico.

Pérez-Peraza, J., Leyva-Contreras, A., Libin, I. Ya, Formichev, V., Guschina, R.T., Yudakhin, K., Jaani, A., 1997. Simulating the mechanism of the action of heliophysical parameters on atmospheric processes. Geofísica Internacional 36 (4), 245–280.

Pérez-Peraza, J., Leyva-Contreras, A., Libin, I. Ya., Ishkov, V., Yudakhin, K., Gulinsky, O., 1998. Prediction of interplanetary shock waves using cosmic ray fluctuations. Geofisica International 37 (2), 87–93.

Pérez-Peraza, J., Leyva-Contreras, A., Valdez-Baron, M., Bravo-Cabrera, J.L., Libin, I. Ya., Jaani, A., 1999. Influence of solar activity on the cyclic variations of precipitation in the Baltic region. Geofisica International 38 (2), 73–81.

Pérez-Peraza, J., Libin, I. Ya., Jaani, A., Yudakhin, K., Leyva-Contreras, A., Valdez-Barron, M., 2005. Influence of solar activity on hydrological processes. Hydrology and Earth System Sciences (HESS) 2, 605–637.

Pérez-Peraza, J., Kavlakov, S., Velasco, V., Gallegos-Cruz, A., Azpra-Romero, E., Delgado-Delgado, O., Villicaña-Cruz, F., 2008a. Solar, geomagnetic and cosmic ray intensity changes, preceding the cyclone appearances around Mexico. Advances in Space Research 42, 1601–1613.

Pérez-Peraza, J., Velasco, V., Kavlakov, S., 2008b. Wavelet coherence analysis of North Atlantic hurricanes and cosmic rays. Geofisica Internacional 47, 231–244.

Pérez-Peraza, J., Velasco, V., Kavlakov, S., Gallegos-Cruz, A., Azpra-Romero, E., Delgado-Delgado, O., Villicaña-Cruz, F., 2008c. On the trend of Atlantic hurricane with cosmic rays. In: 30th international cosmic ray conference, Merida, Mexico, vol. 1, (SH). pp. 785–788.

Pérez-Peraza, J., Velasco-Herrera, V., Alvarez Madrigal, M., Libin, I. Ya., 2011. Do cosmic rays influence ozone depletion in the Antarctic ozone hole? In: 32nd international cosmic ray conference, Beijing, vol. 11, pp. 391–394.

Perry, R., 2008. Blue lakes and silver cities. Espadaña Press, Santa Barbara, CA.

Philander, G.S., 1990. El Niño, La Niña and the Southern oscillation, vol. 46. Academic Press, New York (International Geophysics Series).

Piccardi, G., 1962. The chemical basis of medical climatology. Charles C. Thomas, Springfield, IL.

Polygiannakis, J., Preka-Papadema, P., Moussas, X., 2003. On signal-noise decomposition of time-series using the continuous wavelet transform: Application to sunspot index. Astrophysics 343 (3), 725–734.

Pokrovsky, O.M., 2005. The change in surface temperature in the North Atlantic and climate variability in Europe. Study of Earth from Space 4, 24–34.

Porfiriev, B.N., 2005a. The danger of natural and man-made disasters in the world and in Russia. In: Red, N.N., Marfenin, M. (Eds.), Russia in the outside world: 2004, pp. 37–62. Analytical Yearbook.

Porfiriev, B.N., 2005b. Risks and crises: the new direction of social science research. Modern and Contemporary History 3, 231–238.

Porfiriev, B.N., 2005c. Preserve the traditions of salvation. Political Journal 19, 69.

Porfiriev, B.N., 2006a. Public administration in crisis: a global perspective and Russia. Scientific Expert, Moscow, Russia.

Porfiriev, B.N., 2006b. Natural hazards in terms of modern economic growth: theory and practice of state and non-regulation. The Russian Business Magazine 1, 37–48.

Porfiriev, B.N., 2006c. Natural disaster or sustainable development? Strategy for Russia 3, 25–31.

Porfiriev, B.N., 2006d. Transportation of petroleum resources in Asia and the Pacific: methodology and results comparing the effectiveness of options. The Russian Business Magazine 4, 53–62.

Porfiriev, B.N., 2006e. Disaster and crisis management in transitional societies. In: Rodriguez, H., Quarantelli, E.L., Dynes, R. (Eds.), Handbook of disaster research. Springer, New York, pp. 368–387.

Porfiriev, B.N., 2008. Economics of Climate Change. M.: Ankil, p. 168.

Porfiriev, B.N., 2009a. Russia's transition to an innovation strategy: factor climate risks. In: Infra-M., (Ed.), Innovative development: economic, intellectual resources, knowledge management. Moscow, Russia.

Porfiriev, B.N., 2009b. Managing security and safety risks in the Baltic Sea region. In: Avenhaus, R., Sjonstedt, G. (Eds.), Negotiated risks: International talks on hazardous issues. Germany: Springer, Berlin–Heidelberg.

Porfiriev, B.N., 2010a. Economic crisis: management issues and problems of innovative development. Problems of Forecasting 5, 20–26.

Porfiriev, B.N., 2010b. Climate change as environmental and economic hazard. Earthscan, London, England.

Prilutsky, R.E., 1988. Methods and tools of statistical analysis of fluctuations of cosmic rays. Preprints IZMIRAN 41 (795). Ed. IZMIRAN, p. 24 Moscow.

Program of the Scientific Committee on Solar-Terrestrial Physics (STEP), 1997. University of Illinois-Department of Electrical and Computer Engineering, SCOSTEP Secretariat, Champaign, IL. p. 8.

Prospero, J.M., 1999. Long-term measurements of the transport of African mineral dust to the southeastern United States: implications for regional air quality. Journal of Geophysical Research 104 (D13), 15917–15927.

Prospero, J.M., 2006. Case study, Saharan dust impacts and climate change. Oceanography 19 (2), 60–61.

Prospero, J.M., Carlson, T.N., 1981. Saharan air outbreaks over the tropical North Atlantic. Pure and Applied Geophysics (PAGEOPH) 119, 677–691.

Prospero, J.M., Nees, R.T., 1986. Impact of the North African drought and El Niño on mineral dust in the Barbados trade winds. Nature 320 (6064), 735–738.

Prospero, J.M., Schmitt, R., Cuevas, E., Savoie, D.L., Graustein, W.C., Turekian, K.K., Volz-Thomas, A., Diaz, A., Oltmans, S.J., Levy II, H., 1995. Temporal variability of summer-time ozone and aerosols in the free troposphere over the eastern North Atlantic. Geophysical Research Letters 22, 2925–2928.

Prospero, J.M., Ginoux, P., Torres, O., Nicholson, S., Gill, T., 2002. Environmental characterization of global sources of atmospheric soil dust identified with the NIMBUS 7 Total Ozone Mapping Spectrometer (TOMS) absorbing aerosol product. Review of Geophysics 40, 1002.

Prospero, J.M., Lamb, P.J., 2003. African droughts and dust transport to the Caribbean: climate change implications. Science 302, 1024–1027.

Ptitsina, N.G., Villorezi, G., Dorman, L.I., Yucci, N., Tiasto, M.I., 1998. Natural and man-made low-frequency magnetic field as a factors that are potentially dangerous to health. Successes of Physical Sciences, 168, p. 767.

Pudovkin, M.I., 1995. The influence of solar activity on the state of the lower atmosphere and weather. Soros Educational Journal 10, 106–113.

Pudovkin, M.I., Lyubchich, A.A., 1989. Manifestation of the solar and magnetic activity cycles in the variations of the air temperature in Leningrad. Geomagnetism and Aeronomy 29 (3), 359.

Pudovkin, M.I., Raspopov, O.M., 1992a. The mechanism of the solar activity influence on the lower atmosphere and the meteoreological parameters. Geomagnetism and Aeronomy 32 (5), 593–608.

Pudovkin, M.I., Raspopov, O.M., 1992b. The mechanism of action of solar activity on the state of the lower atmosphere. Geomagnetism and Aeronomy 32 (5), 1–22.

Pustilnik, L.A., Yom Din, G., 2004. Influence of solar activity on state of wheat market in medieval England. Solar Physics 223 (1–2), 335–356. doi:10.1007/s11207-004-5356-5. Retrieved from: http://arxiv.org/ftp/astro-ph/papers/0411/0411165.pdf

Ragulskaya, M.V., 2004. Relationship of periodic processes in the body caused by the rhythm of the environment, with variations of the solar magnetic field. Biomedical Technology and Electronics 1–2, 1–6.

Ragulskaya, M.V., 2005. Effect of variations in solar activity on the functional health of people (Abstract for the degree of candidate of physical-mathematical sciences). In: IZMIRAN (Ed.), Physics of the Sun. Moscow, Russia, p. 32.

Ragulskaya, M.V., Khabarova, O.V., 2001. The influence of solar disturbances on the human body. Biomedical Electronics 2, 5–15.

Rahmstorf, S., Cazenave, A., Church, J., Hansen, J., Keeling, R., Parker, D., Somerville, R., 2007. Recent climate observations compared to projections. Science 316 (5825), 709.

Raisbeck, G.M., Yiou, F., 1980. 10Be in polar ice cores as record of solar activity. In: Pepin, R.O., Eddy, J.A., Merrill, R.B. (Eds.), Proceedings of the Conference on the Ancient Sun. Pergamon, New York, pp. 185–190.

Raisbeck, G.M., Yiou, F., 2004. Comment on "Millennium scale sunspot number reconstruction: evidence for an unusually active Sun since the 1940s." Physical Review Letters 92, 199001.

Ram, M., Stolz, M.R., Tinsley, B.A., 2009. The terrestrial cosmic ray flux: its importance for climate. EOS 90 (44), 397.

Rao, C.R., 1971. Estimation of heteroscedastic variances in linear models. J Am Stat Assoc. 65 (329), 161–172. JSTOR 2283583.

Raspopov, O.M., Shumilov, O.I., Kasatkina, E.A., Turunen, E., Lindholm, M., Kolstrom, T., 2001. The non-linear character of the effect of solar activity on climatic processes. Geomagnetism and Aeronomy 41 (3), 407.

Raspopov, O.M., Dergachev, V.A., Kolstrom, T., 2005. Hale cyclicity of solar activity and its relation to climate variability. Solar Physics 224 (1–2), 455–463. doi:10.1007/s11207-005-5251-8. http://adsabs.harvard.edu/labs/2004SoPh..224..455R.

Raspopov, O.M., Dergachev, V.A., Kolström, T., 2004. Periodicity of climate conditions and solar variability derived from dendrochronological and other palaeoclimatic data in high latitudes. Palaeogeography Palaeoclimatology Palaeoecology 209, 127–139.

Reap, A., 1981. Peipsi-Puhkva jarve Veeseisuel Prognoosist Maaparandus. In: Teaduslintehnilisi Urimistulemusi, Tallin; Teadus, (Eds.), 17–24 (in Estonian).

Reap, A., 1986. On the possibility of ultralong-range forecasting water availability Peipsi-Pskov Lake. Problems and ways of managing natural resources and environmental protection.

In: Proceedings of XI conference of the Republican Hydrometeorological, Siauliai, Lithuania, p. 80.

Reid, G.C., 1987. Influence of solar variability on global sea surface temperature. Nature 329 (6135), 142–143.

Revellel, R., Suess, H., 1957. Carbon dioxide exchange between atmosphere and ocean and the question of an increase of atmospheric CO_2 during the past decades. Tellus 9, 18–27.

Rivin, Yu. R., Zvereva, T.I., 1983. The frequency spectrum of quasi-biennial variations of the geomagnetic field. In: Nauka (Ed.), Solar wind, magnetosphere and the geomagnetic field. Moscow, Russia, pp. 72–90.

Rivin, Yu. R., 1985. Spectral analysis of changes in the amplitude of the 11-year cycles of solar activity. Solar Data 9, 78–82.

Rodrigo, F.S., Esteban-Parra, M.J., Pozo-Vazquez, D., Castro-Diez, Y., 2000. On the variability of rainfall in southern Spain in decadal to centennial time scales. International Journal of Climatology 20 (7), 721–732.

Roig, F.A., Le-Quesne, C., Boninsegna, J.A., Briffa, K.R., Lara, A., Grudd, H., Jones, Ph., Villagran, C., 2001. Climate variability 50,000 years ago in midlatitude Chile as reconstructed from tree rings. Nature 410 (6828), 567–570.

Rosemarie, E.C., John, M.E., Veizer, J., Azmy, K., Brand, U., Weidman, C.R., 2007. Coupling of surface temperatures and atmospheric CO_2 concentrations during the Palaeozoic era. Nature 449 (7159), 198–201.

Rosenfeld, D., Rudich, Y., Lahav, R., 2001. Desert dust suppressing precipitation: a possible desertification feedback loop. In: Proceedings of the National Academy of Science U.S.A., 98, pp. 5975–5980.

Rozhkov, V.A., 1979. Methods of probabilistic analysis of oceanographic processes. Gidrometeoizdat, Leningrad, Russia.

Rozhkov, V.A., 1988. Methods of probability analysis of oceanological processes. Gidrometeoizdat, Leningrad, Russia.

Rusanov, A.I., Kuni, F.M., Shchekin, A.K., 2009. Point excesses in the theory of ordinary and micellar solutions. Russian Journal of Physical Chemistry A 83 (2), 223–230.

Russian Academy of Sciences, 2006. National report to the international association of seismology and physics of the Earth's interior of the international union of geodesy and geophysics 2003–2006. Russian Academy of Science, Moscow, Russia.

Sahney, S., Benton, M.J., 2008. Recovery from the most profound mass extinction of all time. In: Proceedings of the Royal Society B: Biological Sciences, 275, pp. 759–765 (1636) doi:10.1098/rspb.2007.1370.

Sassen, K., DeMott, P.J., Prospero, J.M., Poellot, M.R., 2003. Saharan dust storms and indirect aerosol effects on clouds: CRYSTAL-FACE results. Geophysical Research Letters 30, 1633. doi:10.1029/2003GL017371.

Saunders, M.A., 2009. Earth's future climate. Benfield Greig Hazard Research Centre, Department of Space and Climate Physics, University College London, Surrey, U.K.

Savitzky, A., Golay, M.J.E., 1964. Smoothing and differentiation of data by simplified least squares procedures. Analytical Chemistry 36, 1627–1639.

Schwabe, H., 1844. Sonnen-Beobachungen in Jahre 1843. Astronomische Nachricthen 21, 233–236.

Seabrook, V., Wood, R.W., 1946. The modern magician physical laboratory. State Publishing House of Technical and Theoretical Literature, Leningrad.

Seinfeld, J.H., Pandis, S.N., 2006. Atmospheric chemistry and physics—from air pollution to climate change, second ed. John Wiley & Sons, New York.

Seuss, E., 1875. Die entstehung der Alpen [The origin of the Alps]. W. Braunmuller (in German), Vienna, Austria.

Selevin, V.A., 1933. On the oscillations of the level of Lake Balkhash. Nature 7, pp. 39–42 (in Russian).

Shanahan, T.M., Overpeck, J.T., Anchukaitis, K.J., Beck, J.W., Cole, J.E., Dettman, D.L., Peck, J.A., Scholz, C.A., King, J.W., 2009. Atlantic forcing of persistent drought in West Africa. Science 324, 377–380.

Shapiro, L.J., 1987. Month to month variability of the Atlantic Tropical Circulation and its relationships to tropical cyclone formation. Monthly Weather Review 115, 2598–2614.

Shaviv, N.J., 2002. Cosmic ray diffusion from the galactic spiral arms, iron meteorites and a possible climatic connection. Physical Review Letters 89 (5) 051102.

Shaw, S.J., 1976. History of the Ottoman Empire and modern Turkey, vol. 1. Cambridge University Press, Cambridge, U.K, p. 18.

Sherstiukov, B.G., 2007. Long-term prediction of monthly and seasonal air temperatures, taking into account the periodic unsteadiness. Meteorology and Hydrology 9, 14–26.

Shiklomanov, I.A., Georgievsky, V. Yu., Ezjov, A.V., 2003. Probabilistic forecast of the Caspian Sea. In: Gidrometeoizdat, St. Petersburg (Eds.), Hydrometeorological aspects of the Caspian Sea basin. Russia, pp. 323–340.

Shindell, D.T., Schmidt, G.A., Mann, M.E., Rind, D., Waple, A., 2001. Solar forcing of regional climate change during the Maunder Minimum. Science 294, 2149–2152.

Shnitnikov, A.V., 1949. Common features of cyclical fluctuations in lake level and moisture across Eurasia in connection with solar activity. Bulletin of the Research Commission of the Sun, 3–4.

Shnol, S.E., 2001a. Macroscopic fluctuations of discrete forms of distribution as a consequence of the arithmetic and space physics of reasons. Biophysics 46 (5), 775–782.

Shnol, S.E., 2001b. Macroscopic fluctuations—a possible consequence of fluctuations of space-time. Arithmetic and cosmophysical aspects. Russian Chemical Journal 45 (1), 12–15.

Shnol, S.E., 2009. Cosmophysical factors in the random processes. Svenska fysikarkivat, Stockholm, Sweden.

Shnol, S.E., Kolombet, V.A., Pojarsky, E.V., Zenchenko, T.A., Zvereva, I.M., Konradov, A.A., 1998. Realization of discrete states during fluctuations in macroscopic processes. Soviet Physics—Uspekhi 41 (10), 1025–1035.

Shnol, S.E., Zenchenko, T.A., Zenchenko, K.I., Pojarsky, E.V., Kolombet, V.A., Konradov, A.A., 2000. Regular variation of the fine structure of statistical distributions as a consequence of space physics of reasons. Successes of Physical Sciences 403 (2), 205–209.

Shpindler, I.B., Zengbush, A., 1896. Lake Peipsi. In: Proceedings of the Imperial Russian Geographical Society, 32 (4) 47–69. http://tallinn.ester.ee/record=b2478515~S7*est.

Shpindler, I.B., Zengbush, A., 1896. Lake Tchudskoye, Izvestia Imperial Russian Geographycal. Society, 32–34.

Simpson, J.A., 1963. Recent investigations of the low energy cosmic ray and solar particles radiations (Preprint 25) (pp. 323–352). Vatican: Pontifical Academy of Sciences.

Sitnov, S.A., 2009. The effect of 11-year cycle of solar activity on the quasi-biennial variability in ozone and temperature in the Arctic. Izvestiya-Physics of the Atmosphere and Ocean 45 (3), 1–7.

Solanki, S.K., Usoskin, I.G., Kromer, B., Schüssler1, M., Beer, J., 2004. Unusual activity of the Sun during recent decades compared to the previous 11,000 years. Nature, 431 (7012), 1084–1087.

Solntsev, V.N., 1997. A geometric approach to the system of multivariate statistical analysis. In: Danilov, D.L., Zhiglyavskogo, A.A. (Eds.), The main components of time series: the method of "Caterpillar". University House PRESSKOM, St. Petersburg, Russia, pp. 252–256.

Solntsev, V.N., Filatova, T.N., 1999. Multivariate statistical analysis and probabilistic modeling of real processes. In: Aivazian, S.A. (Ed.), Identification of key patterns in the long-term variations of hydrological characteristics. Abstracts of the international jubilee session of the scientific seminar. CEMI, Moscow, Russia, pp. 185–189.

Solntsev, V.N., Nekrutkin, V.V., 2003. Non-traditional method of analyzing the structure and time series prediction "Caterpillar-SSA". Probabilistic ideas in science and philosophy. In: Proceedings of the Regional Science Conference (with participation foreign scientist), September 23–25, 2003. Institute of Philosophy and Law, Russian Academy of Science, Novosibirsk State University, Novosibirsk, Russia, pp. 126–129.

Sokolik, I.N., Toon, O.B., 1996. Direct radiative forcing by anthropogenic airborne mineral aerosols. Nature 381, 681–683.

Sokolik, I.N., Winker, D., Bergametti, G., Gillette, D., Carmichael, G., Kaufman, Y., Gomes, L., Schuetz, L., Penner, J., 2001. Introduction to special section on mineral dust: outstanding problems in quantifying the radiative impact of mineral dust. Journal of Geophysical Research 106, 18015.

Stern, N., 2006. The economics of climate change. Cambridge University Press, Cambridge, U.K.

Stoilova, I., Dimitrova, S., 2006. Solar activity and health. Investigation of influence of geophysical factors and climatic factors on human health in Bulgaria. Bulgarian Academy of Sciences Journal 4, 36–39.

Stoilova, I., Dimitrova, S., Breus, N., Zenchenko, T., Yanev, T., 2008. Human health and solar-terrestrial interactions. Solar-Terrestrial Physics 12 (2), 336–339.

Stozhkov, Yu. I., 2007. What can be extracted from data on the concentrations of Be-10 and C-14 natural radionuclides? Bulletin of the Lebedev Physical Institute 34 (5), 135.

Stozhkov, Yu. I., Charakhchian, T.N., 1969. 11-year modulation of cosmic ray intensity and distribution gelioshirotnoe spots. Geomagnetism and Aeronomy 9 (5), 803–808.

Stozhkov, Yu. I., Okhlopkov, V.P., Svirzhevsky, N.S., 2004. Cosmic ray fluxes in present and past times. Solar Physics 224 (1), 323.

Stozhkov, Yu. I., Tikhomirov, T., 2008. On the current minimum of solar activity. In: SINP Workshop, "Cosmic Ray Astrophysics and Space Physics." FIAN, St. Petersburg, Russia.

Stozhkov, Yu.I., Ermakov, V.I., Okhlopkov, V.P., 2008. Is the forecast of cold weather climate?. In: SINP Workshop, "Cosmic Ray Astrophysics and Space Physics." FIAN, St. Petersburg, Russia.

Sutton, R.T., Hudson, D.R.L., 2005. Atlantic Ocean forcing of North American and European summer climate. Science 290, 2133–2137.

Svalgaard, L., Wilcox, J.M., Duvall, T.L., 1974. A model combining the solar and sector structured polar magnetic field. Solar Physics 37, 157.

Svensmark, H., 2007. Cosmoclimatology: a new theory emerges. Astronomy & Geophysics 48 (1), 1.18–1.24. doi:10.1111/j.1468-4004.2007.48118.

Svensmark, H., 2008. The cosmic-ray/cloud seeding hypothesis is converging with reality. Retrieved from: www.thegwpf.org/the-observatory/3779-henrik-svensmark-the-cosmic-raycloud-seeding-hypothesis-is-converging-with-reality.html.

Svensmark, H., Friis-Christensen, E., 1997. Variations of cosmic ray flux and global cloud coverage—A missing link in solar-climate relationship. Atmospheric and Solar Terrestrial Physics 59, 1225–1232.

Svensmark, H., Calder, Nigel, 2007. The Chilling Stars: A New Theory of Climate Change. Totem Books. ISBN 978-1840468151.

Svensmark, H. 2008. The Cosmic-Ray/Cloud Seeding Hypothesis Is Converging With Reality, http://www.thegwpf.org/the-observatory/3779-henrik-svensmark-the-cosmic-raycloud-seeding-hypothesis-is-converging-with-reality.html.

Svensmark, H., Bondo, T., Svensmark, J., 2009. Cosmic ray decreases affect atmospheric aerosols and clouds. Geophysical Research Letters VOL. 36, L15101. doi:10.1029/2009GL038429.

Svensmark, H., Calder, N., 2010. The Sun, cosmic rays, clouds, and the latest research (real climate science). Retrieved from: www.neuralnetwriter.cylo42.com.

Takata, M., 1941. About a new biologically active component of solar radiation. Contribution to an experimental basis the Heliobiology. Archives for Meteorology, Geophysics, and Bio-climatology-Series B 2 (2), 486–489.

Tanré, D., Haywood, J., Pelon, J., Léon, J.F., Chatenet, B., Formenti, P., Francis, P., Goloub, P., Highwood, E.J., Myhre, G., 2003. Measurement and modeling of the Saharan dust radiative impact: overview of the Saharan Dust Experiment (SHADE). Journal of Geophysical Research 108 (D18), 8574. doi:10.1029/2002JD003273.

Tegen, I., Lacis, A.A., Fung, I., 1996. The influence of mineral aerosols from disturbed soils on the global radiation budget. Nature 380, 419–422.

Tegen, I., Werner, M., Harrison, S.P., Kohfeld, K.E., 2004. Relative importance of climate and land use in determining present and future global soil dust emission. Geophysical Research Letters 31 L05105. doi:10.1029/2003GL019216.

Thurber, C.H., Brocher, T.M., Zhang, H., Langenheim, V.E., 2007. Three-dimensional P wave velocity model for the San Francisco Bay region, California. Journal of Geophysical Research 112, B07313, p. 19. doi:10.1029/2006JB004682.

Tinsley, B.A., 1996. Correlations of atmospheric dynamics with solar wind induced air-Earth current density into cloud tops. Journal of Geophysical Research 101, 29701–29714.

Tinsley, B.A., 2000. Influence of solar wind on global electric circuit and infrared effects on cloud microphysics, temperature and dynamics in the troposphere. Space Science Reviews 94 (231), 231–238.

Tinsley, B.A., Beard, K.V., 1997. Meeting summary: links between variations in solar activity, atmospheric conductivity, and clouds: an informal workshop. Bulletin of the American Meteorology Society 78, 685–687.

Torrence, C., Compo, G., 1998. A practical guide to wavelet analysis. Bulletin of the American Meteorology Society 79, 61–78.

Torrence, C., Webster, P., 1999. Interdecadal changes in the ESNO-Monsoon System. Journal of Climate 12, 2679–2690.

Trenberth, K., 2005. Uncertainty in hurricanes and global warming. Science 308 (5729), 1753–1754.

Tsalikov, Kh. R., 2008. Danger and threat to the northern territories of the Russian Federation due to global climate change. In: Proceedings of the conference "Ensuring integrated safety of the northern regions of the Russian Federation," April 22, 2008. Russian Emergencies Ministry National Emergency Management Center, Moscow, Russia, pp. 8–14.

Usokin, I.G., Schüssler, M., Solanki, S., Mursula, K., 2005. Heliospheric modulation of cosmic rays: monthly reconstruction for 1951–2004. Journal of Geophysical Research 110 A101102. doi:10.1029/2004JAA010946.

Usokin, I.G., Kovaltsov, G., 2007. Cosmic rays and climate of the Earth: possible connection (Review Paper). Comptes Rendus Geoscience 340, 441–450. doi:10.1016/j.crte.2007.11.001.

Valdez-Galicia, J.F., Lara, A., Mendoza, B., 2005. The solar magnetic flux mid-term periodicities and the solar dynamo. Journal of Atmospheric and Solar-Terrestrial Physics 67, 1697.

Vakulenko, N.V., Monin, A.S., Sonechkin, D.M., 2003. The dominant role of the amplitude modulation of precession cycles in the alternating glacial Late Pleistocene. Reports of the Academy of Sciences 391 (6), 817–820.

Vasiliev, S., Dergachev, V.A., 2009. Solar activity over the past 10 thousand years the data on cosmogenic isotopes. Izvestiya Akademii Nauk SSSR—Seriya Fizicheskaya 73 (3), 396–398.

Vasilieva, G.Y., 1984. Information and cybernetic approach to the study of the solar system as a whole, In: Problems of research of the space (No. 9) (pp. 5–21). Ed. SCIENCE, Moscow-Leningrad.

Velasco, V., Mendoza, B., 2008. Assessing the relationship between solar activity and some large scale climatic phenomena. Advances in Space Research 42 (5), 866–878. doi:10.1016/j.asr.2007.05.050.

Velasco Herrera, V., Pérez-Peraza, J., Velasco Herrera, G., Luna González, L., 2010. African dust influence on Atlantic hurricane activity and the peculiar behaviour of category 5 hurricanes (Los Alamos Laboratories Series). Retrieved at arXiv:1003.4769v1 [physics.ao-ph], 1–5.

Venkatesan, D., 1990. Cosmic ray intensity variations in the 3-dimensional heliosphere. Space Science Reviews 52, 121–194.

Veretenenko, S.V., Dergachev, V.A., Dmitriev, P.B., 2005. Long-period effects of solar activity in the intensity of cyclonic processes at midlatitudes. In: Proceedings of the all-Russian conference "Experimental and theoretical studies of the foundations of forecasting heliogeophysical activity, " October 10–15, 2005, Troitsk, IZMIRAN Retrieved from: http://helios.izmiran.rssi.ru/Solter/prog2005/prog/abstracts.htm (in Russian).

Vernov, S.N., Christiansen, G.B., 1982. Cosmic rays of ultrahigh energies. Vestnik Moskovskogo Universiteta—Seriya 3: Fizika Astronomiya 23, 3–15.

Veselovsky, I.S., Persiantsev, I.G., Dolenko, S.A., Shugay, Yu. S., 2007. Investigation of the relationship between coronal holes on the Sun, high-speed flows korotiruyuschimi solar wind and recurrent geomagnetic disturbances in the decay phase of solar cycle. The physical nature of solar activity and the prediction of geophysical phenomena, 2–7 July 2007. GAO Pulkovo, Pulkovo, St. Petersburg (pp. 375–376).

Vitinsky, Yu. I., Kopecky, M.I., Kuklin, G.V., 1986. Statistics of the Sun's spot formation activity. Nauka, Leningrad, Russia.

Vitinsky, Yu. P., 1973. Cycles and forecasts of solar activity. Nauka, Leningrad, Russia.

Vitinsky, Yu. P., Ol'h, A.I., Sazonov, B.I., 1976. The Sun and the Earth's atmosphere. Gidrometeoizdat, Moscow, Russia.

Vitinsky, Yu. P., 1983. Solar activity. Nauka, Moscow, Russia.

Vize, Yu. V., 1925. Fluctuation of the hydrological factors, in particular the fluctuation of water level in the Lake Victoria, in connection with the general atmospheric circulation and solar activity. Izvestiya State Gidrological Institute 13, 52–59.

Vladimirsky, B.M., Sidyakin, V.G., Temuryants, N.A., Makeev, V.B., Samokhvalov, V.P., 1995. Space and biological rhythms Simferopol. In: Libri, Kiev (Eds.), Ukraine, p. 242.

Volfendeyl, A., Dyalai, D., Erlikin, A.D., Kudela, K., Sloan, T., 2009. About the nature of the correlation between the intensity of cosmic rays and clouds. Izvestiya Akademii Nauk - Rossiyskaya Akademiya Nauk, Seriya Fizicheskaya 73 (3), 408–411.

Von Däniken, E.A.P., 1969. "Chariots of the Gods". Souvenir Press Ltd.

Voropinova, E.N., Kiselev, V.V., 2001. Economic cycles. In: The cycles of nature and society: Mater. IX International Conference. Stavropol, Sept. 25–28, 2001. Stavropol University V. D. Chursin, Stavropol, Russia, pp. 198–201.

Voyskovsky, M.I., Gusev, A.A., Lazarev, A.A., Pankov, V.M., Pugacheva, G.I., 2006. Communication tropical cyclogenesis with solar and magnetospheric activity. Astrophysics and Space Physics: IKI, 521 (06), 202–204.

Wagner, G., Beer, J., Laj, C., Kissel, C., Masarik, J., Muscheler, R., Synal, H.A., 2000. Chlorine-36 evidence for the Mono Lake event in the Summit GRIP ice core. Earth and Planetary Science Letters 181, 1–6.

Wald, A., 1960. Sequential analysis (translated from English). Fizmatgiz, Moscow, Russia.

Ward, M.N., 1998. Diagnosis and short-lead time prediction of summer rainfall in tropical North Africa at interannual and multidecadal timescales. Journal of Climate 11, 3167.

Weaver, C.J., Ginoux, P., Hsu, N.C., Chou, M.D., Joiner, J., 2002. Radiative forcing of Saharan dust: GOCART model simulations compared with ERBE data. Journal of Atmospheric Science 59, 736–747.

Webster, P.J., Holland, G.J., Curry, J.A., Chang, H.-R., 2005. Changes in tropical cyclone number, duration, and intensity in a warming environment. Science 309 (5742), 1844–1846. doi:10.1126/science.1116448.

Webster, P.J., Curry, J.A., Liu, J., Holland, J., 2006. Response to comment on "Changes in tropical cyclone number, duration, and intensity in a warming environment." Science 311 (5768) 1713c.

White, W.B., Dettinger, M.D., Cayan, D.R., 2000. The solar cycle and terrestrial climate. In: Proceedings of 1st Solar and Space Weather Euroconference, Santa Cruz de Tenerife, Spain, ESA-SP-463, p. 125.

Wilcox, J.M., Hundhausen, A.J., 1983. Comparison of heliospheric current sheet structure obtained from potential magnetic field computations and from observed polarization coronal brightness. Journal of Geophysical Research 88, 8095.

Willett, H.C., 1957. Meteorological evidence for solar control of weather. Possible responses of terrestrial atmospheric circulation to changes in solar activity. In: High Altitude Observatory, Institute for Solar ± Terrestrial Research Technical Report No. 1. University of Colorado pp. 65 ± 69.

Willet, H.C., 1974. Recent statistical evidence in support of the predictive significance of solar-climatic cycles. Monthly Weather Review 102 (10), 679.

Yagodinsky, V.N., 1987. Alexander Leonidovich Chijzhevsky, 1897–1964. Nauka, Moscow, Russia.

Yiou, F., Raisbeck, G.M., Baumgartner, S., Beer, J., Hammer, C., Johnson, S., Jouzel, J., Kubik, P.W., Lestringuez, J., Stiévenard, M., Suter, M., Yiou, P., 1997. Beryllium 10 in the Greenland Ice Core Project ice core at Summit, Greenland. Journal of Geophysical Research 102, 26783–26794.

Yoshimura, H., 1979. The solar-cycle period-amplitude relation as evidence of hysteresis of the solar-cycle nonlinear magnetic oscillation and the long-term/55 year/cyclic modulation. The Astrophysics Journal 227, 1047.

Yudakhin, K.F., Libin, I. Ya, Prilutzkiy, K., 1991. Autorregressive analysis. Preprint IZMIRAN (No. 16). p. 24.

Yudakhin, K.F., Chuvilgin, L.G., Dorman, L.I., Gulinsky, O.V., Ptuskin, V.S., 1994. Cosmic ray scintillations in random magnetic rield with non-zero helicity. Astrophysics and Space Science 213 (2), 185–196.

Zadde, G.O., Kijner, L.I., 2008. The main stages of numerical methods for analyzing and forecasting the weather (Electronic resource). TSU IDO, Tomsk, Russia.

Zavolokin, A., 2002. Will be a Global Flood? In: Ogoniok, (Eds.), Moscow, Russia. Retrieved from: www.ogoniok.ru/archive/ 2002/4756/28-32-35 (in Russian), p. 24.

Zherebtsov, G.A., Kovalenko, V.A., Molodikh, S.I., Rubtsova, O.A., 2005. Model of the solar activity on climatic characteristics of the Earth's troposphere. Atmospheric and Ocean Optics 18 (12), 1042–1050.

Zherebtsov, G.A., Kovalenko, V.A., Molodikh, S.I., Rubtsova, O.A., Vasilieva, L.A., 2008. Effect of disturbances on the thermobaric heliophysics and climatic characteristics of the Earth's troposphere. Space Research 46 (4), 368–377.

Zhvirblis, V.E., 1987. Cosmophysical origins dissymmetry of living systems. In: Nauka, (Ed.), The principles of symmetry and systematic in chemistry (pp. 1–87). Moscow, Russia.

Zveryaev, I.I., 2007. Climatology and long-period variability of the annual air temperature over Europe. Meteorology and Hydrology 7, 18–24 (in Russian).

Zverinsky, V.V., 1871. Statistical review. In: The list of settlements of the Russian Empire. The Russian Empire, St. Petersburg, Russia, pp. 3–191.

Index

Page references followed by "f" indicate figure and "t" indicate table.

B

C

D

E

F

G

H

L

M

N

X

Y

Z

Printed and bound by CPI Group (UK) Ltd, Croydon, CR0 4YY

08/05/2025

01864883-0001